W9-BBT-160

THE PASSIVE SOLAR CONSTRUCTION HANDBOOK

THE PASSIVE SOLAR CONSTRUCTION HANDBOOK

Featuring hundreds of construction details and notes, materials specifications, and design rules of thumb

by Steven Winter Associates, Inc.

lead author: M. Emanuel Levy, Principal
coauthors: Deane Evans, Senior Associate
Cynthia Gardstein, Senior Associate

Rodale Press, Emmaus, Pennsylvania

The authors are indebted to Edward Mazria and Rodale Press for permission to use portions of *The Passive Solar Energy Book.* The Rules of Thumb and many of the data tables used in this publication are from *The Passive Solar Energy Book,* © 1979 by Edward Mazria, published by Rodale Press, Emmaus, Pa.

The plans and illustrations of the TVA case study houses were prepared by the Tennessee Valley Authority. They are part of the public domain and are not subject to copyright.

Printed in the United States of America on recycled paper containing a high percentage of de-inked fiber.

Book design: Linda Jacopetti
Illustrations: Rita Marks

Library of Congress Cataloging in Publication Data

Steven Winter Associates, Inc.
The passive solar construction handbook.

Bibliography: p.
Includes index.
1. Solar energy—Passive systems. I. Levy, M. Emanuel.
II. Evans, Deane. III. Gardstein, Cynthia. IV. Title.
TH7413.L48 1983 690'.868 83-4608
ISBN 0-87857-455-7 hardcover

2 4 6 8 10 9 7 5 3 1 hardcover

CONTENTS

ACKNOWLEDGMENTS

Although hundreds, and perhaps thousands, of passive solar homes have been built in the United States over the past ten years, the "passive" industry, as far as standardized construction detailing is concerned, is still in its infancy. With only a few exceptions, the existing stock of passive solar residences are one-of-a-kind structures, frequently built by people not in the mainstream of the housing industry. Many of these homes are built using unique construction detailing, difficult to reproduce and expensive to construct.

The few hundred details illustrated in this text are a first attempt at developing standard and accepted construction practices for passive residences. Many of the details developed were based on a handful of examples selected by the authors from a review of nearly 700 sets of working drawings of existing passive homes. Others are standard industry details modified, in some cases substantially, to accommodate the requirements of passive design. Much of the information that accompanies these details is of the same nature and was derived from similar sources. Consequently, the text tends to be conservative in scope. Experimental data and construction procedures have been avoided in favor of techniques and information that are "tried and true."

The Passive Solar Construction Handbook is based, in part, on a document of the same title written by the authors under contract to the U.S. Department of Energy and the National Concrete Masonry Association.

Throughout the development of the *Handbook*, an advisory board provided direction and assistance on many technical and construction issues. The board was comprised of representatives of industry and government groups, including Paul

Lenchuk, Kevin Callahan, and Steven Heibein of the National Concrete Masonry Association (NCMA); Mario Catani of the Portland Cement Association (PCA); Steven Szoke of the Brick Institute of America (BIA); Al Lutz and Milton Morrison of the National Fenestration Council (NFC); and Kalevi Turkia of the Southern Solar Energy Center (SSEC).

In addition, in the early stages of the work, a panel of builders met in Atlanta to review a preliminary draft. The panel included Charles Wagenheim of Centrex Homes (Florida), Edward Schmitt of Bob Schmitt Homes (Ohio), Mark Kelly of Acorn Structures (Massachusetts), and Peter Janopaul of the Graystone Block Company (California). A second panel comprised of Richard Everhart of U.S. Homes (Arizona), Robert Peterson of Peterson Construction Company (Nebraska), John Arbib of Pasadena Homes (Florida), and Edward Bell of SSEC provided a similar review. These panels helped resolve inaccuracies in the construction information contained in the *Handbook*.

The technical and analytical discussions were reviewed by Robert Jones and William Wray of the Los Alamos Scientific Laboratories and Thomas Hartman of SSEC.

Timely and crucial assistance and suggestions were provided by Michael Bell of the National Association of Home Builders; Steven Selkowitz, Marlo Martin, and Richard Johnson of the Lawrence Berkeley Laboratories; Douglas Balcomb, Robert McFarland, and Scott Noll of the Los Alamos Scientific Laboratories; Don Abrams and Robert Erwin of SSEC; George Barney of PCA; Graig Young of NFC; Janice Hamrin of the California Energy Commission; Michael O'Neill of the National Solar Heating and Cooling Information Center; William Freebourne of the U.S. Department of Housing and Urban Development; Harry Gordon and William Fisher of Burt, Hill, Kosar, and Rittelmann Associates; Susan and Wayne Nichols; and Mark Schiff of Ed Mazria and Associates, Architects.

On our staff at Steven Winter Associates, Steven Winter, Alexander Grinnell, and Don Carr provided much needed support and criticism. The drawings of the construction details and concept sketches were produced by Rita Marks with assistance from Dana Tuluca and Ron Vargo.

And, special thanks to our editor, Carol Stoner, for her patience, perseverance, and meticulous review of the text.

Finally, it is our hope that the *Handbook* will serve as a constant construction reference, avoiding bookshelf dust.

M. Emanuel Levy
Deane Evans
Cynthia Gardstein

New York, New York
1983

Chapter 1
INTRODUCTION

Over the past few years passive solar design has enjoyed a growing audience of enthusiasts working in a variety of areas, mostly on the periphery of the housing industry. These researchers, builders, economists, architects, and others from a wide variety of disciplines have generated a significant body of information on passive solar technology. As is true of many innovative building technologies, this body of technical and analytical studies has preceded the development of standard passive solar construction procedures.

Parallel to the development of the technical foundations of passive solar design is a growing interest among builders and designers in incorporating passive concepts into their standard building practices. Unfortunately, the construction details and information that have developed out of the many successful passive solar projects are dispersed within the enormous volume of literature on passive design and are not readily available to the building community at large. In effect, few publications have been widely distributed that focus on the construction detailing of passive solar homes. In most respects, the construction of a passive solar residence utilizes standard construction procedures, although the actual configuration of the materials may not be typical. *The Passive Solar Construction Handbook* represents a comprehensive effort to fill this gap by illustrating methods of adapting standard building practices to passive solar applications.

The information contained in the *Handbook* has been developed primarily for the builder. However, it is hoped that by providing a variety of construction and technical information, others involved in the building industry and owner-builders will find the contents of value.

The *Handbook* focuses specifically on the construction of passive solar, single-family detached residential buildings. It will be evident, however, that much of the construction information can be successfully adapted to other building types.

Discussions of construction, of course, cannot stand apart from considerations of design. Where appropriate in the text, basic concepts in the design of passive homes are discussed and general design guidelines are offered. It should be emphasized that these basic principles are presented only in a general way. More specific design information can be obtained from any of a variety of passive solar design primers currently available (see Bibliography). In fact, it is strongly recommended that the *Handbook* be used as a companion volume to one of these design guides.

The illustrations used throughout the text are intended to aid the reader in interpreting the basic concepts of passive solar design. They are general in nature and are not intended to convey a specific or preferred method of design.

The construction details, on the other hand, provide a specific suggested use of materials and construction procedures. Details are provided for only those areas of the home that are part of the passive system.

It should be emphasized that the *Handbook* is intended to be national in scope. As a result, the construction information and details tend to be general in nature, not focusing on specific regional building practices. Many construction-related concerns that would ordinarily be identified and discussed in a regionally oriented text are, therefore, simply identified as areas of regional concern.

Included in this list of regional issues are the use and location of vapor barriers; the use of termite shields and soil poison; pressure-treating wood; attic ventilation; and various special structural requirements in seismic areas. These and other regional construction considerations will need to be overlaid onto the basic construction information presented in the illustrations and text. The details illustrated in Chapters 3 through 6 will then need to be adapted, conforming to local building practices and codes. Thus, the information presented in the *Handbook* is intended to serve as a general construction guide and to convey the basic principles of passive solar design and construction.

The text and appendices are organized in a manner intended to acquaint the reader with the basic concepts and provide a source of construction information that can be easily referenced. The discussion that follows briefly outlines this organization, the topics covered in each chapter, and methods for quickly and efficiently using the *Handbook* as a construction reference guide.

Chapter 2: Passive Solar Fundamentals— In Chapter 2 many of the basic elements of passive solar design are reviewed. The discussions are intended to: (1) introduce the reader to the unique design constraints present in passive homes; (2) provide a framework for cataloging passive solar home "types"; and (3) briefly describe many of the salient issues influencing design decisions such as economic and energy-saving calculations. The reader familiar with the topics presented may wish to skim the discussions but is advised to read carefully the paragraphs describing the basic passive systems, a section that delineates the organizational framework for the remainder of the text.

It should be reiterated that although many significant considerations in designing passive solar homes are outlined in Chapter 2, the section is not intended to substitute for any of the readily available passive solar design guides.

Chapters 3 through 6: Direct Gain, Thermal Storage Wall, Attached Sunspace, Convective Loop—These four chapters taken together contain all of the basic design and construction information. They will serve as a constant guide in the design and construction phases of building passive homes.

Each chapter discusses one of the basic passive system types introduced in Chapter 2 (except the Thermal Storage Roof, which is discussed briefly in Chapter 2). The chapters are all subdivided into the following sections: (1) a general discussion, or design overview, of the important design and construction issues that apply to the particular system under review; (2) the construction details and notes; and (3) case studies illustrating designed and built examples of the system type.

In reading each chapter, particular attention should be paid to the design overview, which outlines the major advantages and disadvantages of the system type, as well as the more specific rules of thumb for design and construction. These rules will aid in selecting and constructing the passive design.

The actual construction details illustrate both the recommended materials of construction and one suggested design. The precise organization of the details may vary from chapter to chapter. In all chapters, however, a basic set of details is illustrated, typically followed by variations to that detail and/or alternate details, and a set of construction notes. A short introduction summarizes the use of the detail section in each chapter.

The case studies at the end of each chapter describe the design and construction of successful passive solar homes with particular emphasis on the construction detailing. In most instances, the case study demonstrates the use of two or more system types combined in the same home.

Chapter 7: Materials—While quickly thumbing through the details, it may be noted that although material types are indicated, specific products and material dimensions are not. These designations and selections can be made with the aid of the information contained in this chapter. If used in conjunction with the construction chapters, these discussions provide guidance to available and suitable materials for the various passive solar applications. The information in this chapter is particularly intended to be used with the rules of thumb in the construction chapters.

Bibliography—For further reading, an extensive bibliography is provided at the end of the text. Each entry is coded with a category number indicating the primary type of material contained in that entry. As noted, it is suggested that one of the design primers (category [1]) be used as a companion volume to the *Handbook*.

Glossary—Throughout the text, a variety of terms are used that may be unfamiliar. A glossary has been included to clarify these terms.

Appendices—To supplement many of the discussions in the text, a number of appendices have been included at the end of the *Handbook*. The appendices contain technical information valuable in the process of designing and renovating passive homes.

Chapter 2

PASSIVE SOLAR FUNDAMENTALS

Introduction

The implications of dwindling fossil fuel supplies are creating a marked awareness within the building community of the need to conserve energy in the design of residential buildings. Prospective homeowners are, with increasing frequency, buying the "energy-efficient" or "passive solar" model. With each revision, building codes and standards become increasingly stringent with regard to the selection of energy-use-related materials and components. Utilities are providing incentives and assistance and, in some cases, are requiring that the home be built to minimize the amount of energy used before providing service. And, significantly, lending institutions are recognizing the financial attractiveness of energy-conserving designs, which, by reducing operating costs, lower their investment risks.

There are numerous techniques and opportunities for conserving energy in residential design. Most remain virtually untapped in the vast majority of new home construction nationwide. Simple and familiar techniques, such as increased insulation levels or multiple glazings instead of single glazing, are, of course, effective and expedient measures to reduce energy use. These and other energy-conservation techniques are primarily useful as climatic buffers, protecting against the deleterious effects of, for example, cold outdoor temperatures. They reduce the demand for energy in the home.

There are, however, a variety of additional techniques that effectively utilize the forces of nature to provide energy to the home, reducing, and in some cases eliminating, the need for conventional fossil-fuel-fired heating and cooling equipment. In residential design, most of these techniques are based upon the use of solar radiation to provide space heating and water heating, and, occasionally, space cooling. These solar techniques can be subdivided into two basic categories: active and passive. Although the distinction between them is not always clear-cut, in general, passive solar techniques or systems heat and/or cool without the need for energy from other nonrenewable sources, such as electric power to run a pump. On the other hand, the familiar roof-mounted solar panels, which require pumps and fans whose consumption of electricity may be significant relative to the amount of energy generated by the overall system, are classified as active systems.

In the development of passive solar designs, the range of possibilities should not be limited by strict adherence to this definition of solar types. If a small fan will improve the efficiency of a certain passive installation, it should be used even if the resulting system is not a purely passive one. Similarly, the use of passive techniques should not preclude the use of active components. So-called "hybrid" systems, which combine active and passive elements, can be very effective in providing most of the home's energy needs.

It should be emphasized that passive solar techniques are effective in reducing energy use only if they are combined with, not substituted for, standard energy-conservation techniques. Any energy gains that a passive system might generate can be easily offset by the energy losses that will occur in a poorly designed and constructed building. It is for this reason that the incorporation of energy-conservation measures, in both the design and the construction phases of the building, is an absolute prerequisite for the passive solar home to work effectively. A partial list of recommendations for improving the energy-conserving capability of a building includes the following:

- Insulate the house as well as possible.
- Reduce windows and other openings on the north side of the building.
- Orient the building to maximize the effects of summer breezes and to minimize those of winter winds.
- Utilize landscape elements to increase thermal performance by providing shade and blocking unwanted winds.
- Take advantage of overhangs and projections to shade the building during the cooling season.
- Use earth berming where possible to increase the insulating capacity of walls and/or roofs and to block unwanted winds.
- Ventilate the roof properly to avoid condensation problems and consequent loss of insulating capability.
- Provide adequate vapor barriers.
- Insulate openings by using double or triple glazing or storms on windows and doors.
- Use insulating shades and shutters instead of standard uninsulated units.

- Reduce air infiltration by properly weatherstripping and caulking *all* openings and by using proper sill sealers.

- Use air-lock entries.

All of the above recommendations will help create an energy-conserving structure that will reduce the demand for supplemental energy sources. The next step in the design process is to include passive systems that will actually allow the home to provide for its own energy requirements. The result will be a truly "energy-conscious" home.

The chapters that follow contain some basic guidelines for the design of such a home but concentrate primarily on the materials and methods of constructing passive solar features. For more precise information regarding overall design strategies, any of the variety of passive solar design primers listed in the Bibliography can be consulted.

Principles of Passive Solar Design

Passive solar systems rely on the intelligent design and organization of the spaces in a home and on the careful selection of building materials to derive heating and cooling benefits from the free and abundant energy available in the natural environment. Such systems depend on two basic material properties for their effect: (1) the ability of certain materials to store large amounts of heat and to release that heat slowly to the living spaces of the home; and (2) the ability of glass and many other glazing materials to transmit solar radiation (light) but to remain opaque to thermal radiation (heat). When a passive system admits sunlight through its south-facing windows, this light strikes objects and surfaces within the space and is transformed into heat. Much of this heat is prevented from passing back out through the windows because the glazing is opaque to radiant heat. This phenomenon, known as the "greenhouse effect," will be familiar to anyone who has opened the door of a car that has been sitting in the sun with its windows closed on a summer day.

To date, most research and development in the area of passive solar has dealt with providing heating rather than cooling. Thus, the basic passive systems developed and presented in the *Handbook* have been designed primarily for heating and only secondarily for cooling. In fact, the essential processes of a passive system (collection and storage) are specifically oriented toward gaining heat rather than rejecting it. This is not to say that aspects of a system cannot be adapted for cooling. Storage elements *can* be used to store "coolth" (coolth is to cooling as warmth is to heating) and to absorb heat from the living space (thereby cooling it) during summer days.

Basic Passive Systems

The various strategies for designing and building a passive home can be grouped into five basic system types: Direct Gain, Thermal Storage Wall, Attached Sunspace, Thermal Storage Roof, and Convective Loop. Each of these systems (discussed in detail later in this chapter) can be visualized as an assembly of *components*, each of which performs a unique function and all of which are required for the effective operation of the overall system.

The basic passive systems are each comprised of five mutually dependent basic components: collector, absorber, storage, distribution, and control.

Collector—The solar collector component is composed of transparent or translucent glazing, sealed in a frame and located on the south-facing side of the home. The collector glazing surface can be positioned vertically, as in windows, or sloped, as in a skylight on the roof.

Absorber—The absorber is a solid surface, usually dark colored, that is exposed to the sunlight entering through the collector glazing. The absorber "converts" the solar radiation into heat that is either delivered directly into the space for immediate use or stored in a storage component.

Storage—The storage component is composed of material or materials that have the capacity to retain heat for a period of hours or perhaps days. The mass of a material is an important measure of its capacity to store heat, and residential construction materials that are high in mass, such as masonry, will be appropriate for storage components. The slow rate of heat discharge from the storage materials helps maintain a steady, comfortable temperature within the spaces to be heated. In some cases, the storage materials can be used to store coolth by absorbing heat from the living spaces and effectively cooling them. The storage material is sized according to the amount of solar heat the collector is intended to provide and is usually located in, or adjacent to, the rooms it is intended to heat and/or cool. The absorber and storage components are often one and the same, as in a masonry floor or wall.

Distribution—The distribution component delivers the collected and/or stored heat or coolth into the living spaces to be conditioned. Distribution can be by natural means, such as the radiation of heat from a wall or the movement of air by natural convection, or it can be assisted by small pumps or fans that blow heat away from the absorber to the living spaces or to remote storage.

Control—Control components regulate the heat loss or gain into and out of the passive system. There are three basic types of control components: (1) shading devices, which reduce the amount of radiation allowed to pass through the collector glazing; (2) reflectors, which increase the amount of solar radiation passing through the collector glazing; and (3) movable insulation, which reduces heat flow through the collector and into adjacent spaces.

In summary, all these components interact in the following manner:

- The collector admits solar radiation onto the absorber;
- the absorber "converts" the radiation into heat;
- the storage medium retains the heat not immediately used;
- the distribution component transfers heat between absorber and/or storage and the living spaces; and

- the controls reduce heat loss and increase solar radiation gain in the heating season and/or shade the collector during the cooling season, reducing heat gain.

The precise configuration and appropriate materials for each of the components vary depending upon the system type. Each of the basic passive systems is described below, as well as the arrangement of components that characterizes each.

Direct Gain

Of all the passive solar types, a Direct Gain system is perhaps the easiest to envision and construct. A Direct Gain design is one in which solar radiation *directly* enters and heats the living spaces. The home itself can be visualized as the collector of solar heat. Direct Gain is primarily a heating-type system used mainly in mild and moderate climates.

In the heating season during the daylight hours, sunlight enters through south-facing windows, glass doors, clerestories, and skylights, all common *collector* components in Direct Gain homes. The solar radiation strikes and is *absorbed* and *stored* by elements within the space. The most common or primary *absorber/storage* components within a Direct Gain home are the floors and walls (see Figure 2.1).

Although ceilings and other elements, such as furniture, within the space can be designed to store heat, common Direct Gain storage component materials are most readily incorporated in the floors and walls. Frequently the storage component also serves a structural function, as a bearing wall, for example. The *distribution* of heat is generally not crucial to the operation of a Direct Gain system, since the heat is stored in the same space in which it is used. Heat flow that does occur is by natural convection, usually from one room to another through doorways and stairwells.

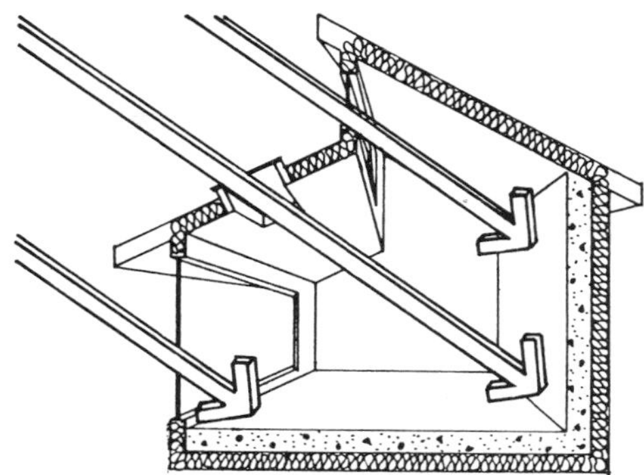

Figure 2.1

In a Direct Gain design, the amount of solar heat that is collected and retained can be *controlled* by three types of components. During the daylight hours, components that reflect sunlight can be placed on the exterior of the home, usually adjacent to the collector components, to increase the amount of solar radiation entering the Direct Gain space (see Figure 2.2).

Essentially, the reflecting component increases the effective collector area without increasing the actual square footage, and the accompanying heat losses, of the collector.

At night, insulating components can be moved into place to prevent heat flow back out through the collectors. This movable insulation can be mounted on either

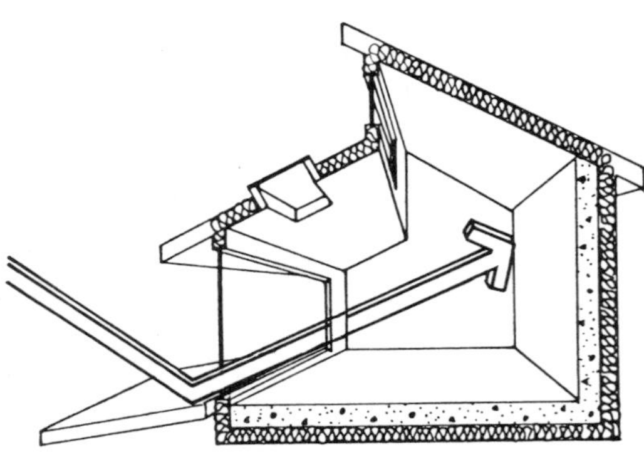

Figure 2.2

the outside or the inside of the collector. Frequently, in cases where a reflecting component is employed, the reflector serves a dual role by both covering and insulating the collector in the evening (see Figure 2.3).

During times of the year when heating is not required, the Direct Gain collector should be shaded to control the amount of solar radiation reaching it. Failure to shade the collector adequately may result in overheating of the Direct Gain spaces, increasing the cooling load. The most effective shading components are mounted outside, intercepting the sunlight before it passes through the collector. Interior shades (such as venetian blinds) can also be effective, reflecting some or most of the direct and diffuse sunlight, but they will allow some heat buildup to occur. The movable insulating component can also be used to block solar radiation and reduce heat gain during the cooling season.

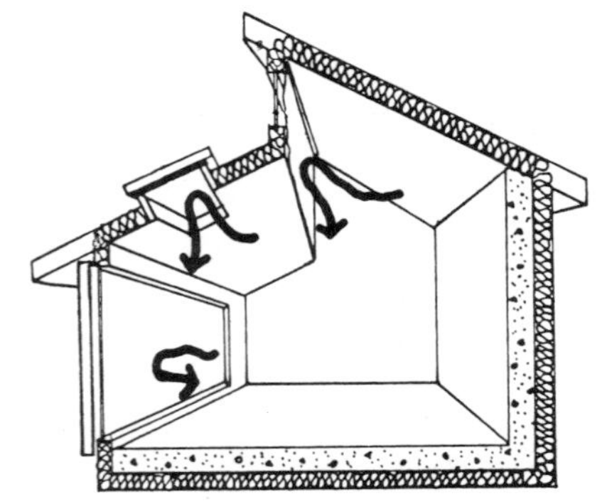

Figure 2.3

Thermal Storage Wall

A Thermal Storage Wall system is one in which the storage component is positioned between the collector and the living space. The storage wall typically provides structural support in addition to serving as the system storage component. Although occasionally used for cooling, Thermal Storage Walls are primarily used to supplement the space-heating requirements. In general, Thermal Storage Walls should be considered for areas that experience mild to severe winters.

In the heating season, solar radiation is allowed to pass through the collector glazing and is *absorbed* and *stored* by the wall. The wall heats up, transferring some of its heat to the living space and some to the column of air between the collector and the wall. This column of air can be exhausted directly into the living space, in which case the wall is referred to as a vented, or Trombe, wall (see Figure 2.4), or the heated air can simply remain trapped between the glazing and the wall, in which case it is an unvented, or stagnating, wall (see Figure 2.5). In both cases, the heat in the wall will radiate to the adjacent living space throughout the day and evening hours.

Beyond the action of the vents in the Trombe wall, distribution components are

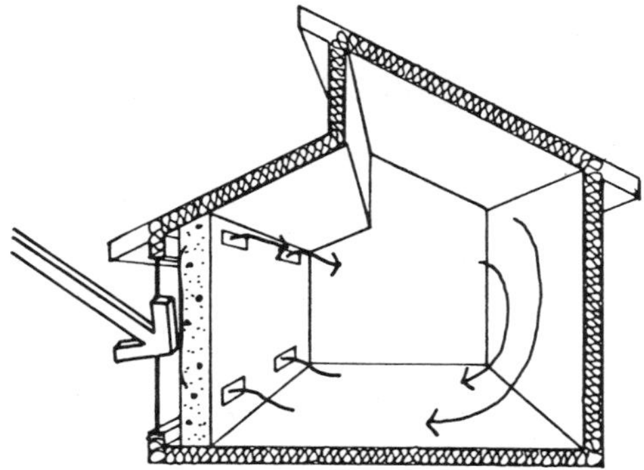

Figure 2.4

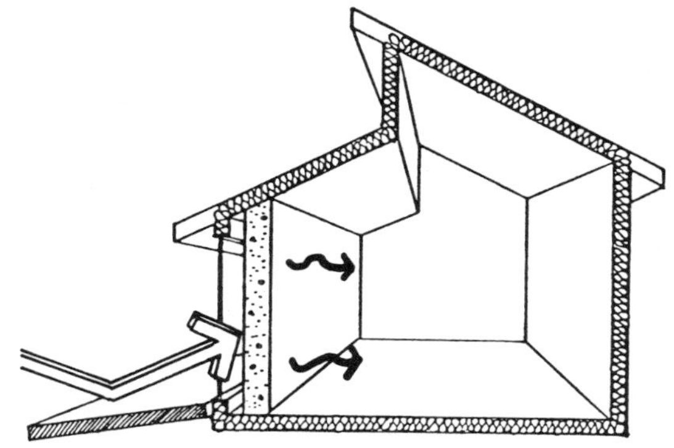

Figure 2.6

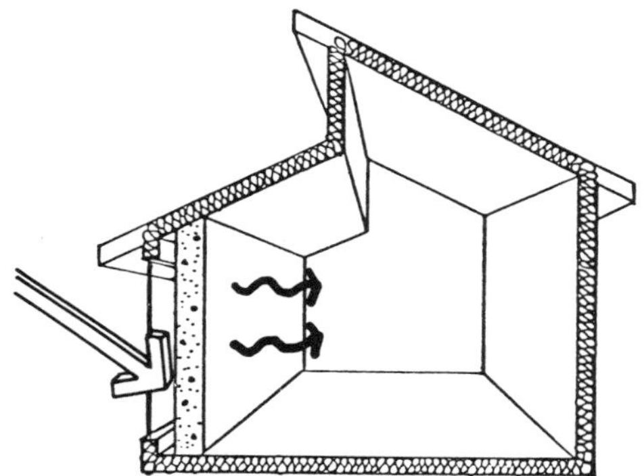

Figure 2.5

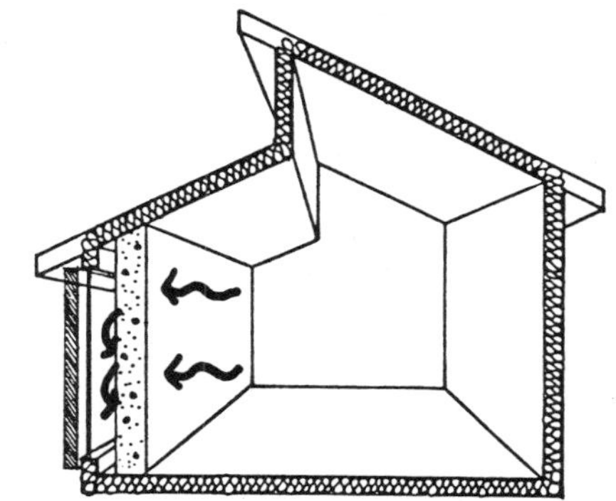

Figure 2.7

not fundamental to either type of Thermal Storage Wall. In some cases, warmed air is ducted from the airspace between the collector and the storage wall to remote living spaces, but this is not characteristic of the system operation and should be avoided if possible.

In a manner similar to the operation of the Direct Gain system, the performance of the Thermal Storage Wall can be regulated by *control* components. During the heating season, a reflector can enhance the amount of solar radiation striking the wall, and in the evening hours, an insulating component can be placed either between storage and collector or outside the collector component to reduce heat losses (see Figures 2.6 and 2.7).

In the cooling season, the impact of solar radiation should be controlled by shading the collector and storage components. Shading components are most effective when placed on the exterior but can also be installed between the collector and the storage wall (see Figure 2.8).

If designed properly, the movable insulation used to control nighttime heat flow out of the home during the heating season can be operated in the daytime during

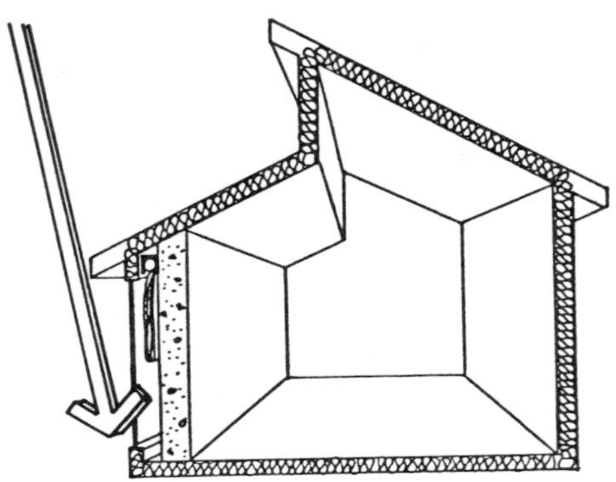

Figure 2.8

the cooling season to control unwanted solar heat gain. To further promote cooling, the airspace between collector and storage components can be vented to the outside. This is particularly effective in areas where the nighttime temperatures drop below the indoor temperatures. Venting the airspace will cool the storage wall in the evening, allowing it to absorb heat from the space during the day and reducing the space-cooling requirements.

Attached Sunspace

Attached Sunspaces are typically built as glass-enclosed spaces, frequently constructed as extensions to homes. They are generally considered secondary-use spaces in which heat is collected and either stored for later use or exhausted directly into the living space. The solar energy collected and absorbed by an Attached Sunspace is used to heat both the Sunspace itself and the adjacent living areas. The ways in which this energy is stored and distributed form the basis for distinguishing five separate Attached Sunspace subsystems, which differ significantly from each other in terms of their operation, construction, and configuration of basic components. These are: open wall subsystem, direct gain subsystem, air exchange subsystem, thermal storage wall subsystem, and remote storage subsystem.

During the heating season, sunlight enters the Attached Sunspace through the south-facing *collector* (an expanse of glass or plastic), is *absorbed* by elements within the Sunspace, and is converted to heat. This basic process is the same for all the Attached Sunspace subsystems. It is only in the operation and location of the *storage, distribution,* and *control* components that these subsystems differ significantly from each other. (For further information, see Chapter 5: Attached Sunspace.)

All Attached Sunspaces will be designed for one of two basic modes of operation. In the first, the Sunspace is thermally isolated from the living area (see Figure 2.9). As a result of this isolation, the Sunspace cannot be used year-round, since the temperature within the space is allowed to fluctuate outside the comfort range.

In the second mode, the Sunspace is *not* thermally isolated but is treated as an additional living space (see Figure 2.10). In this case, temperatures are not allowed to fluctuate but are maintained, by a conventional heating system when necessary, within the comfort range.

Similarly, the design and selection of the control components will vary with the selection of subsystem and operational characteristics. To prevent heat loss, a movable insulating component is typically placed along the collector glazing or between the Sunspace and adjacent living spaces, depending upon the type of oper-

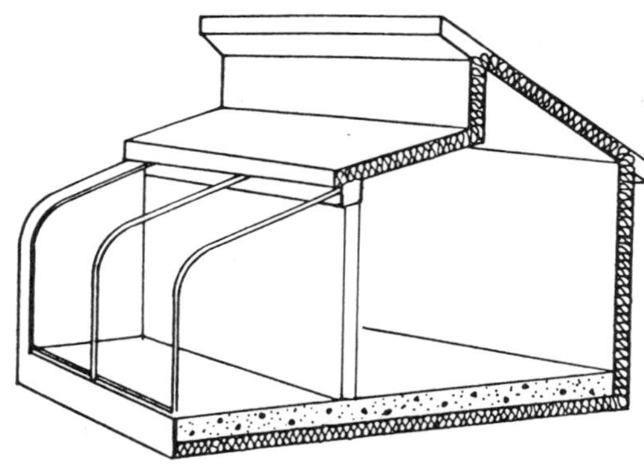

Figure 2.9

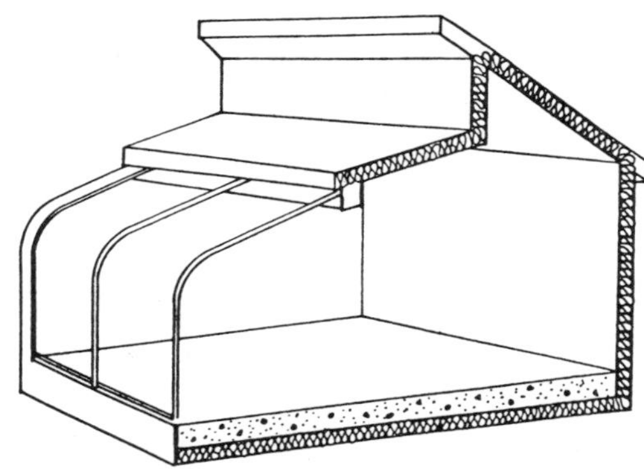

Figure 2.10

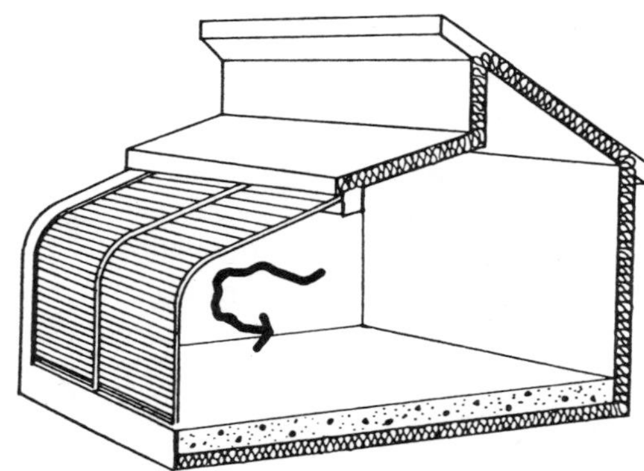

Figure 2.11

ation (see Figure 2.11). It is also possible to enhance solar heat gain by the use of a reflecting component.

In the cooling season, the Attached Sunspace is particularly susceptible to overheating. Control of solar heat gain by shades and ventilation is critical to proper system functioning. In climates where the nighttime temperatures drop below comfort levels, it is often possible to use the Attached Sunspace mass for storing coolth, thus reducing daytime cooling loads.

Thermal Storage Roof

As the name implies, the *storage* mass in a Thermal Storage Roof system (usually water or masonry) is located on the roof of the home. It is typically supported by some form of roof decking (metal or concrete) that serves as the ceiling for the space below and that provides for the efficient and even transfer of heat from the mass above. The location and operation of the storage component make the Thermal Storage Roof an appropriate passive system type for desert climates, where it can provide both effective heating and cooling.

The Thermal Storage Roof is unique in that it does not require a collector component. During the day, the storage component is simply exposed to the sky and heated by direct radiation from the sun. This heat is stored and slowly transferred through the ceiling deck to the living spaces below (see Figure 2.12).

At night and during prolonged overcast periods, some form of movable insulation is provided to cover the storage component and reduce heat losses to the cool night air (see Figure 2.13).

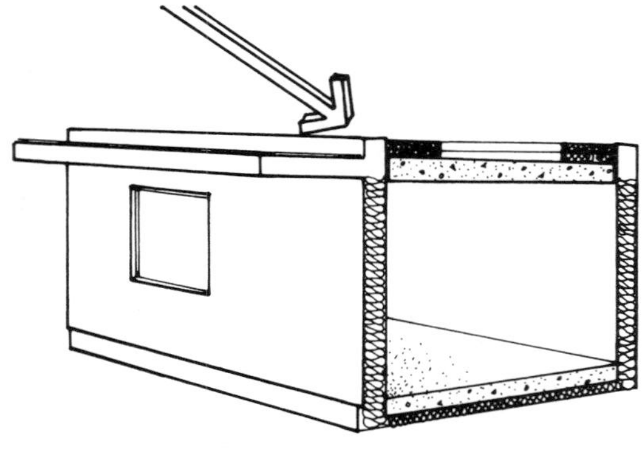

Figure 2.12

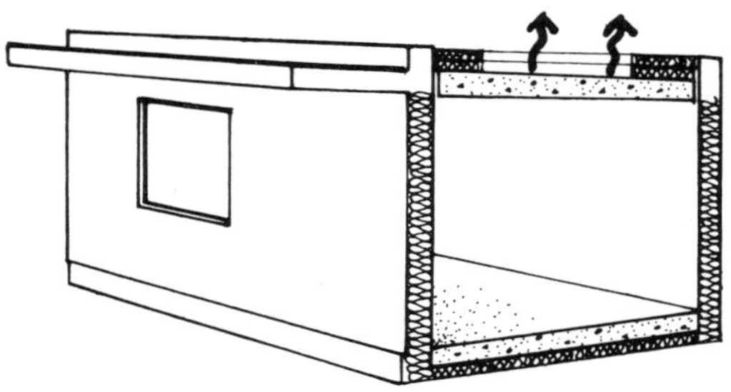

Figure 2.14

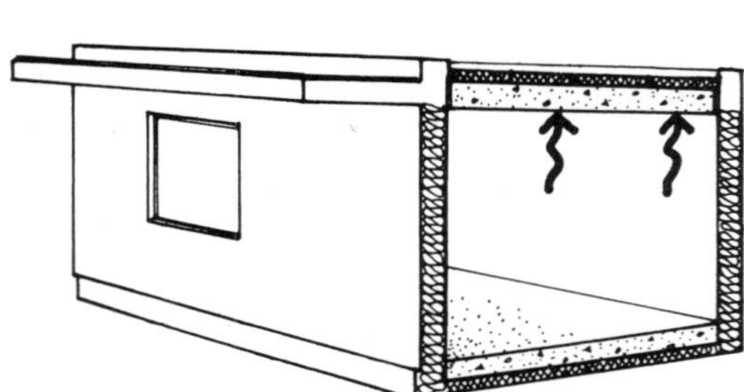

Figure 2.13

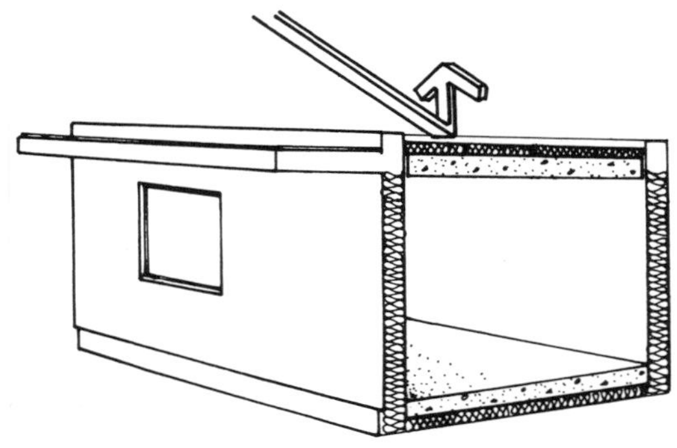

Figure 2.15

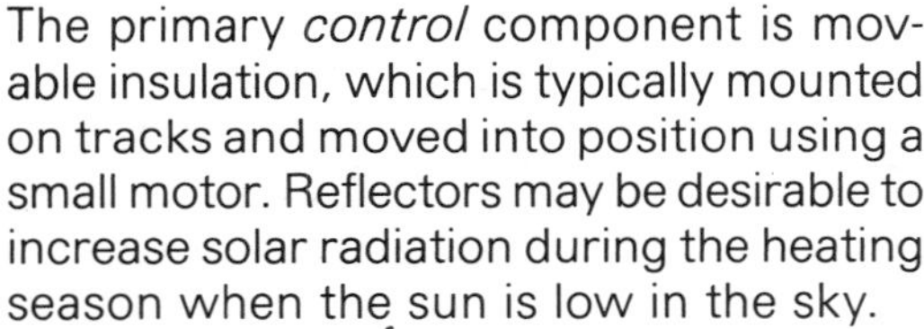

The primary *control* component is movable insulation, which is typically mounted on tracks and moved into position using a small motor. Reflectors may be desirable to increase solar radiation during the heating season when the sun is low in the sky.

Of all the basic passive systems, the Thermal Storage Roof is perhaps the most effective technique for providing passive cooling. During the cooling season, in areas with high diurnal temperature swings and clear night skies, the exposed storage component will be cooled at night. Loss of the collected coolth can be controlled by movable insulation that is placed over the storage component during the daytime to provide insulation and shading. The storage component will absorb heat from the living area during this period and will effectively cool the living space. At night, the insulation is removed, and the diurnal cycle is repeated (see Figures 2.14 and 2.15).

The location of the storage component, however, also presents substantial structural considerations. Particularly in designs where heavyweight materials, such as masonry or water, are specified as storage, the load-bearing capacity of the roof and walls must be increased. Such modifications can be costly—a fact that has limited the popularity of this passive system type.

To date, only a few Thermal Storage Roof designs have been constructed. Due to this lack of practical construction experience, high cost, and structural considerations, details for this system are not included in a subsequent chapter.

Convective Loop

The Convective Loop is primarily a heating-type system most appropriate in areas experiencing moderate to severe winters. It is based on the principle that a fluid, such as air, will rise when heated.

During the heating season, the Convective Loop *collector* admits sunlight, which in turn strikes an *absorber* surface and is converted to heat. This heat is transmitted to the air in the space between the absorber surface and the collector. As the heat is collected at the lowest point in the system, the heated air rises up, passing out through a duct located at the top of the absorber, to be replaced by cooler return air drawn in from below. This warm air can be dumped directly into the living spaces, or it can be diverted into a remote *storage* area, such as a rock bed, and used to provide heat as required (see Figure 2.16).

The major *control* component required by this system is some method for preventing the loop from reversing itself at night or during prolonged cloudy periods, drawing heated air out of storage. Simple backdraft or automatic thermostatically controlled dampers can be used to prevent this back flow. If such controls are provided, it will not be necessary to insulate or shade the collector during the heating season.

It may be desirable to provide some form of reflector to enhance the amount of solar radiation striking the collector. Such reflectors are particularly appropriate for a Convective Loop because the system is thermally isolated, and often physically separated, from the living spaces. This isolation avoids the glare problems that occur in Direct Gain or Attached Sunspaces equipped with reflectors.

During the cooling season, the collector should be well shaded (see Figure 2.17) or covered to prevent possible damage to the absorber surface as a result of excessive heat buildup in the system.

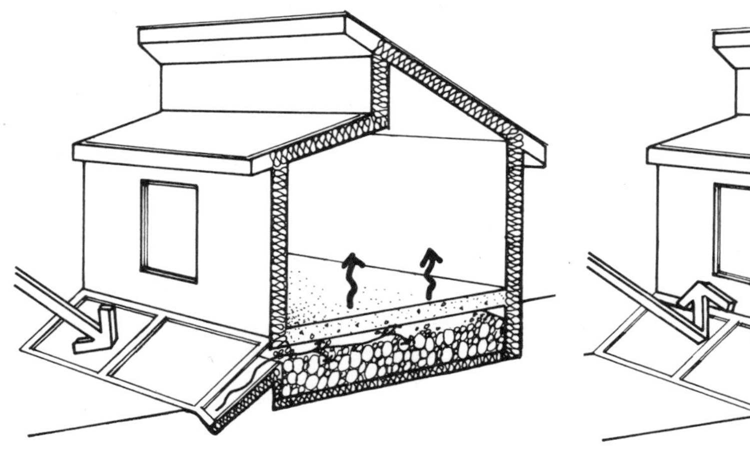

Figure 2.16

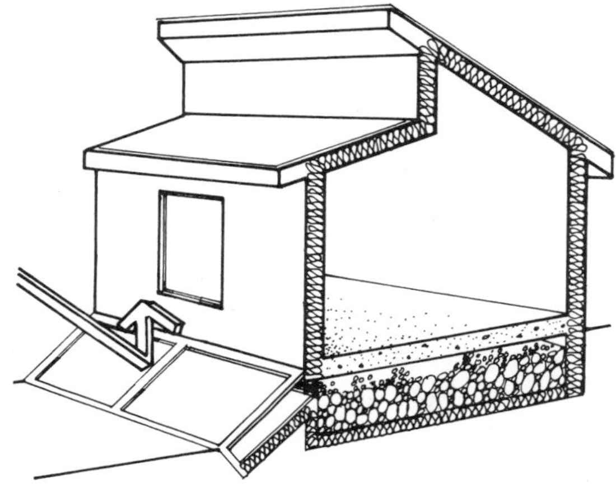

Figure 2.17

Typically, a Convective Loop system is constructed directly on the south wall of a home or is a separate structure placed slightly in front of and below the first floor. In both designs, a small fan is usually employed to assist the natural flow of heat throughout the system.

Passive Solar Design

A well-designed passive home should not be considered as simply an extension of conventional design principles with the addition of "mass and glass." On the contrary, the desire to create a home that is attractive while also saving energy passively will require serious consideration of some basic principles of passive solar and of energy planning.

While the *Handbook* is a useful guide to building passive homes, the discussions assume that the reader has some prior knowledge of passive design principles, such as room layouts that take full advantage of opportunities for passive heating and cooling.

For example, in a well-designed passive home, the living spaces should be placed on the south. Although this planning principle is implied throughout the text, there is no comprehensive discussion on passive solar space planning, a topic that only indirectly influences the construction of the home. On the other hand, the sizing of, for instance, the storage component in a Thermal Storage Wall *is* a design issue that will impact construction and is, therefore, discussed later in a design rule of thumb.

There are, in fact, four design-related topic areas that influence both the design and the construction of a passive home: selecting the appropriate system; specifying the components; estimating the energy savings; and assessing the economics.

Selecting a System

Much of the experience of builders and researchers to date indicates that each of the individual systems described earlier in this chapter will perform best and tend to be most cost-effective in a particular climate region. In defining the five passive system types earlier, five climate types are mentioned: mild, moderate, moderate to severe, severe, and desert. One or more of these regions is identified as being most appropriate for each of the systems.

The recommendations used throughout the text can be summarized as follows:

System	Appropriate Climate(s)
Direct Gain	Mild to moderate
Thermal Storage Wall	Mild to severe
Attached Sunspace	Moderate to severe
Thermal Storage Roof	Desert
Convective Loop	Moderate to severe

The map in Figure 2.18 indicates the boundaries for each of these climate regions (excluding the desert areas). It should be strongly emphasized that these recommendations are intended to serve only as a general guide to system selection and should not supersede specific economic and energy analysis, prior experience, or particular preference. However, it should be kept in mind that these recommendations do indicate, to some extent, the climates to which the inherent operating characteristics of the systems are best suited.

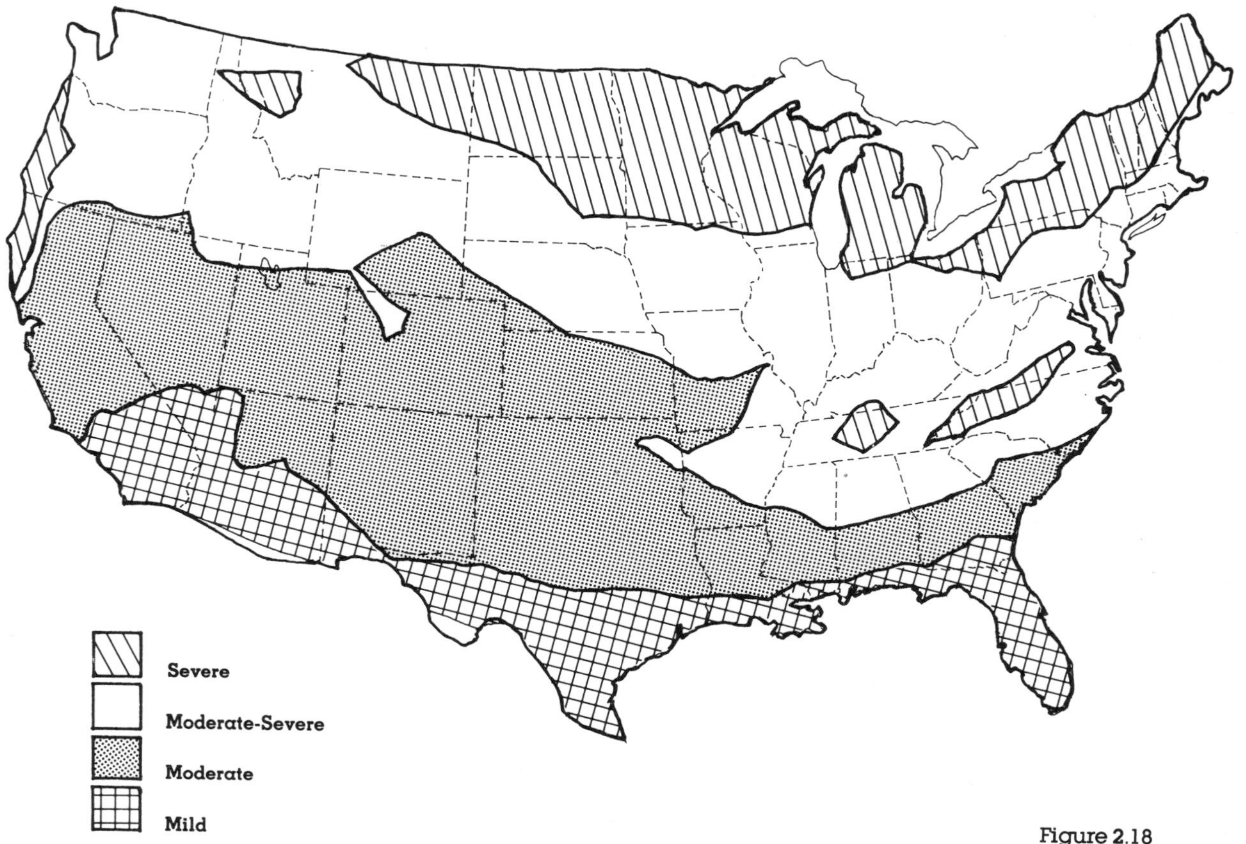

Figure 2.18

Specifying the Components

Once a system is selected, the components that comprise the system need to be specified. As discussed briefly in the *Handbook's* introduction, once a construction detail is chosen, the exact selection and sizing of the materials comprising that detail will require the information and data contained in Chapters 3 through 7.

In most cases, the specification procedure is a simple and straightforward one. For example, if in selecting the Thermal Storage Wall system the storage component is to be specified, the first step is to consult the Storage Rules of Thumb in Chapter 4 (see page 76). These rules of thumb are basic guidelines that will assist in the design of the system. They also indicate criteria for material selection and sizing. Once such criteria are established, Chapter 7: Materials is consulted and particular materials are selected for use.

Continuing with the example above, the detail illustrates a Thermal Storage Wall with masonry selected as the storage component, with sizing criteria obtained from the Storage Rules of Thumb (see page 76), and particular masonry materials specified (see page 201).

In some cases, such as the specification of the collector component, options for materials selection are too numerous to be covered in one book. In these instances, the information in Chapter 7 is of a general nature. For instance, glazing types (such as double glazing) are discussed, but particular manufacturers' products are not presented. In these cases, a check of the products locally available that conform to the rules of thumb will need to be made before a particular material is selected.

Energy-Use Analysis

At many points in the design process, it is useful to have access to design and energy-use analysis methods for the purpose of sizing the passive solar components and estimating the energy use of the home. Depending upon the complexity of the design and required accuracy of results, a variety of methods are available for design and energy-use analysis. These methods can be grouped into four categories reflecting the tools required to use the method and, to a limited extent, the degree of difficulty in using the method, the flexibility of the method to evaluate a range of design and material types, and the accuracy of the results. The four categories are: (1) manual procedures; (2) programmable calculator programs; (3) microcomputer programs; and (4) computer programs. Each category is discussed briefly below. A description of specific programs can be found in Appendix C.

Manual Methods

The first category, manual procedures, includes methods in which the projected energy use can be calculated using paper and pencil and, frequently, preprinted charts, tables, and graphs.

The manual methods are often simplified codifications of computer programs. In the process of simplification, however, some of the flexibility to change building characteristics, allowed in the computer program, is lost. For example, in estimating the performance of a Thermal Storage Wall, most manual methods assume that the wall is 18'' thick. Later it will become evident that such a wall can have a range of thicknesses. Unfortunately, the user of these manual methods may not have the flexibility to vary the material dimension and must, therefore, expect that the projected energy use will be in slight error if a thickness other than 18'' is used. The computer program from which the manual method was derived may have allowed the user to vary the thickness, but the manual method may be applicable only if characteristics such as the wall thickness of the home to be analyzed are similar to the assumptions.

This loss of flexibility may also prohibit the evaluation of innovative designs. The manual methods are, therefore, only applicable for a limited number of design cases (usually the most typical types of designs), yielding only gross estimates of estimated building energy use. Most require a one- to two-day familiarization period. A single analysis can take from a few minutes to a few hours depending upon the method.

Appendix B contains an easy-to-use manual procedure for estimating the performance of a passive solar heating design.

Programmable Calculator Programs

The programmable calculator programs are typically magnetic cards prepared for use with hand-held calculators. Frequently, the accuracy and applicability are similar to the manual methods described above for the same reason: The programs are simplified versions of computer programs requiring numerous fixed characteristics. The advantage of using the calculator programs is the speed with which the analysis can be performed. A single analysis rarely requires more than about 15 minutes and frequently requires less. For most passive solar home designs, the accuracy of the programmable calculators should suffice.

Microcomputer Methods

The next level up from the programmable calculator in both ability to perform energy analyses quickly and degree of flexibility in describing the home is the microcomputer. The use of microcomputers begins to free the designer from the limitations of the manual and programmable calculator methods described above. The results of microcomputer analyses tend, therefore, to be more refined and more accurate for relative comparisons of energy use, and microcomputer methods are able to accommodate wider variations in design than either the manual or programmable calculator methods. A single energy-use analysis of a designed home will typically require about 15 minutes.

Computer Programs

By far the most flexible, complex, and detailed analysis methods are the computer programs. These programs compute, in a few moments, a very accurate profile of the energy-usage patterns of the designed home. A similar analysis, performed manually, would require months of calculation by a knowledgeable mathematician. This high degree of accuracy in computing energy may, in fact, be well out of proportion to the design being evaluated. To program a single home may require several days.

For most residential designs, one of the three methods previously described should be sufficient. Computers are, perhaps, more applicable to commercial or multifamily structures whose complexity demands a sophisticated analysis method. When using a computer program, it is often most economical to purchase computer time from a so-called "timesharer." A single analysis will vary in cost depending upon the program desired and timesharers' rates.

One frequently overlooked aspect of all the energy-use analysis methods discussed above and in Appendix C is the accuracy of such techniques in estimating the actual energy use of the home when constructed. Because of the uncertainties involved in the operation of the home, any correlation of projected and actual energy use is coincidental. This fact should not cloud the real value of employing one of the large variety of energy-use analysis methods available. The value of these tools lies in their relative accuracy of projected energy use.

For example, two homes are analyzed: one poorly insulated with little solar contribution, the other well insulated and designed to provide a large portion of the space-heating and cooling requirements from passive solar. The energy use, projected by one of the methods discussed above, would indicate to the designer the relative merits of the two solutions. If the poorly insulated design requires a projected 500 gal of oil per year for heat and the second home 250 gal, the designer should consider that the solar design will reduce energy use 50%, but will not necessarily save exactly 250 gal of oil.

Again, this should not dissuade the reader from selecting and using one of the analysis techniques. After spending a short time analyzing various designs and passive systems, the designer will develop an intuitive sense of the relative performance of design alternatives. It is in helping to develop this intuition that the analysis tools have greatest value.

Assessing the Economics

Whether or not a home sells has traditionally been a function of a number of factors, not the least of which is cost, specifically first cost. In the past, other costs associated with home ownership, specifically operating costs, rarely entered into home purchase discussions. Consequently, most economic decisions relating to design, material selection, and site planning tended to be based upon first-cost criteria. By far the major costs of home ownership were monthly mortgage payments, a reflection of the first cost and current interest rates. Due to low fuel cost, energy efficiency was seldom a factor in home design.

At present, as energy-cost increases continue to outstrip inflation, the first-cost approach to decision making is being replaced by an economic indicator capable of reflecting the impact of energy costs. Such an indicator combines both first cost and energy cost into a single value familiar to all homeowners: total monthly cost. The total monthly cost includes both mortgage payments (interest and principal) and energy costs.

As a simple example, suppose a home built in New York State costs $100,000. The buyer makes an initial payment of 20%, or $20,000, and takes an $80,000 mortgage. Assuming a 15%, 30-year term rate, the monthly mortgage payments are $1,015. An analysis of the energy use indicates an average monthly expenditure of $124, or $1,488 per annum on fuel. The total monthly cost to the homeowner is, therefore, 1,015 + 124 = $1,139.

Another buyer purchases an equivalent home except that the design has been modified to incorporate passive solar and energy-conservation features. The passive home uses an Attached Sunspace and additional insulation to reduce the home heating requirements (see Figures 2.19 and 2.20).

The following example summarizes the significant criteria for comparing the homes:

Description:
4-bedroom home in New York State

Heated Area:
2,124 sq ft

Solar Features
Attached Sunspace with Movable Insulation

Manufactured Sunspace	$1,800
Additional Masonry	500
Exterior Shades	200
Movable Insulation	800
Added Insulation	300
Total	$3,600

Monthly Expenses:

Base Case	Mortgage	$1,015
	Fuel	124
	Total	$1,139
Passive Home:	Mortgage	$1,051
	Fuel	74
	Total	$1,125

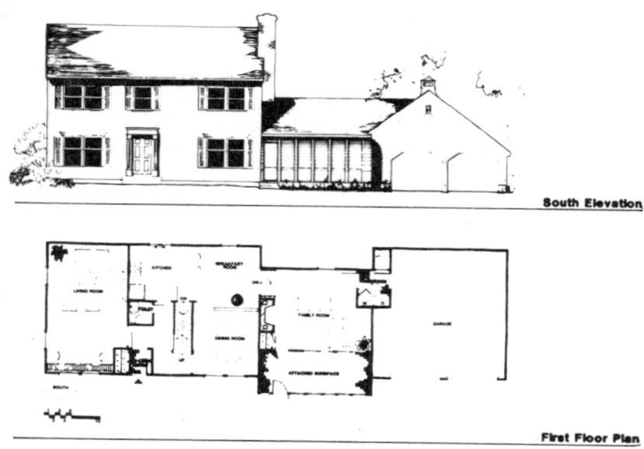

Figure 2.19

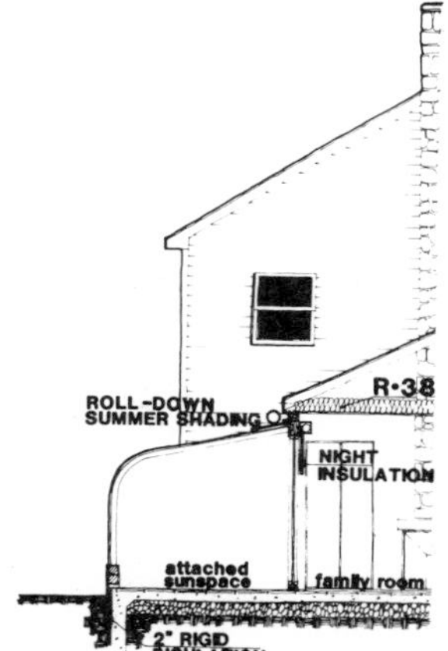

Figure 2.20

It is evident from this example that the solar features add $3,600 to the first cost of the base-case home. Again, assuming that 80% of the total cost ($103,600) is financed, the initial cash outlay is $20,720, or $720 higher than the base case. With the 15%, 30-year term mortgage, the monthly payments are increased to $1,051, $36 above the base-case home. However, due to the energy-saving features, the estimated costs for fuel are now $881 per year, or about $74 per month average. The total monthly homeowner expenditures equal $1,125, or $14 less than the base home. With this $168-per-year savings, the initial additional outlay of $720 is recovered in just over four years. In addition to continually accruing these savings (relative to the less energy-efficient home), the owner of the passive solar home has the additional amenity of the Sunspace!

Ultimately, the decision as to whether or not to invest in a passive solar technique should be an economic one. Throughout the text, numerous design recommendations and rules of thumb are presented to aid in the design process. Design guides should be checked and, if required, superseded by design decisions based upon specific economic analysis using actual local costs.

Chapter 3
DIRECT GAIN

Design Overview

Direct Gain systems are designed to be used primarily for heating and are most effective in areas with mild or moderate winter climates. In a Direct Gain system, the collection, absorption, and storage of solar energy occur directly within the living space. The conceptual simplicity of this arrangement is compelling. As a result of intelligent planning and minor modifications in standard building practice, the home itself becomes the passive solar system. Living within the system can, however, pose certain problems, and attention must be paid, during the design of the building, to questions of space planning, privacy, natural lighting, glare, overheating, and fading and deterioration of fabrics exposed to direct sunlight.

Seasonal Operation—Heating

During the heating season, sunlight enters the living space through an expanse of south-facing collector (usually glass). This solar radiation is in turn absorbed by elements in the space (e.g., floor, walls, ceiling, and furnishings) and converted to heat, which either warms the air in the space or is stored in interior storage elements for later use (see Figure 3.1).

These storage elements are generally made of high-mass materials, typically masonry, which may serve as basic structural components (e.g., walls and floors), or water in suitable containers.

Interior storage mass should be as evenly distributed as possible throughout the spaces to be heated, rather than concentrated in one area. For example, 400 sq ft of 4''-thick exposed mass is more effective than 200 sq ft of 8''-thick material. It is also recommended that as much mass as possible be located so that sunlight strikes it directly.

It is often difficult, however, when distributing mass throughout a Direct Gain space, to provide direct sunlight over the entire surface area. Mass not located in direct sun will absorb heat from warmed air but will not absorb and store as much heat as mass in direct sun. Therefore, in cases where storage mass is evenly distributed throughout the Direct Gain space, some method for distributing the sunlight to this mass should be investigated.

Reflecting sunlight from light-colored nonstorage surfaces to the dark surface of the storage material is one technique. Another effective, although less common, technique is to use a glazing that diffuses and scatters the sunlight. Standard glazing materials and construction procedures can be used in either approach (see Figure 3.2).

Seasonal Operation—Cooling

During the cooling season, the large areas of south-facing collector should be well shaded to prevent excessive solar heat gain (see Figure 3.3).

Shading elements, such as overhangs, should be sized so as to maximize *overall* system performance. If they are too large, overhangs may completely prevent unwanted solar gain during the cooling season, but they will also block a significant amount of radiation during the later part of the heating season. Conversely, if they are too small, solar gain will be

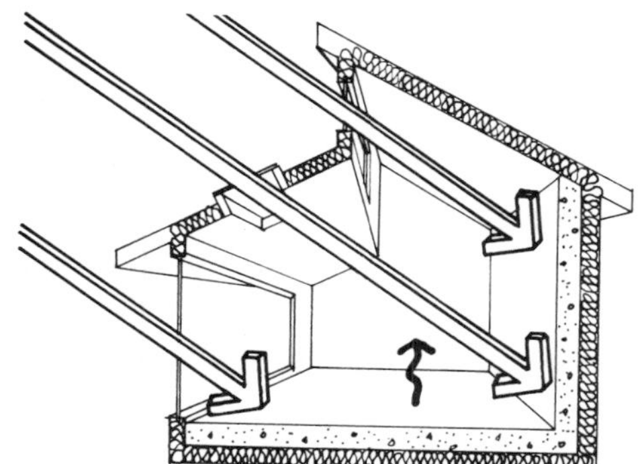

Figure 3.1

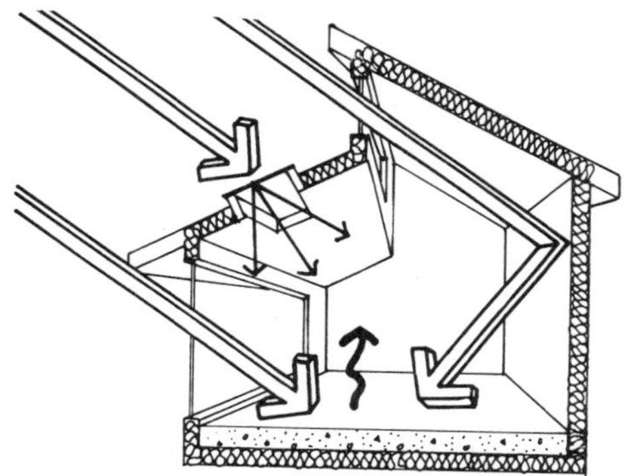

Figure 3.2

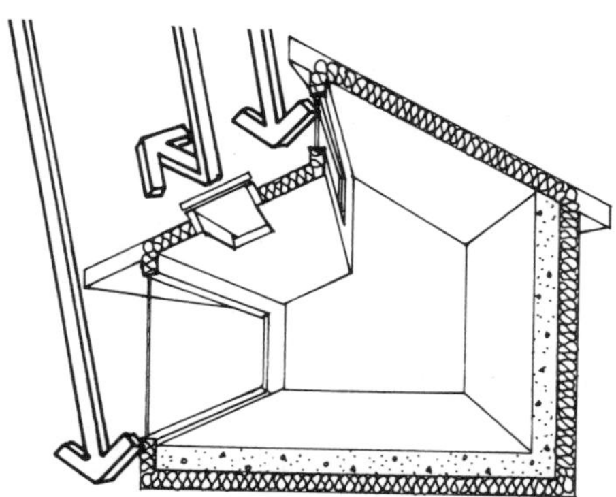

Figure 3.3

maximized when heat is needed, but overheating will occur during the late part of the cooling season, adding to the overall load on the cooling equipment.

Advantages

- Conceptually the most straightforward, Direct Gain can also be the easiest to build using standard construction materials and methods.

- The system does not radically alter the appearance of the home.

- Considerations of privacy make it more likely that movable insulation will be utilized, at least at night, because the Direct Gain living space is often directly visible from the exterior.

- The system provides natural light and view in addition to energy savings and can, therefore, reduce lighting energy consumption.

- The system can have a very low additional cost relative to identical buildings without solar features and to other passive systems.

Disadvantages

- Large expanse of south-facing glazing can cause glare and privacy problems.

- Ultraviolet radiation in sunlight can fade and degrade fabrics.

- Manual operation of movable insulation, which is often required for effective system operation, may meet with homeowner resistance, thus decreasing performance and reducing marketability.

Collector

The principal function of a Direct Gain collector is to admit and trap solar energy so that it can be absorbed and stored by elements within the Direct Gain space. Although generally perceived as energy losers, the large expanses of south-facing collector used in Direct Gain applications can, if properly designed, gain significantly more energy than they lose. There are three basic types of Direct Gain collectors: solar windows, clerestories, and skylights (see Figure 3.4).

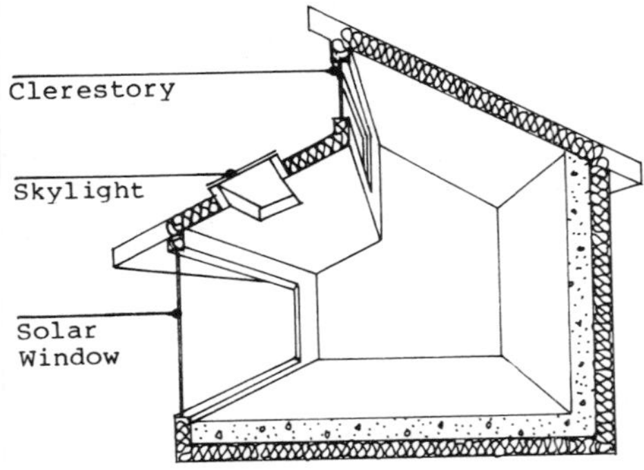

Figure 3.4

Solar Window

A solar window is simply a standard off-the-shelf window or patio door installed in a south-facing Direct Gain wall using conventional construction methods. The collector frame can be wood or metal and should be well caulked and weatherstripped. If metal is used, the frame should contain thermal breaks.

The solar window can be vertical, or it can be slightly tilted so as to maximize the amount of incoming solar radiation. A tilted collector is often difficult to install and may require tempered glass. It can also be hard to shade and may pose significant overheating problems during the cooling season.

Glazing materials other than glass can be used in solar windows, although they are not generally available in preframed, off-the-shelf configurations. Such materials include various forms of fiberglass, acrylics, and polycarbonates (possible local building-code restrictions on the use of these alternate materials should be investigated before they are specified). Materials that will diffuse and scatter the incoming light should receive special consideration, since they can improve the overall performance of a Direct Gain system by increasing the amount of radiation reaching the distributed mass.

Double glazing is recommended for most climates. Triple glazing should be used in severe climates, especially if some form of movable insulation is *not* provided. Single glazing can be used in mild climates if movable insulation *is* provided.

Clerestory

Clerestories serve the same function as solar windows but can avoid problems with privacy, glare, and fading of fabrics. They also allow light to penetrate deeper into a space than may be possible with windows alone, often enabling direct sunlight to strike the north wall of the home. In situations where site obstructions, such as trees, prevent light from reaching standard solar windows, clerestories can be used to admit solar radiation to a Direct Gain space. As in solar windows, clerestories can be vertical or tilted, and the problems associated with the latter configuration are the same for both collector types. Special attention should be paid to properly sealing and caulking the clerestory, especially in tilted installations.

The full range of collector materials available for solar windows can also be used in clerestories. If glass is chosen, double glazing is again recommended, and special glazing may be necessary in tilted configurations to prevent breakage. Several clerestories can be used together in a sawtooth pattern in order to increase the total amount of solar radiation entering a building (see Figure 3.5).

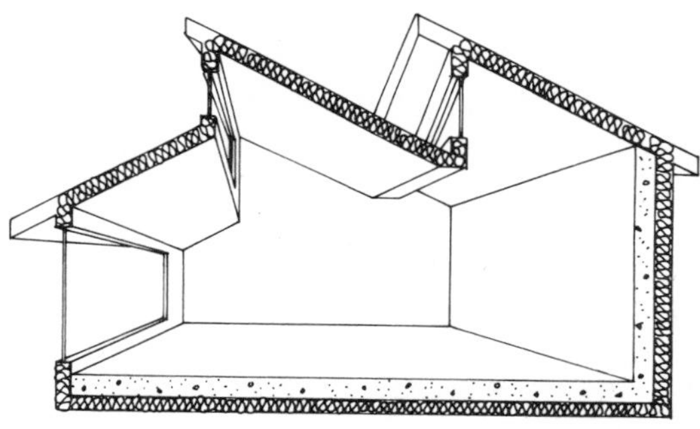

Figure 3.5

Skylight

Skylights serve the same function as the other Direct Gain collectors and can be placed flat on a flat roof or tilted in a south-facing sloped roof. Skylights are the least effective collector type. Their horizontal position receives the least sunlight during the heating season, when the sun is low in the sky and the heating requirements are highest, and maximum sunshine in the cooling season, when the solar heat should be avoided. It is recommended, therefore, that a reflector be used in this configuration to enhance the amount of solar energy hitting the collector in the heating season and that a shade be provided to reduce solar heat gain in the cooling season (see Figure 3.6).

As in clerestory installations, special attention should be paid to properly sealing and caulking a skylight because of the increased potential for water leakage in a flat or tilted configuration. Once again, the full range of collector materials can be used, subject to local building-code restrictions. In many cases, lighter, translucent materials may be preferable to glass because of their optical and handling characteristics, and a variety of plastic "bubble" skylights are available off-the-shelf. If glass is used in a skylight application, it should be tempered.

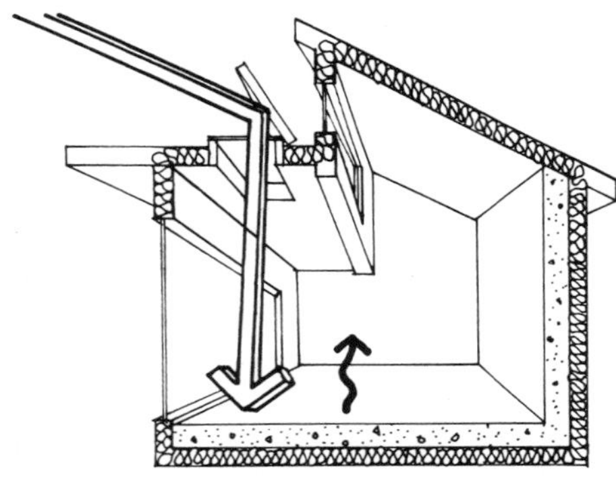

Figure 3.6

Collector Rules of Thumb

All Collectors

- **Orientation**:
  Collector wall must face south. Variations up to 20° east or west of true south, however, will not significantly affect thermal performance. A slight east-of-south orientation may be desired to allow the sun to "wake up" the living space in the morning.

- **Collector Area**:
  A preliminary collector area size per square foot of floor area to be heated can be determined from Table 3.1. Appendix B provides a simple procedure for fine-tuning the size of collector area.

  For example, for Madison, Wisconsin, at 43°NL with an average January temperature of 21.8°F (data available from local and/or national weather bureau), the intersection of the closest latitude (44°NL) and the closest "Average Winter Outdoor Temperature" (20°F) indicates that 0.29 sq ft of glazing per square foot of floor area is required. For a living space with a floor area of 200 sq ft to be heated, 58 sq ft (200 × 0.29) of south-facing glazing will be required.

- **Number of Glazings**:
  Double glazing is recommended for most installations. Triple glazing should be used in severe climates, especially if movable insulation is *not* provided. Single glazing can be used in mild climates if movable insulation *is* provided.

- **Operation**:
  All collectors should be operable to provide natural ventilation during the cooling season.

Clerestory

- **Color**:
  The ceiling adjacent to the clerestory should be light in color to reflect sunlight into the Direct Gain space and onto interior storage elements.

- **Location**:
  Clerestories should be placed in front of a Direct Gain storage wall at a dis-

Table 3.1

Average Winter Outdoor Temperature (Clear Day)*	Glazing/Floor Area**			
	36°NL	40°NL	44°NL	48°NL
COLD CLIMATES				
20°F	.24	.25	.29	.31 (w/NI)
25°F	22	.23	.25	.28 (w/NI)
30°F	.19	.20	.22	.24
TEMPERATE CLIMATES				
35°F	.16	.17	.19	.21
40°F	.13	.14	.16	.17
45°F	.10	.11	.12	.13

Note: NI indicates Night Insulation.

* Temperatures listed are for December and January, usually the coldest months.

** These ratios apply to a well insulated space with a heat loss of 8 btu/degree day/square foot of floor area/°F. If space heat loss is more or less than this figure, adjust the ratios accordingly.

tance of roughly 1.0 to 1.5 times the height of the wall (see Figure 3.7).

- Sawtooth Clerestories:
  The angle of a sawtooth clerestory (as measured from the horizontal) should be equal to, or less than, the altitude of the sun at noon on December 2l, the winter solstice. Angle A can be determined from the following formula: A = 66.5° − latitude. For example, at latitude 36°, angle A will equal 66.5° − 36° or 30.5° (see Figure 3.8).

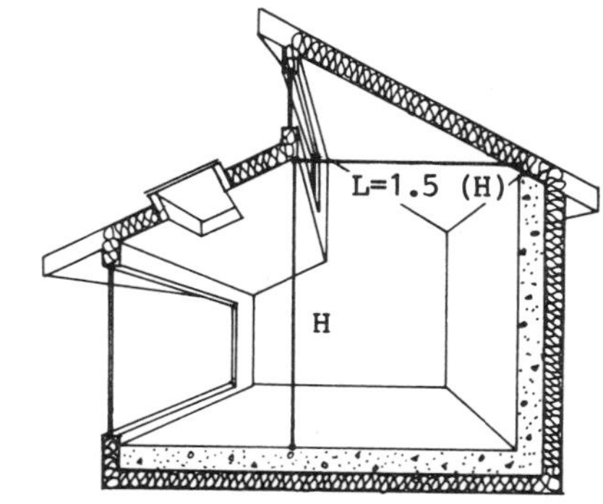

Figure 3.7

Skylight

- Reflector:
  A reflector is recommended for horizontal skylights to increase winter solar gain (see Control later in this chapter).

Storage

In Direct Gain systems, solar energy can be stored in the floor, walls, ceiling, and/or furnishings of the living space if these components have sufficient capac-

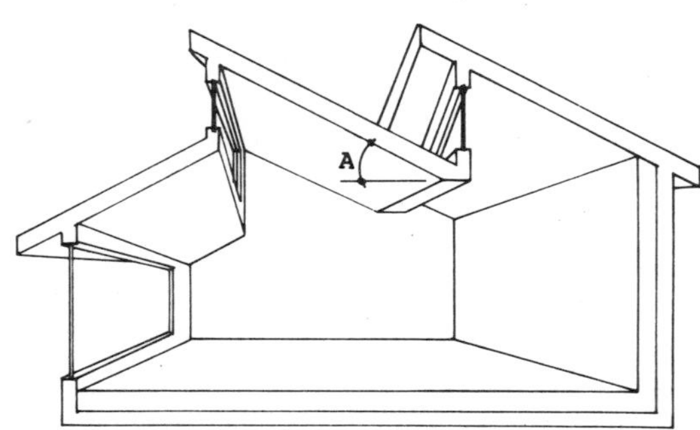

Figure 3.8

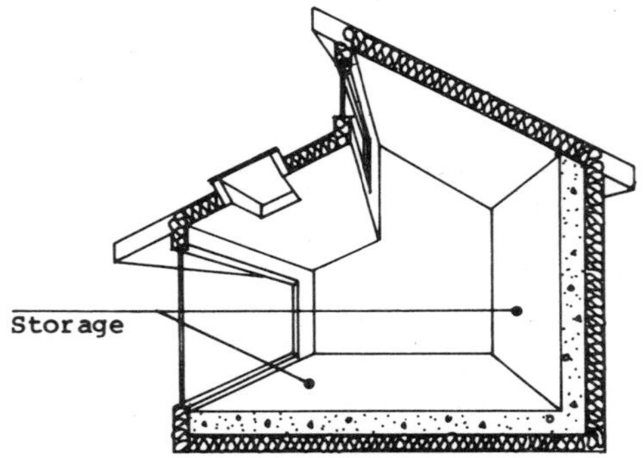

Figure 3.9

ity to absorb and store heat and to reradiate this heat to the living space slowly (see Figure 3.9). Most common high-mass materials (concrete, brick, and water) and phase-change materials (PCMs—materials such as salt or wax that store thermal energy when they melt and release heat when they solidify) have this capability and can be used effectively in Direct Gain applications.

The amount of distributed mass required for effective performance will vary depending upon the requirements of each specific installation. In general, within the thickness ranges covered in the Storage Rules of Thumb, thin material spread over a large area will perform better than thick material concentrated in one part of the space. Care should be taken to ensure a balanced distribution of mass throughout the Direct Gain space.

Storage materials should be located so that they are struck by direct or diffuse sunlight for as much of the day as possible. Storage mass that is heated only indirectly by warm air from the living space requires roughly four times as much area as the same mass in direct sun to provide the same thermal effect. Storage mass that *is* located in direct sunlight will perform slightly better in a vertical (wall) rather than in a horizontal (floor, ceiling) configuration.

If masonry pavers are used in a storage floor, they should be laid so that no voids exist between adjacent units. Good thermal contact should be maintained (by grouting or actual physical contact) between all units to ensure efficient transfer and even distribution of heat.

Storage floors, walls, and ceilings in direct sunlight should be dark in color to increase the absorption of solar energy. All adjacent nonmass surfaces should be light colored in order to reflect as much solar radiation as possible onto the storage surfaces.

If walls and floors are both used for storage, walls can be light if the floor is dark. Interior finishes over the storage materials (other than dark paint) will reduce the ability of the mass to absorb and reradiate heat. Plaster and wallpaper will reduce performance only slightly, and gypsum board can be acceptable (although it is not recommended) if care is taken during installation to apply a very thick coat of construction adhesive in order to increase the thermal connection between the wall and the gypsum board. Wall-to-wall carpet over a storage floor is emphatically *not* recommended.

Storage Rules of Thumb

- **Area:**
  It is recommended that for every square foot of south-facing glazing area, 3 sq ft of storage mass, designed to receive direct sun, be provided.

- **Thickness (Floor):**
  Recommended thickness of the storage floor material is 2-6'' (generally 4'').*

  Dimensions for water containers and phase-change materials can vary but are typically 6-18'' thick. If prefabricated units are used, consult the manufacturers' literature regarding these products.

- **Thickness (Wall):**
  Recommended thickness of storage wall materials is 2-4''.*

  Dimensions for water containers and phase-change materials can vary significantly. Consult the manufacturers' literature regarding these products.

- **Color:**
  Storage mass exposed to direct sunlight should be a dark color.

- **Masonry Grouting:**
  Masonry units should be solid or have grout-filled cores. A full mortar bedding is recommended.

*Dimensions cited are *actual*, not nominal. If restrictions on local availability of materials require using thicknesses outside the ranges listed, it is generally more thermally efficient to employ thicker rather than thinner units. (At floor thicknesses greater than 4'', performance will increase, but not significantly. At thicknesses greater than 8'', performance can decrease.)

Control

Shading

To avoid excessive heat gain in the cooling season and to increase overall system performance, it is recommended that some method be provided for shading the Direct Gain collector. The most thermally effective shading devices are those placed on the exterior of the home, such as simple overhangs (fixed or adjustable), trellises, vegetation (deciduous trees, etc.), awnings, louvers (horizontal or vertical, fixed or adjustable), and wing walls (see Figure 3.10).

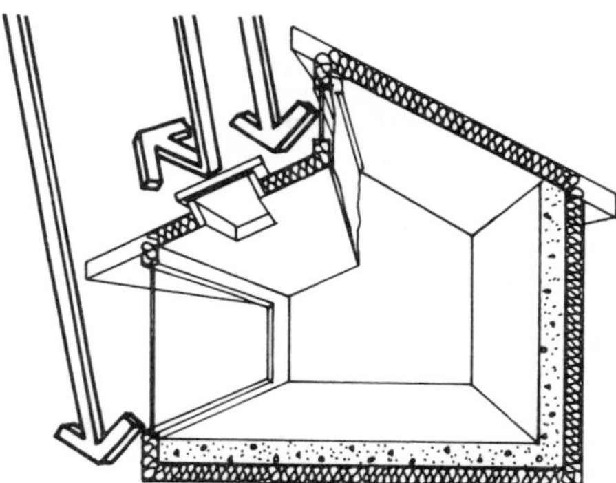

Figure 3.10

Interior shading devices, while often not as thermally effective as exterior units, are generally easier to operate and maintain. Common interior shading devices include roller shades, blinds, drapes, and movable panels. For optimum overall performance of the system, these interior shading elements should also be designed to provide insulation during the day in the cooling season and at night in the heating season (see Insulating—Movable Insulation later in this chapter).

Reflecting

Reflectors placed horizontally above or below a collector can help increase overall system performance by increasing the amount of sunlight reaching the collector (see Figure 3.11).

In certain cases where physical obstructions (e.g., trees or other buildings) on or around the building site shade the collector, the provision of reflectors can often increase solar collection 30-40%. Reflectors are also recommended for skylights placed in a flat roof that may not receive sufficient winter sunlight in the horizontal position.

Reflectors are usually panels, coated on one side with a material of high reflectance, that are placed directly in front of the vertical south-facing collector, or behind a horizontal collector (skylight). When the collector extends all the way to the ground (e.g., a patio door), the reflector is simply laid on the ground in front of it. In the case of collector openings higher up in the wall, or in cases where the reflector is placed above the collector, some form of support will be needed. Reflectors for horizontal skylights can be placed vertically on the roof on the north side of the skylight.

All horizontal panels should be placed so that they slope slightly away from a vertical collector to increase the amount of reflected sunlight and to facilitate drainage (5° is recommended). To be economically and/or aesthetically justifiable they should also be insulated so that they can serve as movable insulation when not in the reflecting mode. It should be noted that reflecting panels may cause glare and/or overheating problems within the Direct Gain living spaces. Light-colored exterior landscape elements, such as patios or terraces, can also serve as reflectors. They will not perform as efficiently as panels with high reflectance, but they will reduce the possibility of glare with overheating (see Figure 3.12).

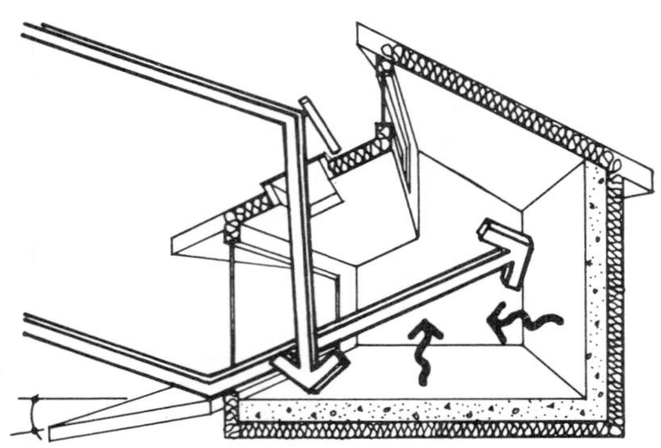

Figure 3.11

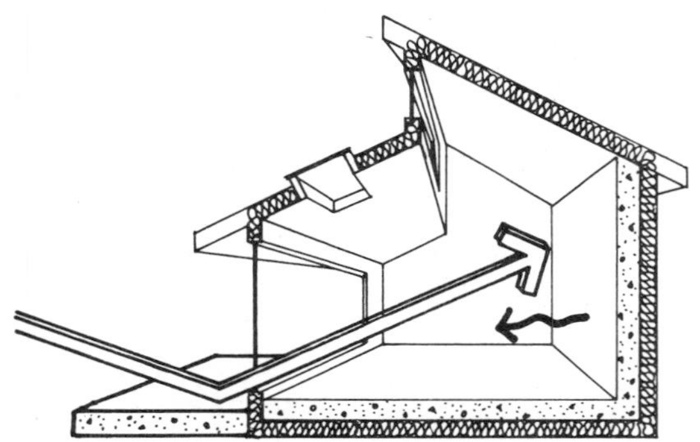

Figure 3.12

Insulating—Fixed Exterior Insulation

Direct Gain storage walls and floors that are exposed to the outside should be insulated on their *exterior* surfaces. Insulating the interior surface of a Direct Gain storage wall effectively nullifies any thermal storage capability of the wall, since it prevents solar energy from being absorbed by the wall and then reradiated back into the Direct Gain space. Therefore, insulation should be placed on the outside of any exterior wall, above and below grade, that is used for Direct Gain storage (see Figure 3.13).

The case of Direct Gain storage floors is a bit more complicated. Storage floors supported by wood frame construction over vented crawl spaces *should* be insulated, and standard batt insulation will work well. The effectiveness of insulating under a slab-on-grade will depend on factors such as the moisture content and minimum winter temperatures of the ground under the slab. If the moisture content is high or the minimum temperature low (less than 50°F), the tendency for heat to be lost through the floor to the ground will increase. In these instances, insulating under the slab, for a minimum of 2′ along the entire perimeter and preferably under the entire slab, is recommended. Two inches of rigid insulation (R-11) is appropriate for Direct Gain slabs-on-grade.

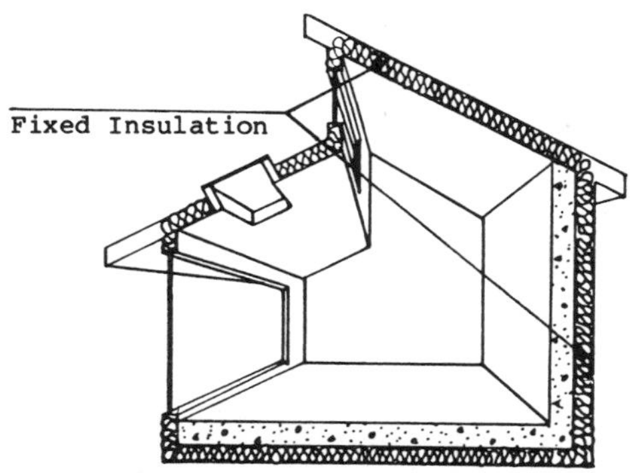

Figure 3.13

Insulating—Movable Insulation

Although Direct Gain collectors can admit and trap a great deal of solar energy during clear, sunny days, they can also lose a great deal of heat during prolonged overcast periods and at night. Providing some form of movable insulation for use during these periods of heat loss can result in very significant increases in overall thermal performance. In fact, in the more severe climates, certain Direct Gain designs may be net energy losers if movable insulation is *not* provided. Movable insulation can also be used to prevent excessive heat gain during the cooling season.

As in the case of shading devices, there are two basic types of movable insulation: those applied to the outer face of the collector, and those applied on the inside. Both can be effective at reducing heat loss during the heating season and, when used like shades, at preventing excessive heat gain during the cooling season.

Insulation on the inside is generally easier to operate and maintain and is not subject to degradation due to weathering. Exterior insulation, on the other hand, is a more efficient shading device, reducing heat gain during the cooling season by intercepting the solar radiation before it can penetrate the collector and be converted to heat (see Figure 3.14).

There are a variety of possible insulating devices, including rigid panels (if placed on the exterior, these can also be used as reflectors—see Reflecting earlier in this chapter), insulating shutters, insulating drapes and shades, and even double glazing with insulation integrated into the airspace. These devices can be hand oper-

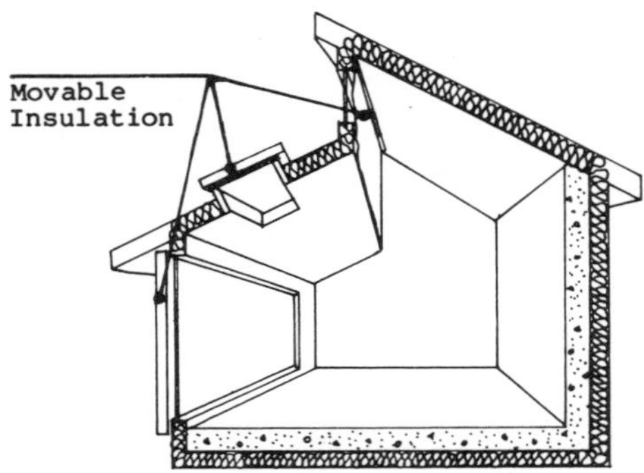

Figure 3.14

ated or motor driven. In choosing the appropriate technique, the willingness of the homeowner to become involved in the day-to-day management of these elements must be carefully weighed. No matter what technique or product is finally chosen, care must be taken to ensure a very tight seal between the insulation and the collector to avoid heat loss around the edges of the insulation.

Control Rules of Thumb

Shading

- **Overhang:**
  Generally, it is not cost-effective to construct a south-facing overhang with a projection exceeding 2′. Overhangs of this dimension can be built using standard truss construction. However, an overhang sized to provide a specified level of shading has the advantage of further reducing both the home's cooling energy requirements and potential glare in the cooling season.

  The projection of the overhang that will be adequate (that will provide 100% shading at noon on June 21) at particular latitudes can be quickly calculated by using the following formula:

$$\text{Projection (L)} = \frac{\text{window opening, i.e., collector (H)}}{\text{F (see table below)}}$$

North Latitude	F Factor*
28°	5.6-11.1
32°	4.0- 6.3
36°	3.0- 4.5
40°	2.5- 3.4
44°	2.0- 2.7
48°	1.7- 2.2
52°	1.5- 1.8
56°	1.3- 1.5

*Select a factor according to your latitude. The higher values will provide 100% shading at noon on June 21, the lower values on August 1.

A slightly longer overhang may be desirable at latitudes where this formula does not provide enough shade during the later part of the cooling season (see Figure 3.15).

For example, for Madison, Wisconsin, at 43°NL, select an F factor between 2.0 and 2.7 (values for 44°NL). Assuming shade is desired on August 1, choose the lower value of 2.0. If the window opening (H) is 8′, the projection (L) of the overhang placed at the top of the collector will be: $\frac{8}{2.0} = 4'$.

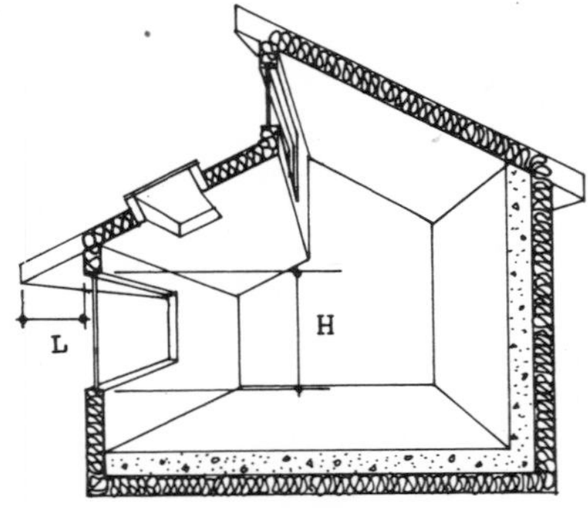

Figure 3.15

Reflecting

- **Sizing:**
  If reflectors are used, they should be sized according to the following:

  1. Solar windows and clerestories: same width as the collector and roughly 1 to 2 times the height.

  2. Skylights: roughly equal in size to the collector.

- **Slope:**
  A downward slope (away from the collector) of roughly 5° is recommended for drainage (see Figure 3.16).

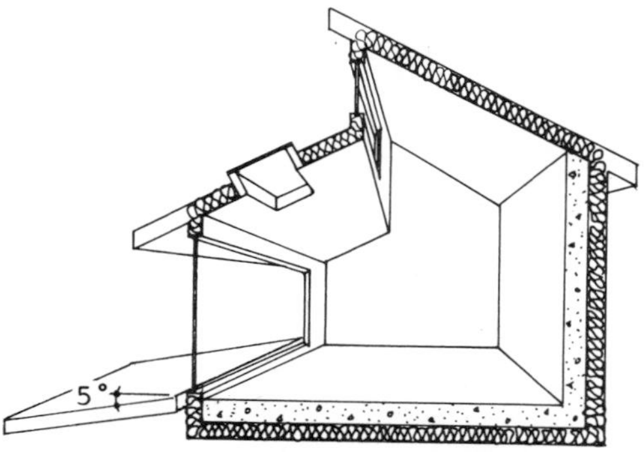

Figure 3.16

Insulating

- **R-Values (Fixed Insulation):**
  Fixed exterior insulation R-values should meet or exceed the levels that would be maintained by the same element in a nonsolar building, which should in turn meet or exceed local building-code requirements.

- **R-Values (Movable Insulation):**
  Movable insulation should have an R-value between R-4 and R-9.

- **Seals:**
  All insulation should provide a tight seal to avoid problems with heat loss at its edges.

Direct Gain Details

The drawings on the following pages illustrate common construction detailing for Direct Gain systems. The details are divided into four sections.

The first three sections illustrate construction options where an interior wall of the home is designed as the storage component. Each of the three individual sections focuses on one of the primary interior wall storage material options: masonry, water, and phase-change materials. The fourth section illustrates the construction of the storage component on an exterior wall. Details in this section illustrate the use of masonry for thermal storage.

The first page in each section consists of two to four individual details that together suggest one possible configuration for all the system components. A page of variations follows, highlighting alternate construction methods and the use of materials other than those suggested on the preceding page. The next page, common to all sections, contains the construction notes referenced from the details. The notes provide guidelines, troubleshooting tips, and other information useful in building a Direct Gain home.

In addition, the first section covering interior storage walls using masonry materials contains a page of collector variations. These options for detailing the collector and the associated construction notes can be used with any of the details in all four sections.

Further, at the end of both the first and fourth sections, component material options are illustrated. These constructions can be substituted for those shown in the sample details on the first page of the section.

DIRECT GAIN

Interior Wall

(Masonry)

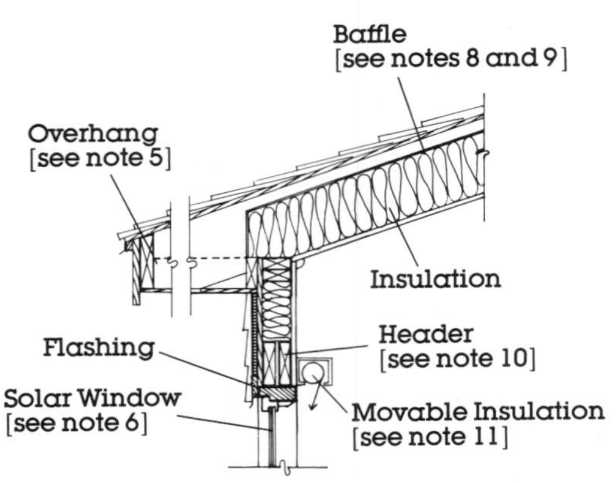

Solar Window at Roof

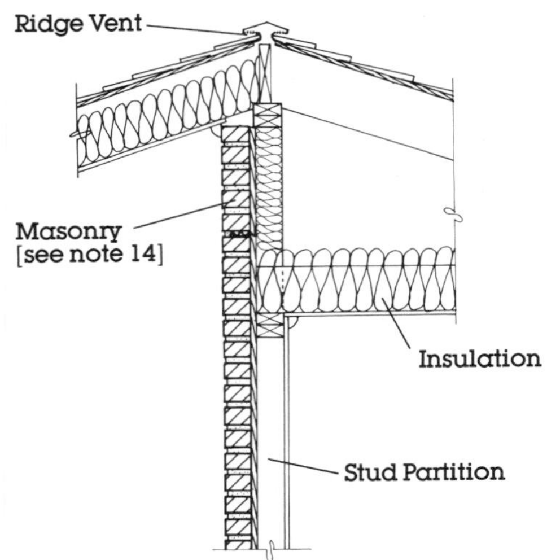

Masonry Veneer at Roof

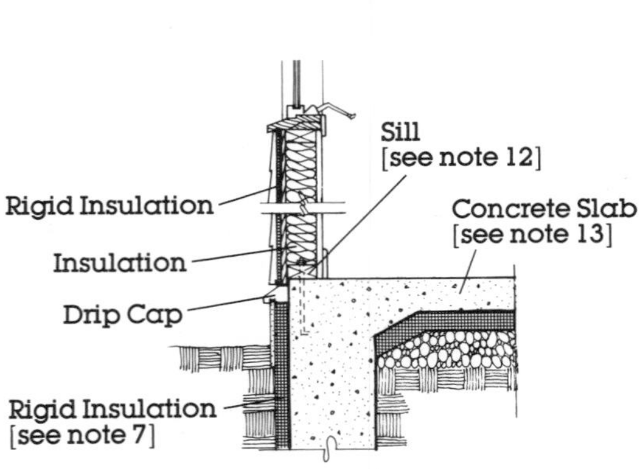

Solar Window at Foundation

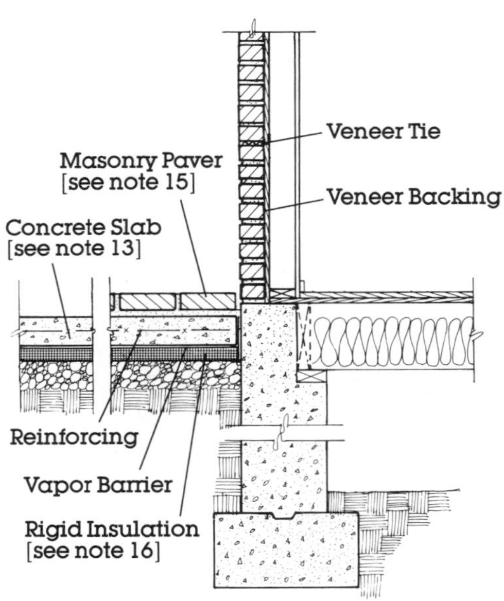

Masonry Veneer at Foundation

Interior Wall Variations
(Masonry)

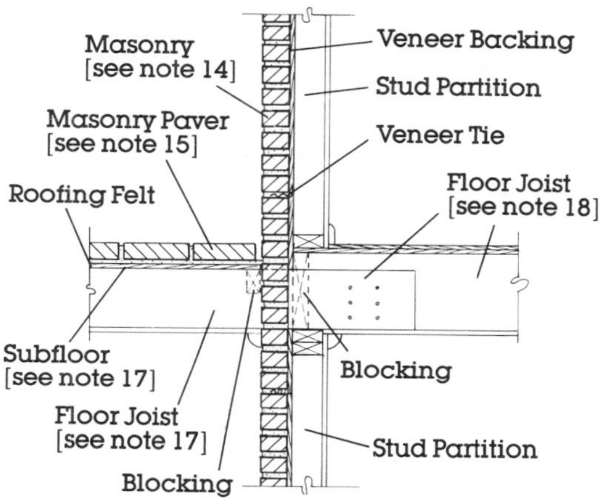

Masonry Veneer at Second Floor

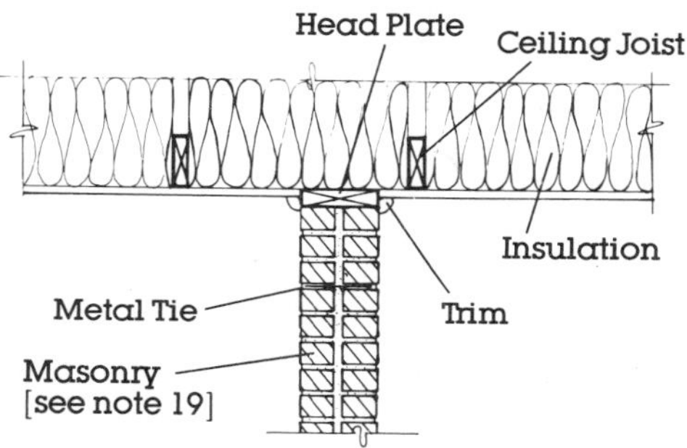

Solid Masonry at Roof

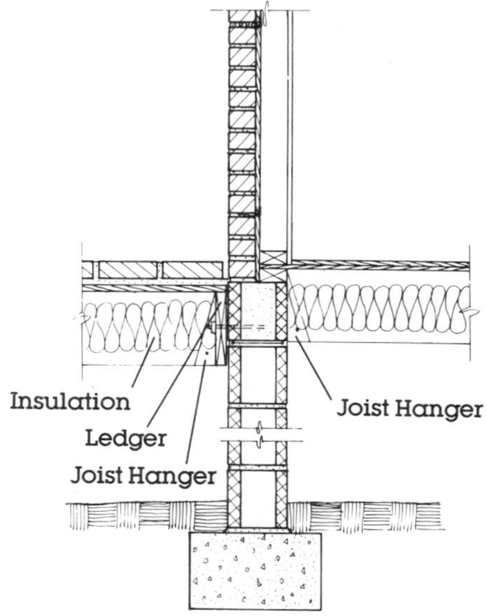

Masonry Veneer at First Floor

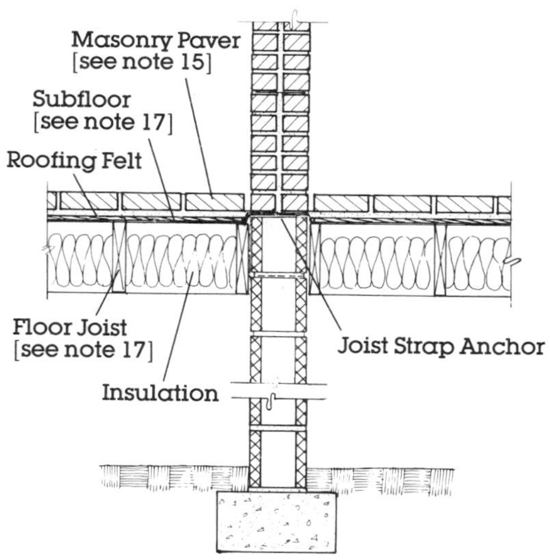

Solid Masonry at First Floor

Collector Variations

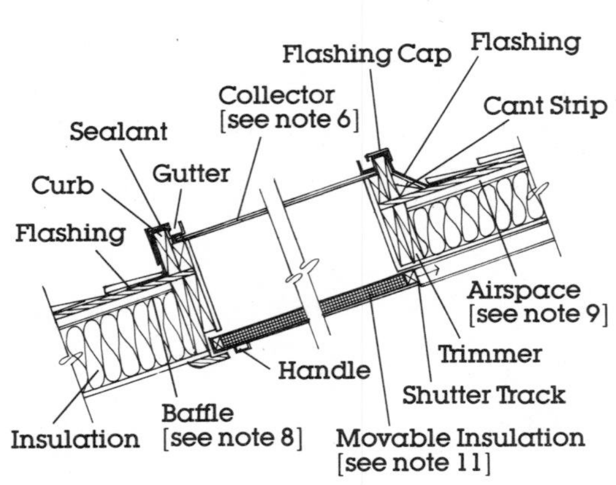

Skylight at Pitched Roof

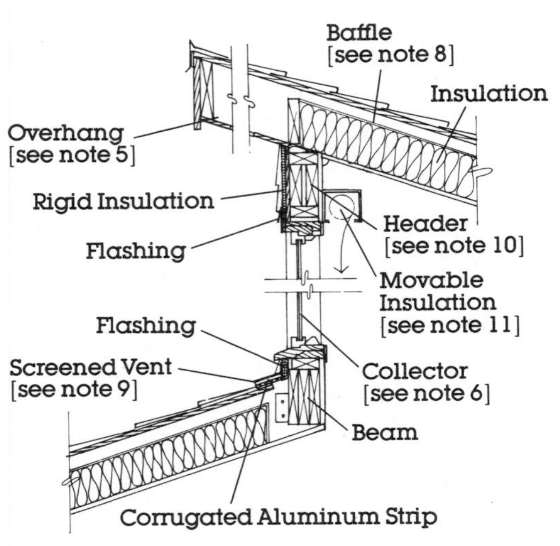

Clerestory at Pitched Roof

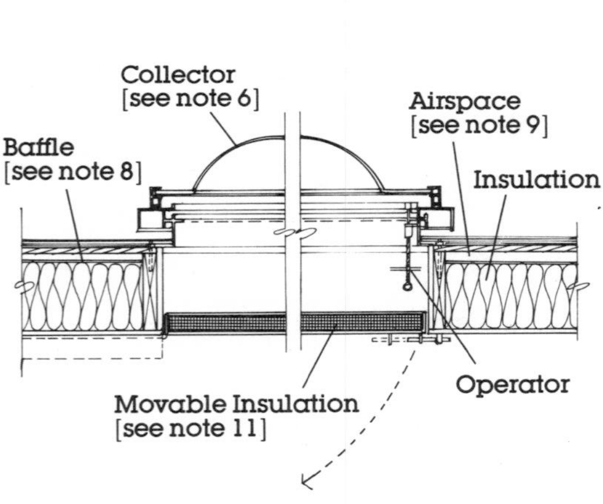

Skylight at Flat Roof

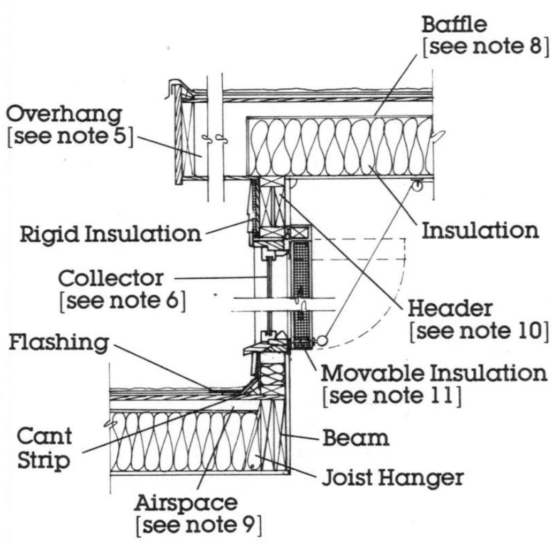

Clerestory at Flat Roof

Interior Wall Notes

(Masonry)

GENERAL NOTES:

1. All footings must bear on undisturbed soil. Adjust footing size, reinforcing, and depth below grade as required by site conditions and/or local building-code restrictions. Placement of all reinforcing, metal ties, and anchor bolts is to be determined by local codes and/or accepted practice. Reinforcing in concrete slab-on-grade construction is not recommended except where required by local conditions and/or code. Expansion joints should be provided, as required, at the connection between concrete or masonry floors, and stud, concrete, or masonry walls to prevent cracking.

2. Insulation levels in foundation/floor, walls, and ceiling must meet or exceed requirements of local building/energy codes. Floor/foundation insulation may be provided between floor joists over crawl space or unheated basement, or on foundation/basement walls and/or under slab.

3. Other roof configurations may be used at builder's option. Consult local building codes for restrictions and special requirements concerning roof/wall connections, especially in seismic and/or high-wind areas.

4. Provide gutters and downspouts as required.

DETAIL NOTES:

5. Overhang at collector should be sized to optimize solar gains during heating season and prevent overheating during cooling season. For further information, see Control Rules of Thumb in this chapter.

6. For optimum thermal performance, it is generally recommended that double glazing be provided. Glass or plastic material may, however, be specified (see Collector Components in Chapter 7). In either case, a ½'' airspace is recommended between glazing layers where feasible. Collectors can be standard preframed wood or metal frame windows, patio doors, prefabricated "bubble" skylights, any of a wide variety of off-the-shelf glazing units currently available, or site-built custom units. Where metal frames are specified, it is generally recommended that only those fabricated with thermal breaks be specified. All installations of prefabricated configurations must be per manufacturers' specifications to ensure proper performance.

7. Insulation must be protected where exposed above grade. Cement plaster over wire lath or other methods may be employed. Where dampproofing is used, allow for complete curing before applying insulation. Consult insulation manufacturer for recommended installation methods.

8. Provide baffles where necessary to maintain 1'' minimum airspace for venting attic.

9. The opaque roof must be vented to ensure proper performance of the insulation. Vents may be located at the

sides of the roof or at the ridge, depending upon the roof construction. An airspace is provided to allow for circulation.

10. Provide insulation at header. Rigid insulation may be specified, or other insulation may be used at builder's option.

11. Movable insulation should be provided to improve the system's thermal performance. Insulating shades, drapes, and shutters may be specified for interior applications. Insulated shutters or weather-resistant panels may be specified for exterior use. Such exterior insulating devices may be coated with a reflective surface so that when open they act as reflectors, increasing the effective collector area. Care must be taken to ensure a tight fit between any insulation and the collector to reduce heat loss at the edges. For further information, see Control Components in Chapter 7.

12. Continuous sill sealer is recommended to provide protection against infiltration.

13. Concrete slab-on-grade may be used as storage component. Alternate storage floor materials may be specified at the builder's option. For further information, see Interior Floor Types in this chapter.

14. Interior masonry veneer may be used for Direct Gain storage. For information on finishes, material variations, and thermal performance characteristics, see Interior Wall Types in this chapter and Storage Components in Chapter 7.

15. For information on masonry pavers used for Direct Gain storage, including recommended installation procedures, see Storage Components in Chapter 7.

16. In cooler climates, install continuous insulation under the slab used for Direct Gain storage. For further information, see Control in this chapter.

17. Consideration must be given to the additional weight of pavers and concrete used as Direct Gain storage. Consult structural design loading tables and local codes to determine suitable grade and spacing of joists, and grade and thickness of subfloor.

18. Joists are framed through continuous masonry veneer wall into structural framing. Joists are lapped and nailed together and bear on the double plate of the stud partition below. The joists are unequal in depth to maintain the floor level.

19. Solid masonry interior partition used for Direct Gain storage wall is perpendicular to south glazing (collector). For information on finishes, material variations, and thermal performance characteristics, see Storage Components in Chapter 7.

DIRECT GAIN

Interior Wall Types

(Masonry)

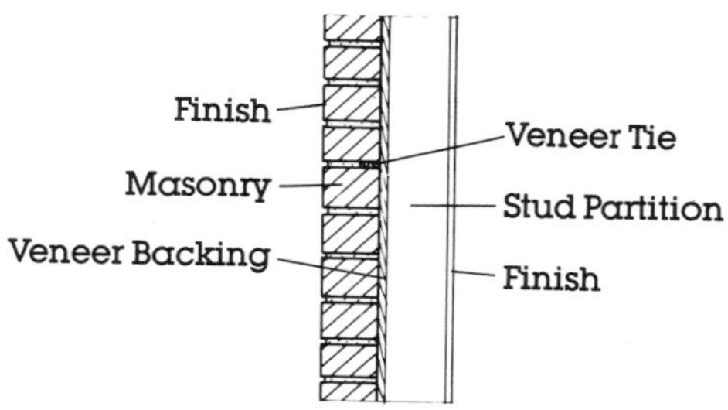

Masonry Veneer

GENERAL NOTES

- Veneer may be attached to the frame wall using either anchored or adhered construction methods. In anchored construction (as shown), metal ties are used to attach the masonry to a veneer backing (e.g., plywood or building paper) or directly to the frame wall studs. In adhered construction, the masonry is secured to a veneer backing material, using a binding material such as mortar. Units specified for adhered construction are generally limited to a maximum weight of 15 lb/ft^2 of surface area.

FINISH

- For optimum thermal performance, masonry should be a dark color and should be left exposed wherever feasible. Certain finishes (e.g., plaster and gypsum board) can be acceptable if properly applied. Particular colors can also be specified from the manufacturer, or the masonry can be stained or painted. For further information on finishes and their effect on thermal performance, see Storage in this chapter and Storage Components in Chapter 7.
- Wall finish on the nonveneer side of wall construction has no effect on thermal performance and may be specified per builder's option.

MASONRY

- Any of the wide variety of standard brick, concrete block, concrete brick, or split, slump, fluted, scored, or recessed block are suitable for thermal storage. To ensure proper thermal performance, masonry must be solid, or hollow with grout-filled cores. For further material information, see Storage Components in Chapter 7. For recommended thicknesses, see Storage Rules of Thumb in this chapter.

- Veneer is assumed to support no load other than its own weight.

VENEER BACKING

- In anchored veneer construction (as shown), veneer backing is not required, although some form of backing (e.g., plywood, gypsum board, or building paper) can make wall construction easier and may be desirable.

- In adhered veneer construction (not shown), some form of veneer backing is required. Plywood, gypsum board, metal lath over building paper, or other similar materials are suitable.

VENEER TIE

- Corrosion-resistant, corrugated metal ties, of a minimum 22-gauge thickness and ⅞'' width, are typically specified. Corrosion-resistant 8d nails, located at the bend in the tie and having a minimum penetration of 1½'' into stud, are recommended. The tie should be embedded a minimum of 2'' into the mortar joint at the veneer. Spacing of the ties is to be determined by local building-code requirements and/or standard practice.

STUD PARTITION

- The interior stud partition walls can be of standard wood or metal frame construction.

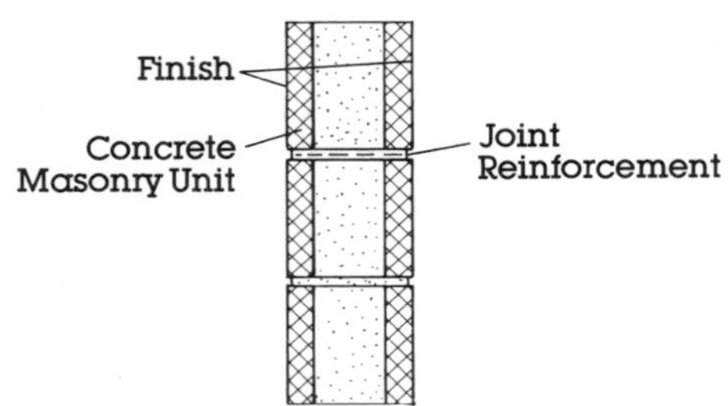

Concrete Masonry Unit

GENERAL NOTES

- Single-wythe masonry units may be built with conventional mortar joints.

- Openings in masonry walls may be spanned by arches, steel angle lintels, or reinforced masonry lintels constructed of standard and/or specially shaped masonry units, mortar, grout, and reinforcing steel. Specification for all three types should conform to local building-code requirements with regard to load, span, and type of wall construction.

FINISH

- For optimum thermal performance, masonry should be a dark color and should be left exposed wherever feasible. Certain finishes (e.g., plaster and gypsum board) can be acceptable if properly applied. Particular colors can also be specified from the manufacturer, or the masonry can be stained or painted. For further information on finishes and their effect on thermal performance, see Storage in this chapter and Storage Components in Chapter 7.

- Wall finish on nonveneer side of wall construction has no effect on thermal performance and may be specified per builder's option.

- For further information on finishes and their effect on thermal performance, see Storage in this chapter and Storage Components in Chapter 7.

CONCRETE MASONRY UNIT

- Concrete block, concrete brick, or architectural facing units, such as split, slump, fluted, scored, or recessed block, may be used for thermal storage. To ensure optimum thermal performance, units must be solid, or hollow with grout-filled cores. For further material information, see Storage Components in Chapter 7. For recommended thicknesses, see Storage Rules of Thumb in this chapter.

JOINT REINFORCEMENT

- Joint reinforcement should be provided to control cracking caused by thermal movement and shrinkage. Truss, ladder, or x-tie configurations are suitable and can be embedded in the wall at vertical intervals per local code requirements or industry recommendations. Splices in reinforcing should be lapped a minimum of 6'', and the width of the reinforcement should be 2'' less than the nominal width of the storage wall.

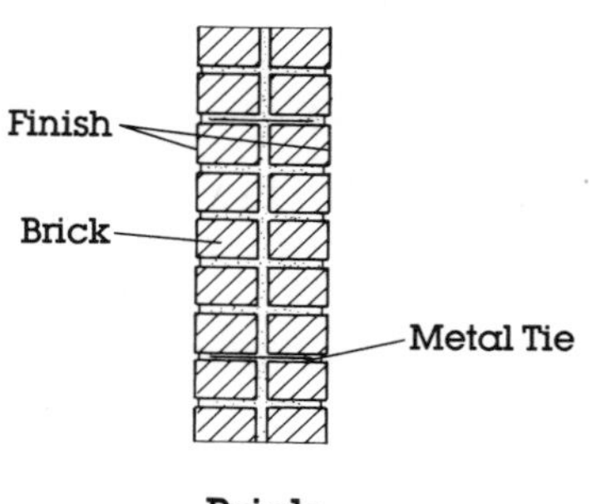

Brick

GENERAL NOTES

- Openings in brick walls may be spanned by arches, steel angle lintels, or reinforced brick lintels constructed of standard and/or specially shaped brick units, mortar, grout, and reinforcing steel. Specifications for all three should conform to the Brick Institute of America recommendations and local building-code requirements with regard to load, span, and all type of wall construction.

FINISH

- For optimum thermal performance, brick should be a dark color and should be left exposed wherever feasible. Certain finishes (e.g., plaster and gypsum board) can be acceptable if properly applied. For further information on finishes and their effect on thermal performance, see Storage in this chapter and Storage Components in Chapter 7.

BRICK

- Any of the wide variety of standard bricks are suitable for thermal storage. To ensure proper performance, bricks must be solid, or hollow with grout-filled cores. For further information, see Storage Components in Chapter 7. For recommended thicknesses, see Storage Rules of Thumb in this chapter.

DIRECT GAIN

Interior Floor Types

(Masonry)

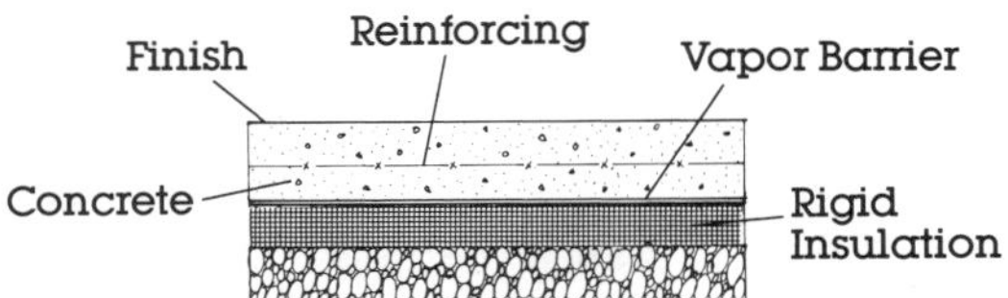

Concrete Slab-on-Grade

FINISH

- For optimum thermal performance, the concrete slab should be a dark color and should be left exposed wherever feasible. Certain finishes (e.g., exposed aggregate, embossed surface, and thin vinyl tile, ceramic tile, and quarry tile) will not significantly affect thermal performance and can be specified. Covering the concrete with carpet is emphatically *not* recommended. For further information on finishes and their effects on thermal performance, see Storage in this chapter and Storage Components in Chapter 7.

CONCRETE SLAB

- Concrete slab-on-grade construction can be used for Direct Gain storage. For material information see Storage Components in Chapter 7. For recommended thicknesses, see Storage Rules of Thumb in this chapter.

REINFORCING

- Reinforcing is generally not recommended except where required by soil conditions and/or local building codes.

VAPOR BARRIER

- A vapor barrier should be placed directly under the slab.

RIGID INSULATION

- In the cooler climates, it is generally recommended that rigid insulation be placed under slabs-on-grade that are acting as Direct Gain storage components. For further information, see Control in this chapter.

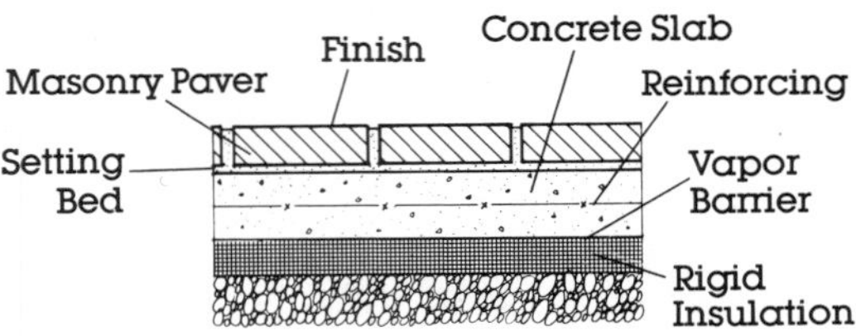

Masonry Paver on Slab

FINISH

- For optimum thermal performance, pavers should be a dark color and should be left exposed wherever feasible. Covering the pavers with carpet is emphatically *not* recommended. Sealers can be used to provide an impervious finish for the pavers, making them easier to clean, without affecting thermal performance. For further information on finishes and their effect on thermal performance, see Storage in this chapter and Storage Components in Chapter 7.

PAVER

- Any of the wide variety of standard brick and concrete masonry pavers are suitable for thermal storage. For material information, see Storage Components in Chapter 7. For recommended thicknesses, see Storage Rules of Thumb in this chapter.

SETTING BED

- Pavers are generally laid in a mortar setting bed. Type N mortar is suitable for interior applications.

- Pavers can also be set in a dry mixture of sand and cement. After the pavers are laid, the surface is wetted down with water to set the mixture. In this application, pavers are typically laid without mortar joints.

- Pavers can be laid without mortar. Care must be taken to maintain continuous contact between units to ensure optimum thermal performance.

CONCRETE SLAB

- Concrete slab-on-grade construction can be used for Direct Gain storage. For material information, see Storage Components in Chapter 7. For recommended thicknesses, see Storage Rules of Thumb in this chapter.

REINFORCING

- Reinforcing is generally not recommended except where required by soil conditions and/or local building codes.

VAPOR BARRIER

- A vapor barrier should be placed directly under the slab.

RIGID INSULATION

- In the cooler climates, it is generally recommended that rigid insulation be placed under slabs-on-grade that are acting as Direct Gain storage components. For further information, see Control in this chapter.

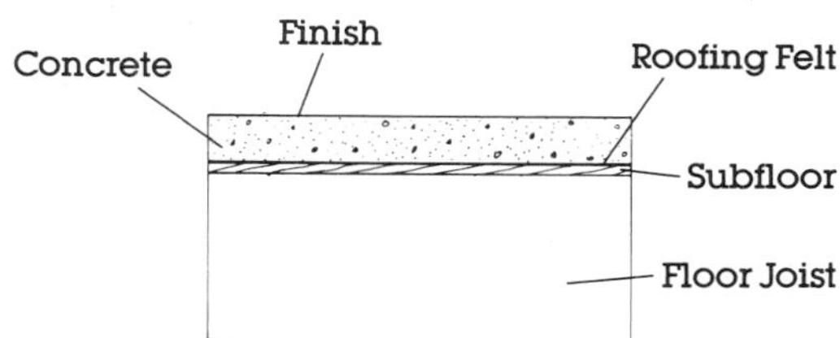

Lightweight Concrete on Wood Floor

FINISH

- For optimum thermal performance, the concrete slab should be a dark color and should be left exposed wherever feasible. Certain finishes (e.g., exposed aggregate, embossed surface, and thin vinyl tiles) will not significantly affect thermal performance and can be specified. Covering the concrete with carpet is emphatically *not* recommended. For further information on finishes and their effect on thermal performance, see Storage in this chapter and Storage Components in Chapter 7.

CONCRETE

- A structural lightweight concrete slab over floor joists can be used for Direct Gain storage. For further material information, see Storage Components in Chapter 7. For recommended thicknesses, see Storage Rules of Thumb in this chapter.

ROOFING FELT

- The installation of two layers of 15-lb roofing felt between the setting bed and the subfloor is recommended. The felt serves as a cushion for the masonry as well as a moisture barrier for the floor assembly. In mortarless applications, it aids in reducing the effects of minor surface irregularities in the base material.

- The felt also serves as a bond break between the setting bed and subfloor.

- Individual sections of roofing felt should be lapped 6''.

SUBFLOOR
- Consideration must be given to the additional weight of the masonry when selecting the subfloor.

FLOOR JOIST
- Consideration must be given to the additional weight of the masonry when selecting and sizing the floor joists.

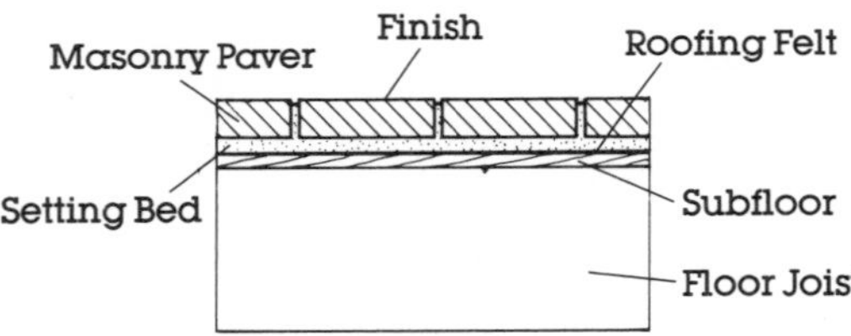

Masonry Paver on Wood Floor

GENERAL NOTES
- The pavers and the setting bed act together as the thermal storage mass. The recommended overall thickness of the storage floor can be attained by increasing or decreasing the depth of the setting bed, depending on the thickness of the pavers specified. For recommended floor thickness, see Storage Rules of Thumb in this chapter.

FINISH
- For optimum thermal performance, pavers should be a dark color and should be left exposed wherever feasible. Covering the pavers with carpet is emphatically *not* recommended. Sealers can be used to provide an impervious finish for the pavers, making them easier to clean, without affecting thermal performance. For further information on finishes and their effect on thermal performance, see Storage in this chapter and Storage Components in Chapter 7.

PAVER
- Any of the wide variety of standard brick and concrete masonry pavers are suitable for thermal storage. For material information, see Storage Components in Chapter 7. For recommended thicknesses, see Storage Rules of Thumb in this chapter.

SETTING BED
- Pavers are generally laid in a mortar setting bed. Type N mortar is suitable for interior applications.

- Pavers can also be set in a dry mixture of sand and cement. After the pavers are laid, the surface is wetted down with water to set the mixture. In this application, pavers are typically laid without mortar joints.

- Pavers can be laid without mortar. Care must be taken to maintain continuous contact between units to ensure optimum thermal performance.

ROOFING FELT

- The installation of two layers of 15-lb roofing felt between the setting bed and the subfloor is recommended. The felt serves as a cushion for the assembly. In mortarless applications, it aids in reducing the effects of minor surface irregularities in the base material.

- The felt also serves as a bond break between the setting bed and subfloor.

- Individual sections of roofing felt should be lapped 6''.

SUBFLOOR

- Consideration must be given to the additional weight of the masonry when selecting the subfloor.

FLOOR JOIST

- Consideration must be given to the additional weight of the masonry when selecting and sizing the floor joists.

DIRECT GAIN

Interior Wall

(Water)

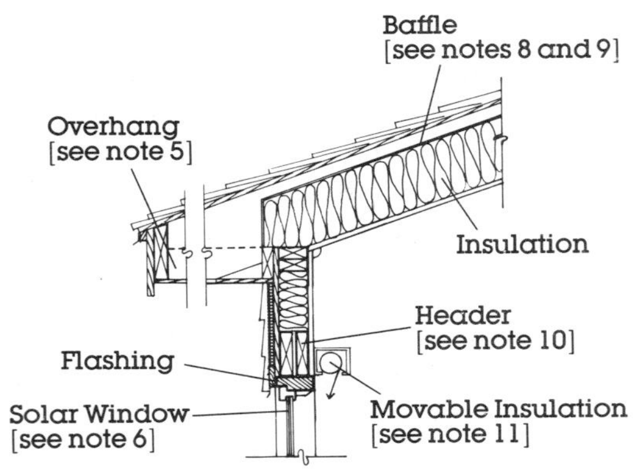

Solar Window at Roof

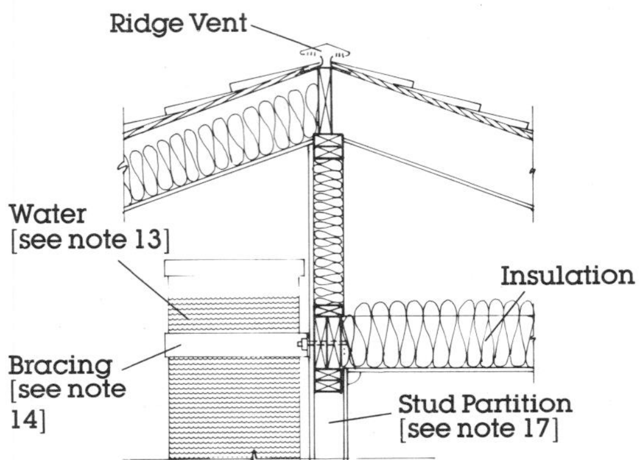

Water Tube at Roof

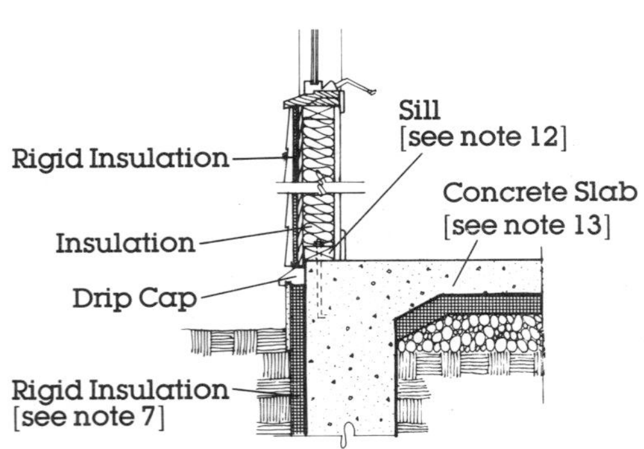

Solar Window at Foundation

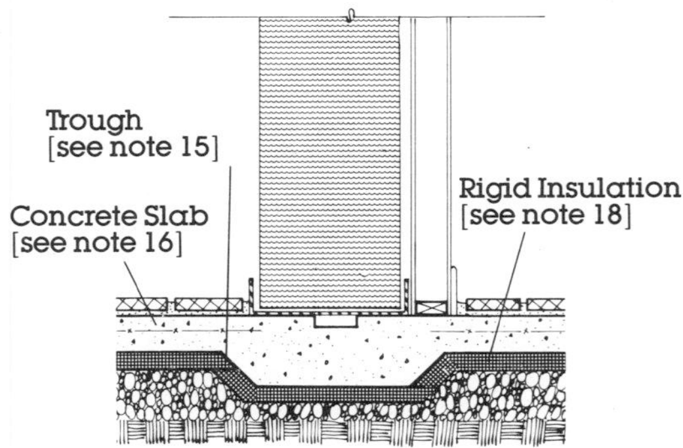

Water Tube at Foundation

Interior Wall Notes

(Water)

GENERAL NOTES:

1. All footings must bear on undisturbed soil. Adjust footing size, reinforcing, and depth below grade as required by site conditions and/or local building-code restrictions. Placement of all reinforcing, metal ties, and anchor bolts is to be determined by local codes and/ or accepted practice. Reinforcing in concrete slab-on-grade construction is not recommended except where required by local conditions and/or code. Expansion joints should be provided, as required, at the connection between concrete or masonry floors, and stud, concrete, or masonry walls to prevent cracking.

2. Insulation levels in foundation/floor, walls, and ceiling must meet or exceed requirements of local building/energy codes. Floor/foundation insulation may be provided between floor joists over crawl space or unheated basement, or on foundation/basement walls and/or under slab.

3. Other roof configurations may be used at builder's option. Consult local building codes for restrictions and special requirements concerning roof/wall connections, especially in seismic and/ or high-wind areas.

4. Provide gutters and downspouts as required.

DETAIL NOTES:

5. Overhang at collector should be sized to optimize solar gains during heating season and prevent overheating during cooling season. For further information, see Control Rules of Thumb in this chapter.

6. For optimum thermal performance, it is generally recommended that double glazing be provided. Glass or plastic material may, however, be specified (see Collector Components in Chapter 7). In either case, a ½'' airspace is recommended between glazing layers where feasible. Collectors can be standard preframed wood or metal frame windows, patio doors, prefabricated "bubble" skylights, any of a wide variety of off-the-shelf glazing units currently available, or site-built custom units. Where metal frames are specified, it is generally recommended that only those fabricated with thermal breaks be specified. All installations of prefabricated configurations must be per manufacturers' specifications to ensure proper performance.

7. Insulation must be protected where exposed above grade. Cement plaster over wire lath or other methods may be employed. Where dampproofing is used, allow for complete curing before applying insulation. Consult insulation manufacturer for recommended installation methods.

8. Provide baffles where necessary to maintain 1'' minimum airspace for venting attic.

9. The opaque roof must be vented to ensure proper performance of the insulation. Vents may be located at the sides of the roof or at the ridge, depend-

ing upon the roof construction. An airspace is provided to allow for circulation.

10. Provide insulation at header. Rigid insulation may be specified, or other insulation may be used at builder's option.

11. Movable insulation should be provided to improve the system's thermal performance. Insulating shades, drapes, and shutters may be specified for interior applications. Insulated shutters or weather-resistant panels may be specified for exterior use. Such exterior insulating devices may be coated with a reflective surface so that when open they act as reflectors, increasing the effective collector area. Care must be taken to ensure a tight fit between any insulation and the collector to reduce heat loss at the edges. For further information, see Control Components in Chapter 7.

12. Continuous sill sealer is recommended to provide protection against infiltration.

13. Water, stored in appropriate containers, is an effective material for thermal storage. Suitable containers include fiberglass tubes, polyethylene or steel drums, and metal culverts. Prefabricated tank systems designed to be set within or on structural wall framing are also available. Care should be taken to ensure that steel and metal containers are lined with corrosion-resistant materials. In all cases, water should be treated with algae-retardant chemicals. For further information on the use of water as thermal storage, including descriptions of appropriate containers, see Storage Components in Chapter 7.

14. It is recommended that all cylindrical containers that are located within or near a traffic or living area be braced at top and bottom against lateral movement. Taller containers should also be braced at midpoint. Bracing may be in the form of a framed valance, a broad steel band tied into a wall, or any other bracketing device.

15. Troughs should be provided as a precaution against leakage of water from containers or from condensation. Troughs may be formed of corrosion-resistant materials and attached or set into the supporting floor system so as to provide added lateral bracing at the base of the container. Troughs may be designed for a single container or a row of containers. It is generally recommended that each trough be provided with a slight slope of the base to a central drain that is tied to the main plumbing system of the building.

16. Concrete slab should be thickened to take the additional weight of water containers. Consult local codes for requirements.

17. Interior stud partition walls can be of standard construction. Wall finish may be specified per builder's option, but it is generally recommended that lighter colors be incorporated behind water storage, since they provide good reflective capabilities.

18. Rigid insulation should be provided to prevent heat losses from the water storage to the ground.

DIRECT GAIN

Interior

(PCM)

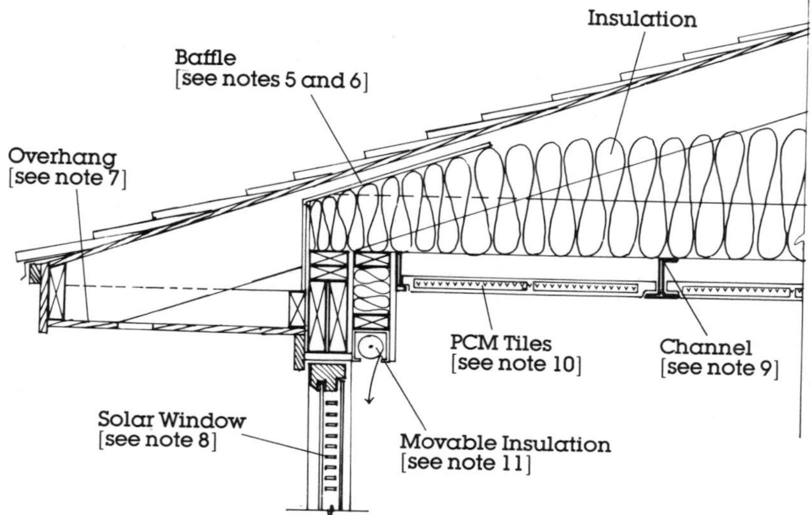

Phase-Change Materials at Ceiling

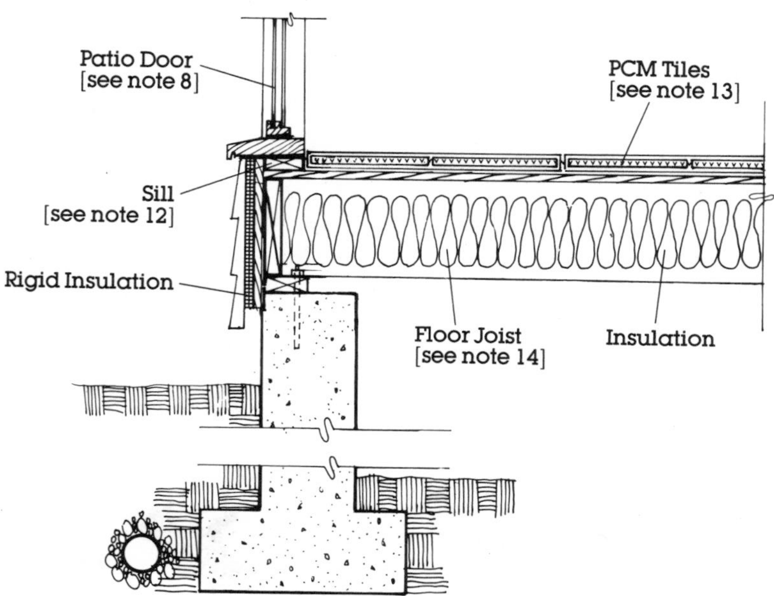

Phase-Change Materials at First Floor

Interior Notes

(PCM)

GENERAL NOTES:

1. All footings must bear on undisturbed soil. Adjust footing size, reinforcing, and depth below grade as required by site conditions and/or local building-code restrictions. Placement of all reinforcing, metal ties, and anchor bolts is to be determined by local codes and/or accepted practice. Reinforcing in concrete slab-on-grade construction is not recommended except where required by local conditions and/or code. Expansion joints should be provided, as required, at the connection between concrete or masonry floors, and stud, concrete, or masonry walls to prevent cracking.

2. Insulation levels in foundation/floor, walls, and ceiling must meet or exceed requirements of local building/energy codes. Floor/foundation insulation may be provided between floor joists over crawl space or unheated basement, or on foundation/basement walls and/or under slab.

3. Other roof configurations may be used at builder's option. Consult local building codes for restrictions and special requirements concerning roof/wall connections, especially in seismic and/or high-wind areas.

4. Provide gutters and downspouts as required.

5. Provide baffles where necessary to maintain 1'' minimum airspace for venting attic.

6. The opaque roof must be vented to ensure proper performance of the insulation. Vents may be located at the sides of the roof or at the ridge, depending upon the roof construction. An airspace is provided to allow for circulation.

DETAIL NOTES:

7. Overhang at collector should be sized to optimize solar gains during heating season and prevent overheating during cooling season. For further information, see Control Rules of Thumb in this chapter.

8. For optimum thermal performance, it is generally recommended that double glazing be provided. Glass or plastic material may, however, be specified (see Collector Components in Chapter 7). In either case, a ½'' airspace is recommended between glazing layers where feasible. Collectors can be standard preframed wood or metal frame windows, patio doors, prefabricated "bubble" skylights, any of a wide variety of off-the-shelf glazing units currently available, or site-built custom units. Specifically designed double-glazing units incorporating louvers within the airspace should be specified with thermal storage systems located in the ceiling. The concave louvers coated with a reflective surface direct sunlight to the storage material. Where metal frames are specified, it is generally recommended that only those fabricated with thermal breaks be specified. All installations of prefabricated configurations must be per manufacturers' specifications to ensure proper performance.

9. Tiles can be mounted in heavy-duty suspended ceiling systems, within light-weight steel framing channels screwed into ceiling joists, or other systems. Tiles weigh approximately 20–30% more than plaster, and consideration must be given to their weight when analyzing the method of suspension. Consult manufacturers' recommendations.

10. Ceiling tiles containing phase-change materials may be used for Direct Gain thermal storage, provided that they receive sunlight. For further information on finishes, material variations, and thermal performance characteristics, see Storage Components in Chapter 7.

11. Movable insulation should be provided to improve the system's thermal performance. Insulating shades, drapes, and shutters may be specified for interior applications. Insulated shutters or weather-resistant panels may be specified for exterior use. Such exterior insulating devices may be coated with a reflective surface so that when open they act as reflectors, increasing the effective collector area. Care must be taken to ensure a tight fit between any insulation and the collector to reduce heat loss at the edges. For further information, see Control Components in Chapter 7.

12. Continuous sill sealer is recommended to provide protection against infiltration.

13. Floor tiles of a reinforced concrete material containing phase-change material can be used for Direct Gain thermal storage. For further information on finishes, material variations, thermal performance on finishes, material variations, thermal performance characteristics, and installation procedures, see Storage Components in Chapter 7.

14. Consideration must be given to the additional weight of tiles used as Direct Gain storage. Consult structural design loading tables and local codes to determine suitable grade and spacing of joists, and grade and thickness of subfloor.

DIRECT GAIN

Exterior Walls

(Masonry)

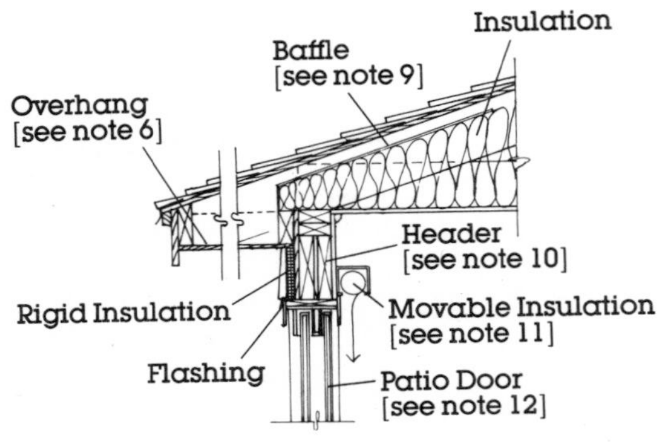

Patio Door at Roof

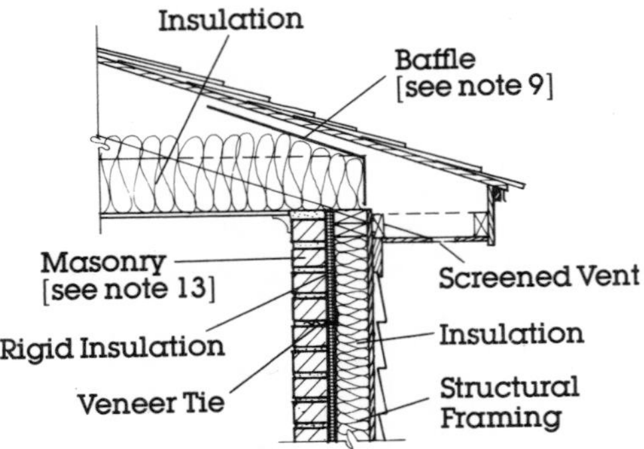

Masonry Veneer at Roof

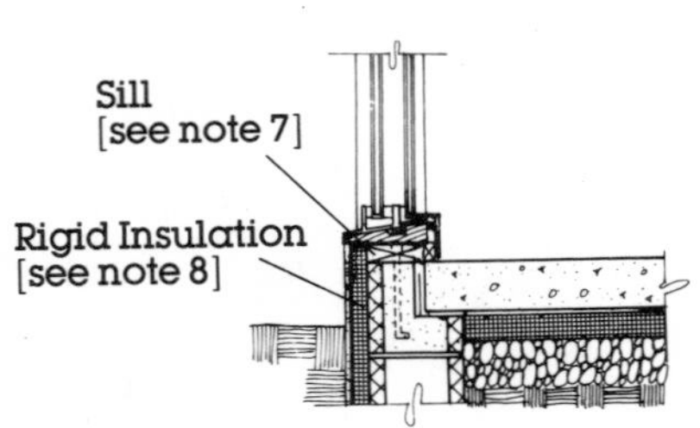

Patio Door at Foundation

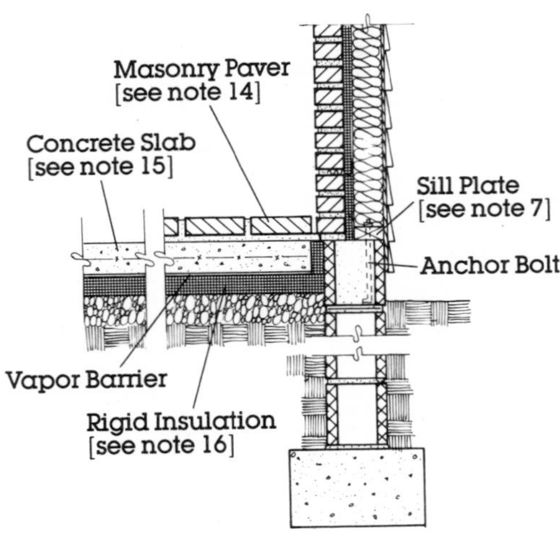

Masonry Veneer at Foundation

Exterior Wall Variations
(Masonry)

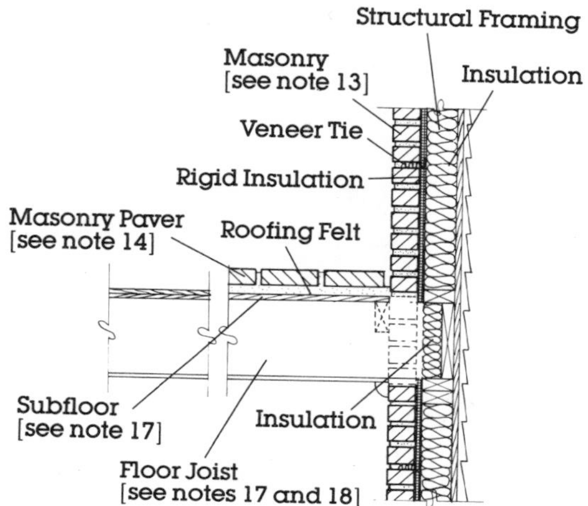

Masonry Veneer at Second Floor

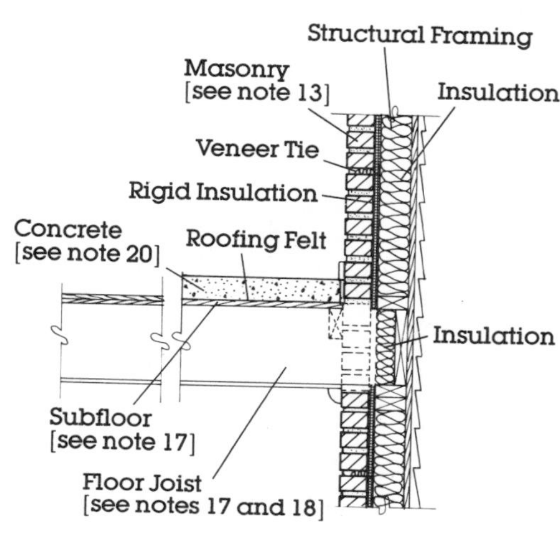

Masonry Veneer at Second Floor

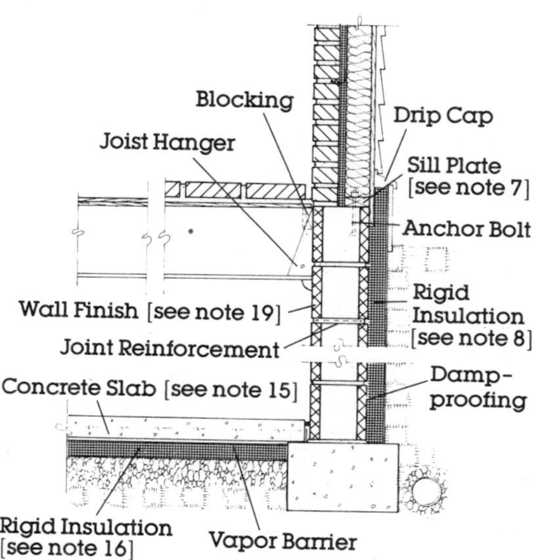

Masonry Veneer at First Floor

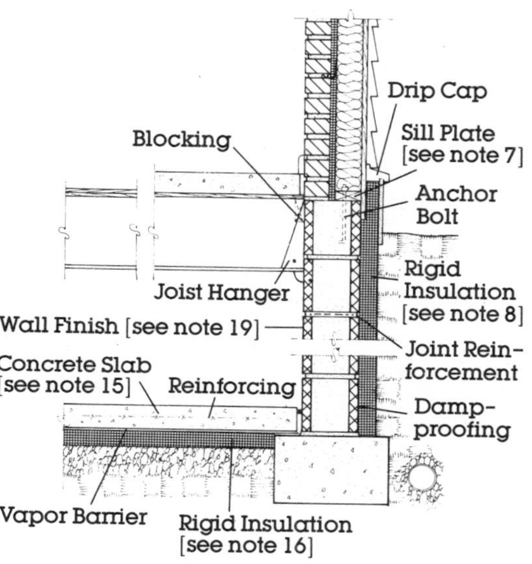

Masonry Veneer at First Floor

Exterior Wall Variations
(Masonry)

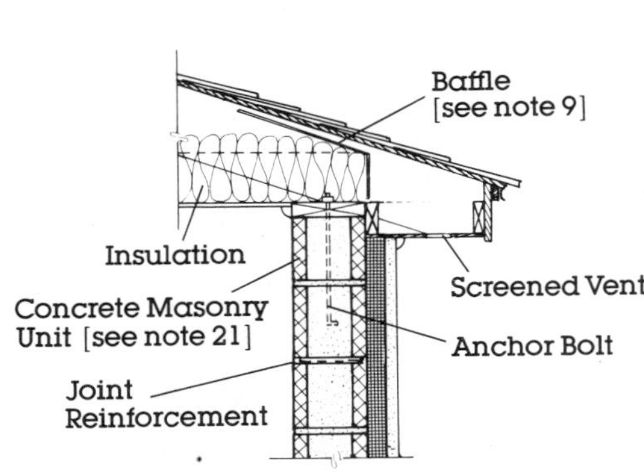

Solid Masonry at Roof

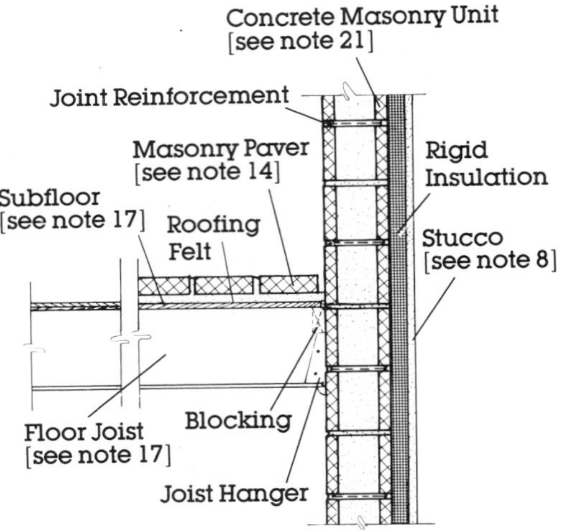

Solid Masonry at Second Floor

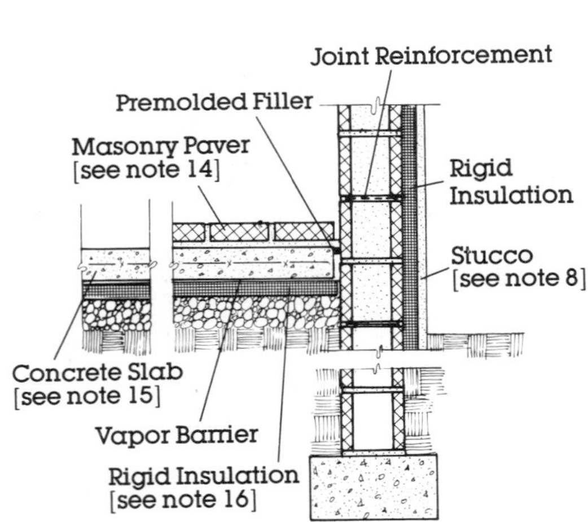

Solid Masonry at Foundation

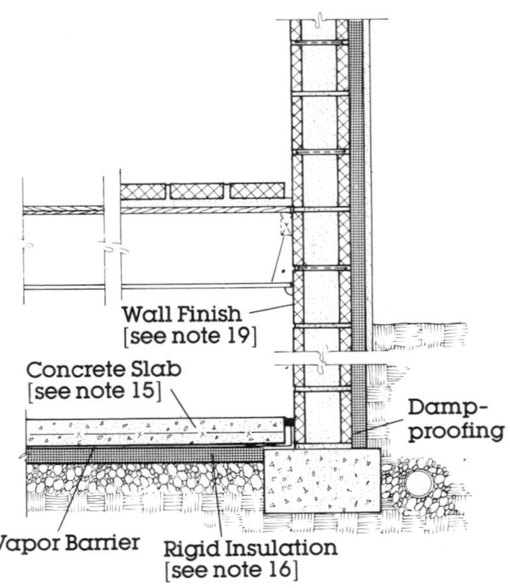

Solid Masonry at First Floor

Exterior Wall Notes

(Masonry)

GENERAL NOTES:

1. All footings must bear on undisturbed soil. Adjust footing size, reinforcing, and depth below grade as required by site conditions and/or local code. Where groundwater problems exist, provide sufficient granular fill to prevent water penetration at slab-on-grade. Reinforcing in concrete slab-on-grade construction is not recommended except where required by local conditions and/or code. Expansion joints should be provided, as required, at the connection between concrete or masonry floors, and stud or masonry walls to prevent cracking.

2. Sizing of headers is to be determined by local building codes and/or accepted practice.

3. Insulation levels at foundation, floors, walls, and ceiling must meet or exceed requirements of local building/ energy codes.

4. Other roof configurations may be used at the builder's option. Consult local building codes for restrictions and special requirements concerning roof/wall connections, especially in seismic and/or high-wind areas. The opaque roof must be vented to ensure proper performance of the insulation. Vents may be located at the sides of the roof or at the ridge, depending upon the roof construction. An airspace is provided to allow for circulation.

5. Provide gutters and downspouts as required.

DETAIL NOTES:

6. Overhang at collector should be sized to optimize solar gains during heating season and prevent overheating during cooling season. For further information, see Control Rules of Thumb in this chapter.

7. Continuous sill sealer is recommended to provide protection against infiltration.

8. Cement plaster (stucco finish applied on wire lath over exterior insulation) is recommended. Other finishes may be specified at builder's option, but rigid insulation must be protected where exposed above grade.

9. Provide baffles where necessary to maintain 1'' minimum airspace for venting attic.

10. Provide insulation at header. Rigid insulation may be used at builder's option.

11. Movable insulation should be provided to improve the system's thermal performance. Insulating shades, drapes, and shutters may be specified for interior applications. Insulated shutters or weather-resistant panels may be specified for exterior use. Such exterior insulating devices may be coated with a reflective surface so that when open they act as reflectors, increasing the effective collector area. Care must be taken to ensure a tight fit between any

insulation and the collector to reduce heat loss at the edges. For further information, see Control Components in Chapter 7.

12. For optimum thermal performance, it is generally recommended that double glazing be provided. Glass or plastic material may, however, be specified (see Collector Components in Chapter 7). In either case, a ½'' airspace is recommended between glazing layers where feasible. Collectors can be standard preframed wood or metal frame windows, patio doors, prefabricated "bubble" skylights, any of a wide variety of off-the-shelf glazing units currently available, or site-built custom units. Where metal frames are specified, it is generally recommended that only those fabricated with thermal breaks be specified. All installations of prefabricated configurations must be per manufacturers' specifications to ensure proper performance.

13. Interior masonry veneer is shown for the Direct Gain storage component. For information on finishes, material variations, and thermal performance characteristics, see Interior Wall Types in this chapter and Storage Components in Chapter 7.

14. For information on masonry pavers used for Direct Gain storage, including recommended installation procedures, see Interior Floor Types in this chapter and Storage Components in Chapter 7.

15. Concrete slab-on-grade is shown as storage component. For further information, see Interior Floor Types in this chapter and Storage Components in Chapter 7.

16. In cooler climates, install continuous insulation under the slab used for Direct Gain storage. For further information, see Control in this chapter.

17. Consideration must be given to the additional weight of pavers and concrete. Consult structural design loading tables and local codes to determine suitable grade and spacing of joists, and grade and thickness of subfloor.

18. Masonry veneer must be continuous. Floor joists are framed through veneer.

19. Below-grade wall may be used for Direct Gain storage if it is designed to be in direct sunlight. For information on interior finishes, see Interior Wall Types in this chapter and Storage Components in Chapter 7.

20. For information on concrete used as Direct Gain storage, including recommended installation procedures, see Storage Components in Chapter 7.

21. Single-wythe concrete masonry unit is shown for the Direct Gain storage component. For information on finishes, material variations, and thermal performance characteristics, see Storage Components in Chapter 7.

DIRECT GAIN

Exterior Walls

(Masonry)

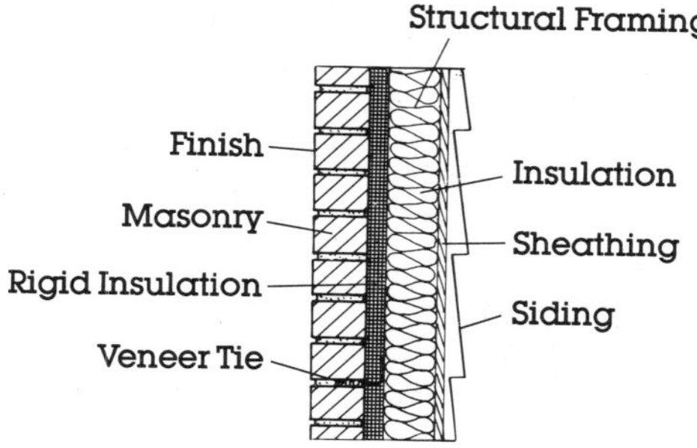

Masonry Veneer

FINISH

- For optimum thermal performance, masonry should be a dark color and should be left exposed wherever feasible. Certain finishes (e.g., plaster and gypsum board) can be acceptable if properly applied. Particular colors can also be specified from the manufacturer, or the masonry can be painted. For further information on finishes and their effect on thermal performance, see Storage in this chapter and Storage Components in Chapter 7.

MASONRY

- Any of the wide variety of standard brick, concrete block, concrete brick, and architectural facing units, such as split, slump, fluted, scored, or recessed block, are suitable for thermal storage. To ensure proper thermal performance, bricks must be solid, or hollow with grout-filled cores. For further material information, see Storage Components in Chapter 7. For recommended thicknesses, see Storage Rules of Thumb in this chapter.
- Veneer is assumed to support no load other than its own weight.

RIGID INSULATION

- Use of rigid insulation between masonry veneer and frame wall may be necessary to develop an acceptable R-value for the exterior wall.

VENEER TIE

- Corrosion-resistant, corrugated metal ties, of a minimum 22-gauge thickness and ⅞'' width, are typically specified. Corrosion-resistant 8d nails, located at the bend in the tie and having a minimum penetration of 1½'' into stud, are recommended. The tie should be embedded a minimum of 2'' into the mortar joint at the veneer. Spacing of the ties is to be determined by local building-code requirements and/or standard practice.

STRUCTURAL FRAMING

- The structural frame walls can be of standard wood or metal frame construction.

INSULATION

- Selection of insulation materials for frame walls is to be per local code requirements and/or standard practice.

SHEATHING

- Plywood, building board, or other similar materials can be specified as sheathing. The use of structural insulative sheathing, as a substitute, will reduce air infiltration and improve overall thermal performance if properly installed.

SIDING

- Material selection and color of siding is to be per builder's preference.

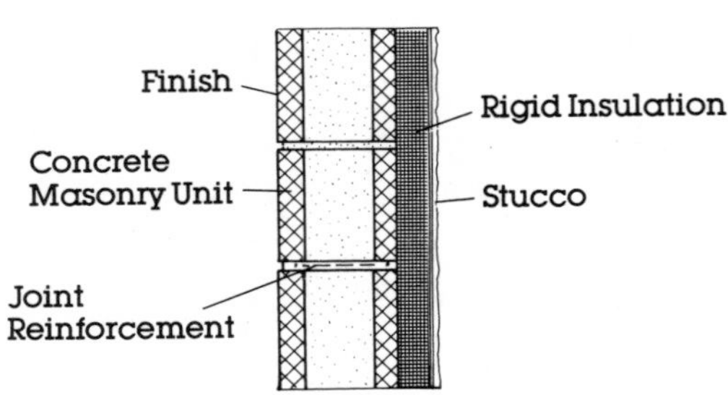

Solid Concrete Masonry

FINISH

- For optimum thermal performance, masonry should be a dark color and should be left exposed wherever feasible. Certain finishes (e.g., plaster and gypsum board) can be acceptable if properly applied. Particular colors can also be specified from the manufacturer, or the masonry can be stained or painted. For further information on finishes and their effect on thermal performance, see Storage in this chapter and Storage Components in Chapter 7.

CONCRETE MASONRY UNIT

- Concrete block, concrete brick, or architectural facing units, such as split, slump, fluted, scored, or recessed block may be used for thermal storage. To ensure optimum thermal performance, units must be solid, or hollow with grout-filled cores. For further material information, see Storage Components in Chapter 7. For recommended thicknesses, see Storage Rules of Thumb in this chapter.

JOINT REINFORCEMENT

- Joint reinforcement should be provided to control cracking by thermal movement and shrinkage. Truss, ladder, or x-tie configurations are suitable and can be embedded in the wall at vertical intervals per local code requirements or industry recommendations. Splices in reinforcing should be lapped a minimum of 6'', and the width of the reinforcement should be 2'' less than the nominal width of the storage wall.

RIGID INSULATION

- Rigid insulation may be applied directly to the masonry with adhesive (type and application per insulation manufacturers' recommendations and/or standard practice).

STUCCO

- Wire mesh or expanded metal lath for stucco finishes is laid over the rigid insulation and mechanically fastened to the concrete masonry. In general, the insulation surface must be free of mud and dust before stucco finish is applied.

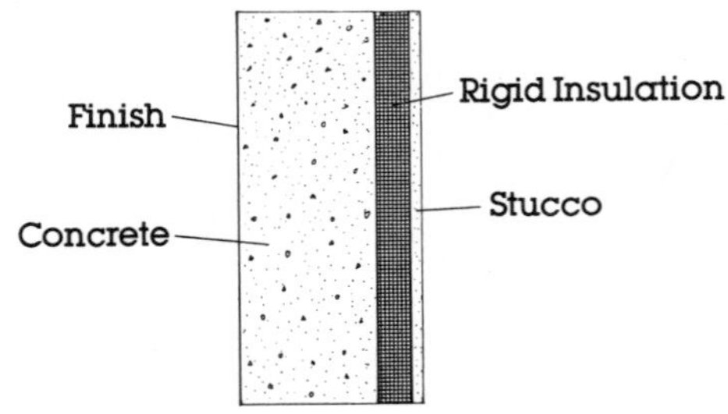

Solid Concrete

FINISH

- For optimum thermal performance, concrete should be a dark color and should be left exposed wherever feasible. Textured finishes can be obtained, using standard construction practice, without significant impact on thermal performance.

- Certain finishes (e.g., plaster and gypsum board) can be acceptable if properly applied. Particular colors can also be specified from the supplier, or the concrete can be stained or painted. For further information on finishes and their effect on thermal performance, see Storage in this chapter and Storage Components in Chapter 7.

CONCRETE

- The use of normal-weight concrete (140–150 lb) is generally recommended. Placement of formwork and pouring of concrete should follow standard practice and/or local building codes. For further information, see Storage Components in Chapter 7. For recommended material thicknesses, see Storage Rules of Thumb in this chapter.

RIGID INSULATION

- Rigid insulation may be applied directly to the masonry with adhesive (type and application per insulation manufacturers' recommendations and/or standard practice).

STUCCO

- Wire mesh or expanded metal lath for stucco finishes is laid over the rigid insulation and mechanically fastened to the concrete masonry. In general, the insulation surface must be free of mud and dust before stucco finish is applied.

TVA Solar House #10
Direct Gain

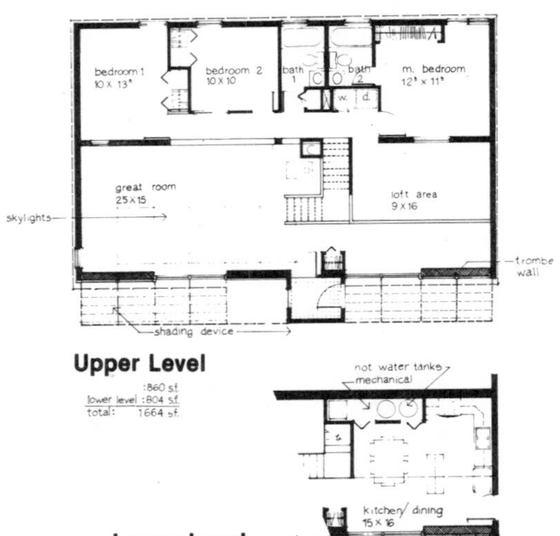

North

Integrated into a south-facing slope, this 1,660-sq-ft two-level house incorporates Direct Gain and Thermal Storage Wall systems for passive solar heating. The site is well suited to conserving energy because it requires minimal exposure of the home on the north, which reduces heat loss, and because a substantial building area is available for solar gain on the south.

Fixed double-glazed skylights illuminate an interior wall composed of grout-filled 12" masonry units. Ceiling-mounted rigid panels insulate the skylights. The panels sit in a track and operate on a counterweight system operated from the first floor.

Thermal Storage Wall panels of fully grouted 8" hollow-core concrete masonry units alternate with double Direct Gain glazing, combining the advantages of both system types. Exterior insulated vents located at the upper and lower edges of the collector frames ventilate the Thermal Storage Wall airspace, reducing heat gain during the cooling season.

A sunscreen spanning the south face of the house affords summer sun protection, while its plywood fins are angled to optimize winter solar heat gain. Designed as a trellis, the sunscreen frames the front porch located on both sides of the air-lock entry.

The hollow-core concrete block foundation is filled with loose-fill insulation, reducing ground heat losses. Additionally, 2" of rigid insulation for the foundation wall, R-30 batt in the ceiling, R-19 batt in the wall, and R-7 duct insulation wrap complete a thermally tight home.

Supplementary heating is provided by a woodburning stove. Solar collectors can be added to reduce the conventional fuel needed to meet domestic hot water heating requirements.

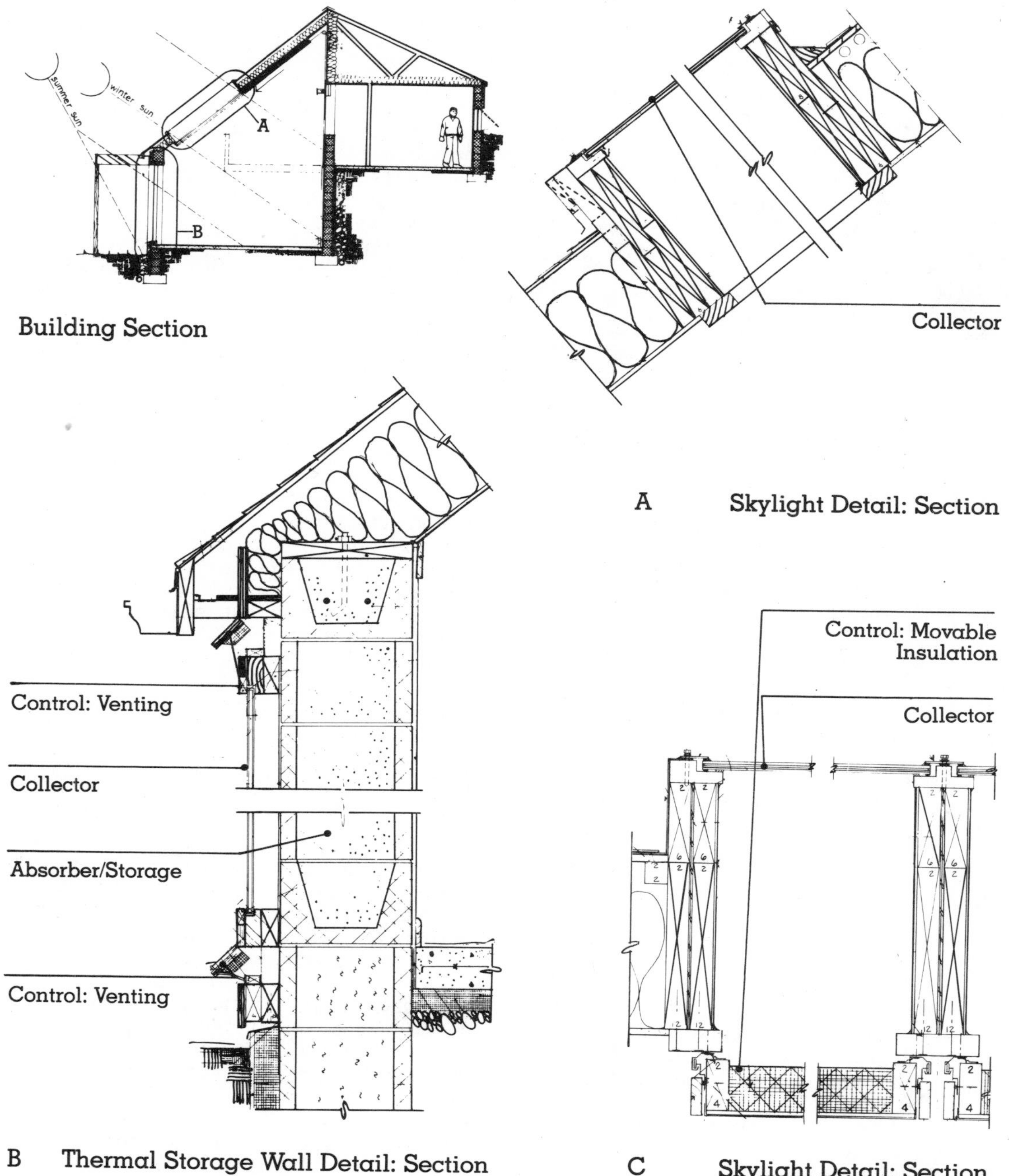

Building Section

A Skylight Detail: Section

B Thermal Storage Wall Detail: Section

C Skylight Detail: Section

TVA Solar House #11
Direct Gain

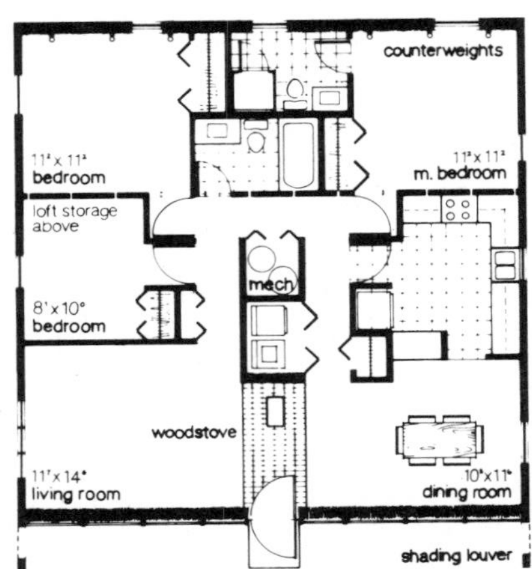

North

Main Floor Plan

This 1,200-sq-ft single-story gabled house provides passive solar heating through both Direct Gain and Thermal Storage Wall systems. Designed for flat or slightly sloping sites, its flexible square floor plan can be rotated and, with only minor changes, can be redesigned to place its passive features on any one of its four facades.

A continuous row of tempered glass skylights lines the upper edge of the south slope of the roof. At their interior, insulated shutters, which have a reflective interior skin and which roll in a track and are operated by a counterweight system, function in several modes to offer climate control. At night and on cloudy days in the heating season, they are closed to reduce heat losses; in the cooling season, they are closed during the day to control heat gain. The insulating panels consist of 3" of rigid insulation sandwiched between ¼" masonry boards. The panels are painted to match the interior walls.

A louvered roof overhang with angled slats positioned to allow the winter sun to strike the collector intercepts the summer sun, protecting the south-facing Thermal Storage Wall from overheating during summer. Its mortar-filled 12" heavyweight concrete masonry units have a cement plaster interior finish. Upper and lower exterior vents in the collector frames may be opened to cool the shaded wall and airspace during the summer.

In addition to warming the interior, the skylights allow sunlight to penetrate to the Direct Gain storage wall on the north side of the home. Its interior veneer of 4" solid concrete masonry units is coated with a dark cement plaster finish for greater heat absorption.

Operable exterior vents high in the east and west gables provide natural ventilation for summer cooling. Adjacent to the entry, a wood-burning stove whose chimney runs through an interior masonry wall offers auxiliary heat.

In addition to loose-fill insulation filling the 8" concrete masonry unit foundation wall, a 2" layer of rigid waterproof R-16 insulation improves the thermal resistance of the vertical foundation. Beneath the slab, another 2" of rigid waterproof R-16 insulation extends 4' in from the perimeter. Roof insulation is specified to a level of R-30.

B
insulating shade
jan 21
loft storage
over private space
A

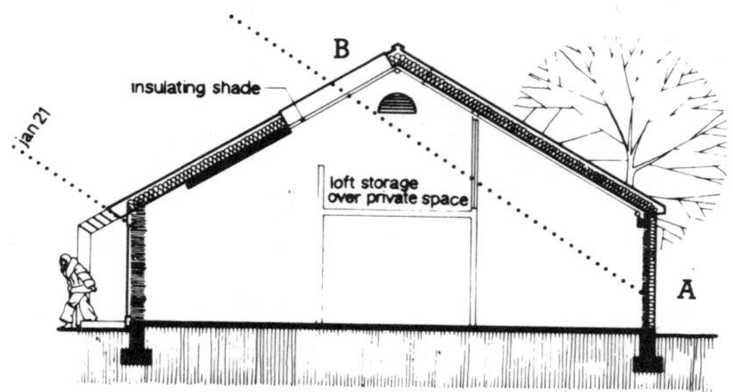

Building Section

Absorber/Storage

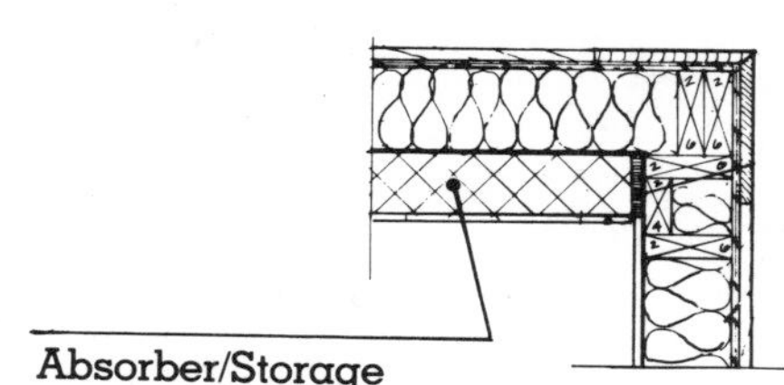

A Corner Detail: Plan

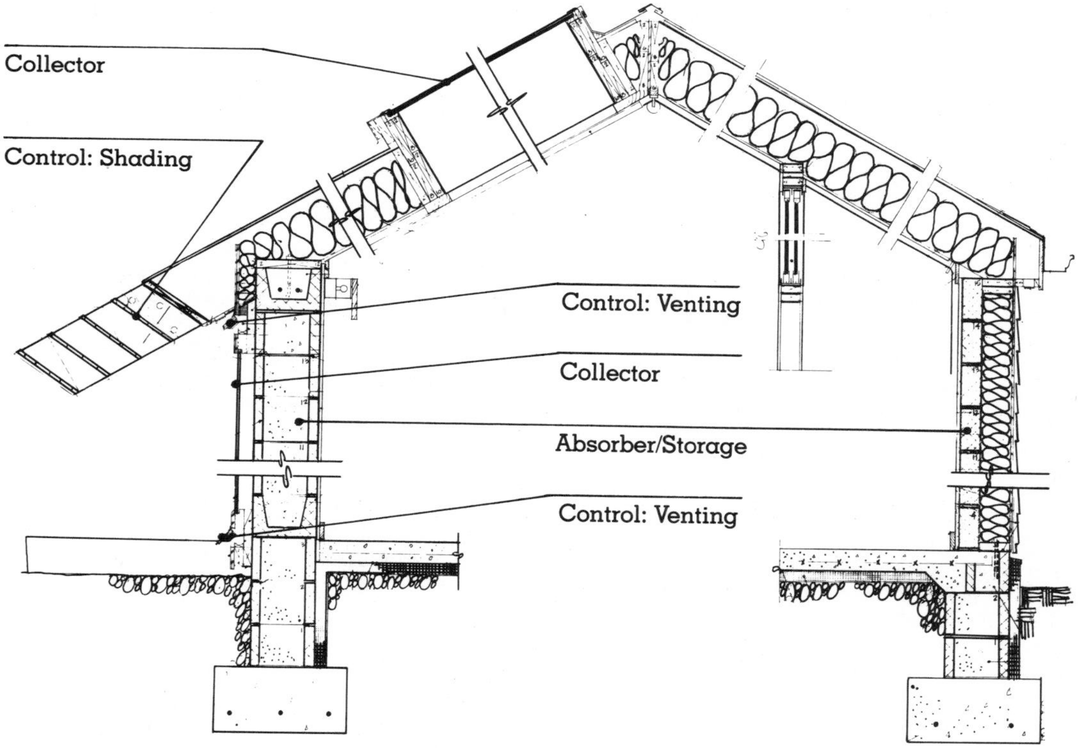

Collector
Control: Shading
Control: Venting
Collector
Absorber/Storage
Control: Venting

B Building Section

Burnside House—Seagroup
Direct Gain

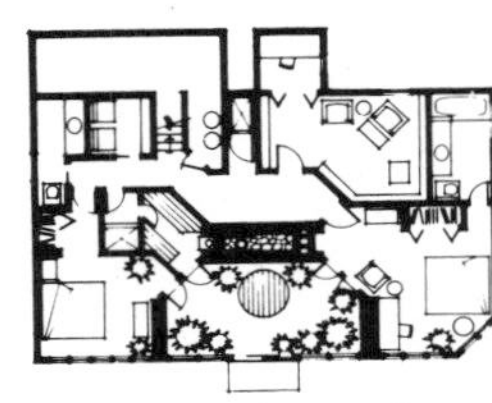

Lower Level

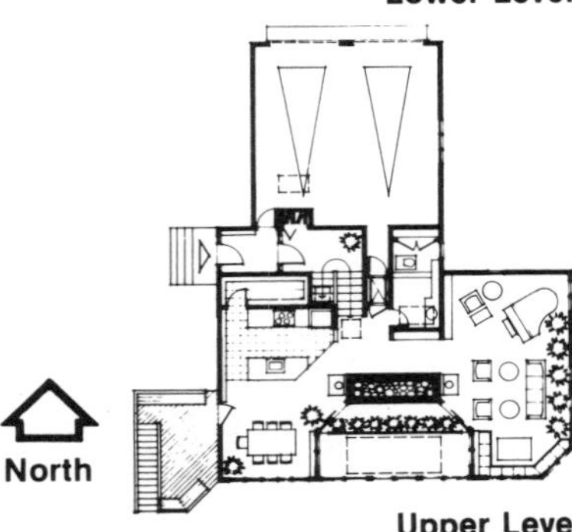

Upper Level

A double-height Attached Sunspace with vertical and horizontal rock bed and masonry thermal storage supplies passive solar heating to the Burnside house in Pawlet, Vermont. The home is vertically zoned with the bedrooms on the lower level and the living areas above. Both share the aesthetic and heating advantages of a double-height Sunspace centrally located on the south side of the home.

Partial earth berming and a north-facing garage shelter the north facade of the house from cold winter winds. Limited fenestration to the east is strategically located to provide morning light into the kitchen and dining room while minimizing heat losses. To the south, large expanses of glazing provide Direct Gain to all the major living spaces.

A skylight with movable insulation increases solar gain into the Sunspace. Thermal mass is provided by brick pavers over concrete slab-on-grade floor in the Sunspace. In adjacent Direct Gain spaces, the slab floors are finished with stone. Both spaces share a brick-veneered concrete masonry wall.

Operable doors and windows into the Sunspace allow heated air to circulate throughout the home. Insulating shades reduce heat losses from all south- and southwest-facing fenestration at night.

On the north side of the Sunspace, a 1½-story vertical rock bed provides central heat for the home. Additionally, a horizontal rock bed has been constructed beneath the floor under the living spaces to the north. During the day, warmed air rises in the Sunspace and is circulated down and across the rock beds with the aid of induction fans. The air releases its heat to the rocks, which in turn warm the adjacent masonry surfaces. This heat slowly dissipates throughout the day and evening hours. When needed for supplemental heat, two woodburning stoves at either end of the vertical rock bed contribute heat to the space and thermal mass.

At the peak of the cathedral ceiling, above the Sunspace and rock beds, pole-operated latched vents exhaust excess heat through the louvered thermal chimney for summer cooling.

Project: Burnside Residence

Builder	**Designer**
Tom Melcher Pawlet, Vermont	David Wright, AIA The Sea Ranch, California

Sponsor
Tom & Elinor Burnside
Pawlet, Vermont

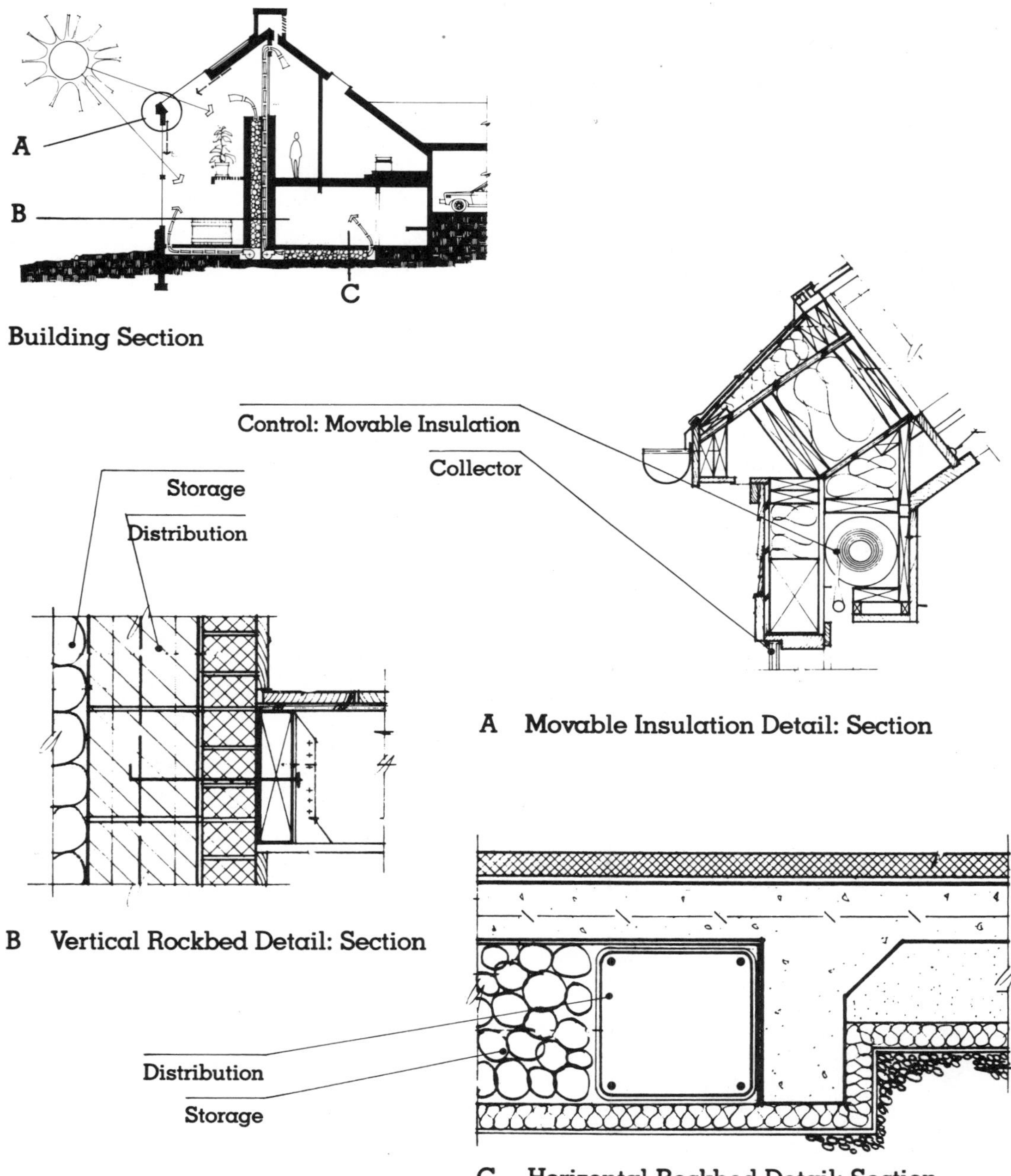

Building Section

A Movable Insulation Detail: Section

B Vertical Rockbed Detail: Section

C Horizontal Rockbed Detail: Section

McCaffrey House
Direct Gain

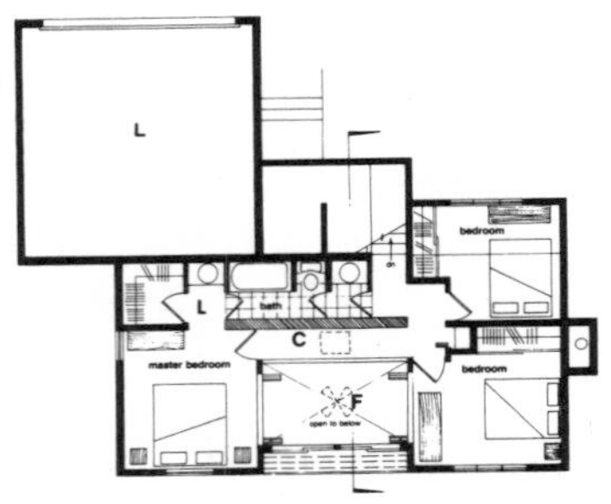

Upper Floor Plan

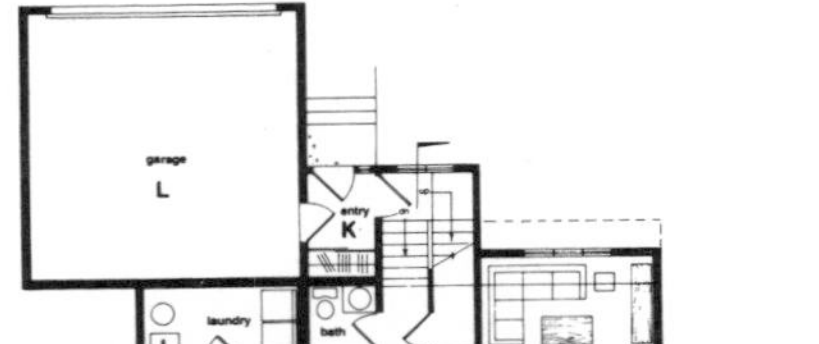

North

Main Floor Plan

A south-facing, two-story Direct Gain dining room with a masonry interior wall beyond combine to provide passive solar heating for the 1,360-sq-ft McCaffrey house in Lafayette, Colorado. Partial earth berming on all sides reduces the home's overall heat loss. A protected entry into an air-lock vestibule adjacent to the garage, and service areas on the north buffer the north, further reducing the home's heating needs. In addition, the absence of glazing along the east and west facades eliminates a major source of summer heat gain.

At the center of the south wall, two sets of double-glazed sliding glass doors on each level illuminate the Direct Gain spaces. On the exterior, a light-colored concrete patio slab acts as a reflector, increasing solar gain into this space. The dining room is open to the kitchen and living room on the first floor, allowing for an even distribution of heat throughout these spaces.

Thermal storage is provided by a two-story-high wall in the center of two layers of 4″ brick units. The wall receives Direct Gain sunlight through a large second collector. A vent at the wall allows heated air to flow into the rooms on the north side of the home. The quarry tile over 5″ concrete slab dining-room and kitchen floor provides additional thermal storage. R-4 quilted rolling shades insulate the glazing against nighttime and cloudy-day heat losses.

A sunscreen trellis at the first floor and a roof overhang above are designed to shade the south facade in the summer. Additionally, the second floor is cantilevered on the south side to shade the first floor. Above the Direct Gain dining area, two thermostatically controlled paddle fans circulate air through the house.

A heat-circulating fireplace and a gas forced-air furnace meet the auxiliary heating needs. The building envelope is constructed with R-38 ceiling and R-19 wall insulation.

Project: SERI #3

Builder
Tradition Homes
Lafayette, Colorado

Designer
WJM Associates
William J. McCaffrey
Ft. Collins, Colorado

Sponsor
Solar Energy Research Institute,
Denver Metro Home Builders Assoc.
Denver, Colorado

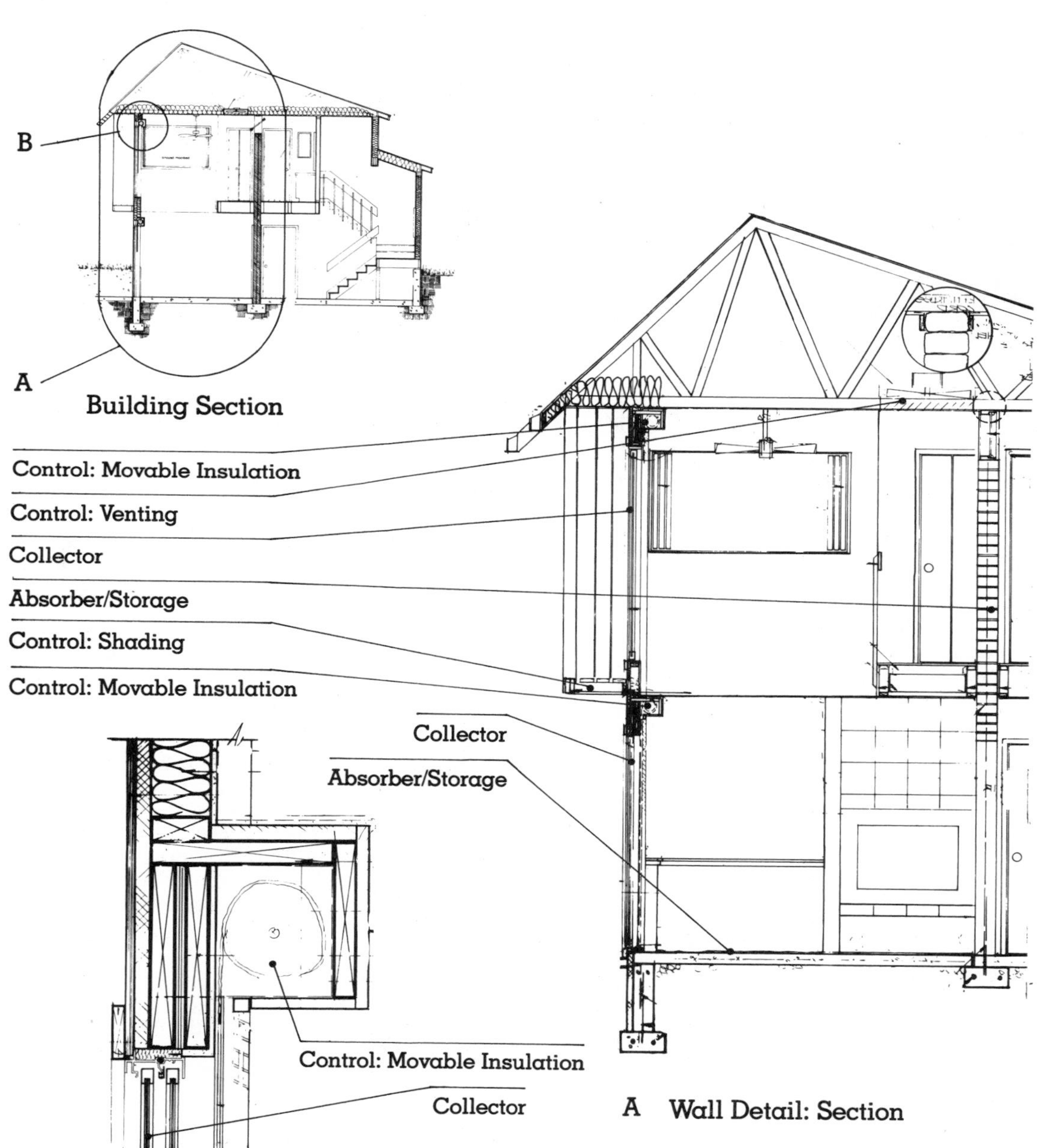

Building Section

A Wall Detail: Section

B Movable Insulation Detail: Section

TVA Solar House #7
Direct Gain

An air-lock entry and garage protect the northern exposure of the TVA Solar House #7 against heat loss. Water-filled tubes lining portions of the fenestrated south wall, and a masonry floor slab provide thermal storage for this two-story 2,300-sq-ft Direct Gain passive house.

Floor-to-ceiling fiberglass water storage tubes alternate with open areas of operable glazing on both floors along the south wall. The water contained in the 12''- and 18''-diameter tubes is treated with dark-colored dyes in order to increase the solar heat absorptivity. Quarry tile over the concrete slab beneath the first-floor tubes and in other selected areas provides additional thermal storage mass. Insulated drapes in the space between these storage tubes and the glazing are essential for the effective operation of the system.

On the upper level, the tubes, which are cantilevered beyond the line of the first floor, are placed between the exterior glazing and pairs of bifold interior doors. In the closed position, these doors enable heat to build up in the water storage tubes. The doors are opened when heat is required in the living spaces.

The house has been designed to accommodate either a flat, or an east- or west-sloping site. Deciduous trees to the southeast and southwest, overhangs, and wing walls provide summer shading. Evergreens have been used as a windbreak on the north. The cantilevered second floor and roof overhang above protect the large expanse of water wall from direct summer sun without interfering with winter solar gain.

The envelope is constructed with R-30 ceiling insulation, R-19 wall insulation, and R-4 insulated window coverings over the predominantly double- and triple-glazed casement windows.

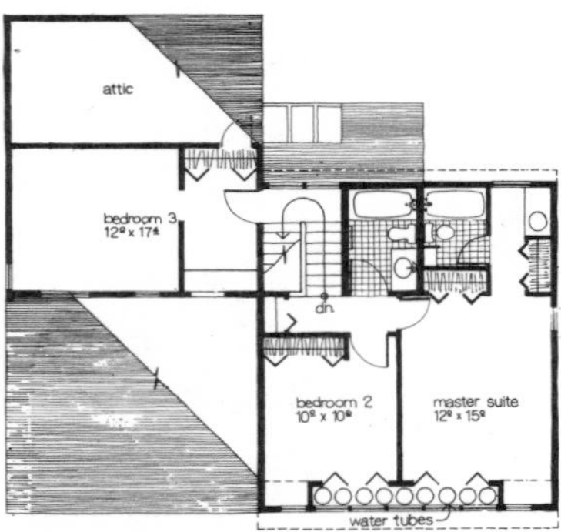

Upper Level

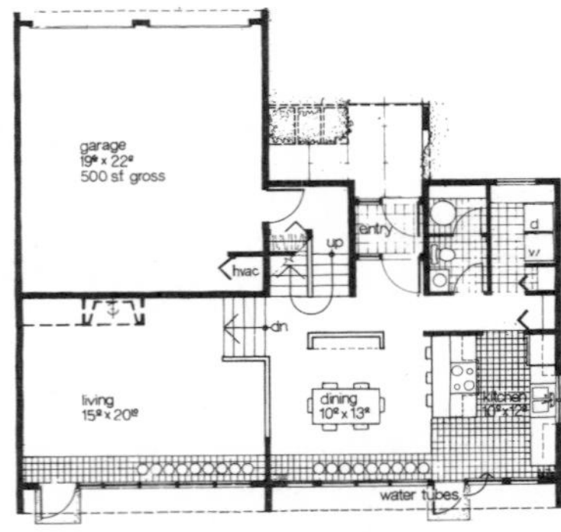

North

Lower Level

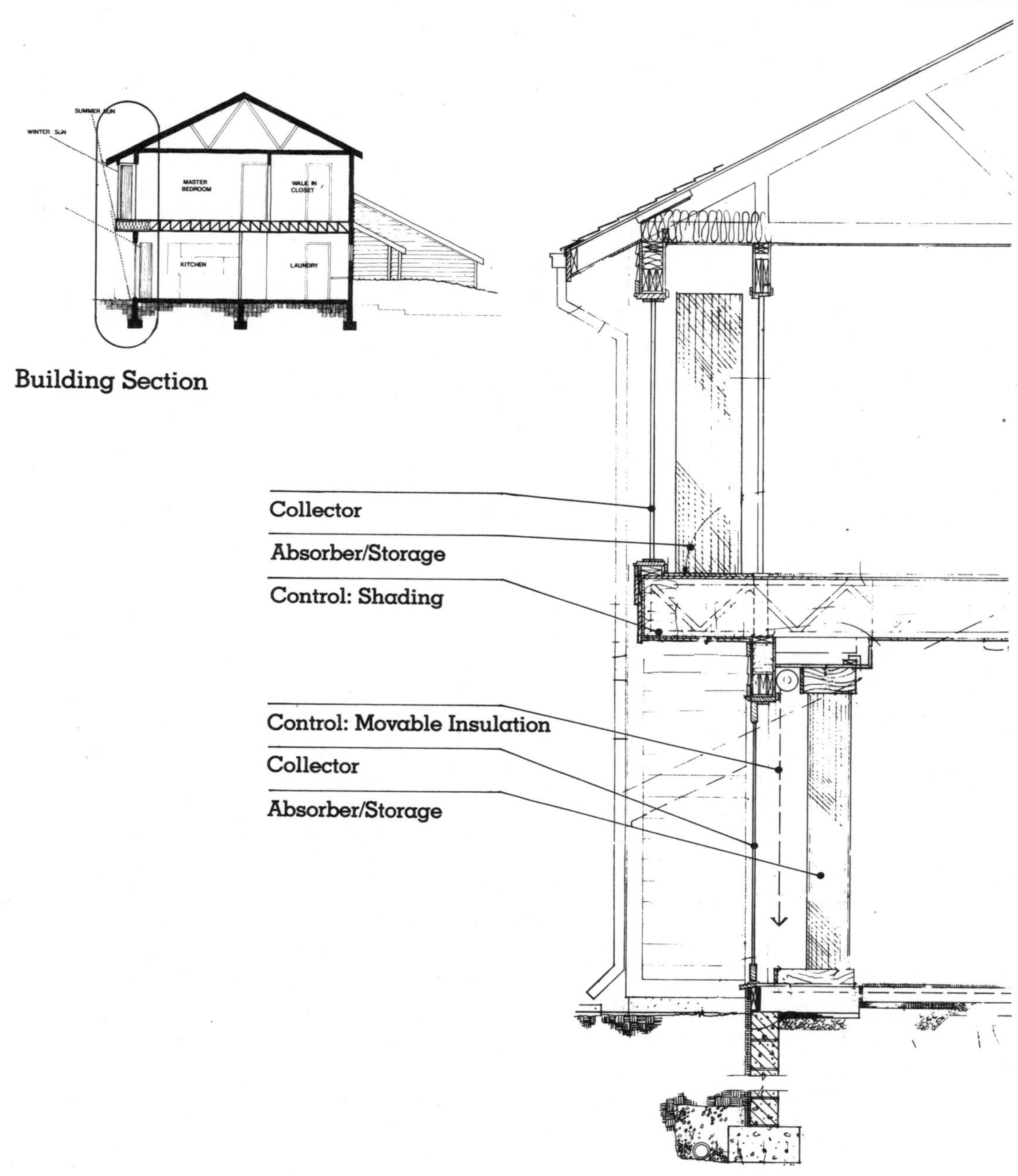

Building Section

Wall Detail: Section

Chapter 4
THERMAL STORAGE WALL

Design Overview

Thermal Storage Wall systems are designed primarily for space heating but can be used in certain climate conditions to provide cooling. For heating, they are effective in areas with mild to severe winters. For cooling, they are best suited to areas with high daily temperature swings (see Seasonal Operation below).

The collector component in a Thermal Storage Wall system is generally a wall of south-facing glass or plastic placed directly in front of a wall that serves as the storage component. Movable insulation control components can be applied outside the collector or in the airspace between the collector and the storage wall. Shading and reflecting devices are typically placed on the exterior.

In a Thermal Storage Wall system, collection, absorption, storage, and control of solar energy occur outside the living space. Heat is transferred to the living space by the storage wall itself, which generally forms one side of the space. The wall can be unvented, or vents can be provided at the top and bottom. Windows can also be integrated into the wall to provide light, view, and some Direct Gain heating.

Seasonal Operation—Heating

During the heating season, sunlight passes through the collector (usually some form of glass or plastic), strikes the storage wall, and heats it. In an unvented Thermal Storage Wall (see Figure 4.1), this heat is stored and slowly migrates to the interior, where it heats the adjacent living space.

If properly designed, the wall can provide adequate heat to the living space throughout the night.

Some of the heat generated in the airspace between the collector and the storage wall is lost back out through the glass or plastic to the outside. The hotter the air in the airspace, the greater this heat loss. These potential heat losses can be reduced by venting the storage wall (see Figure 4.2).

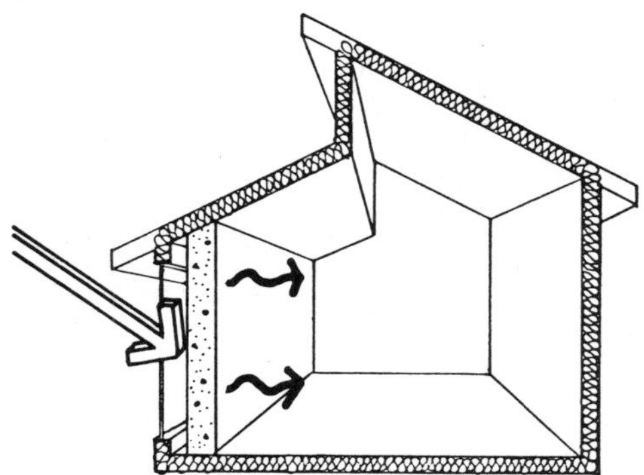

Figure 4.1

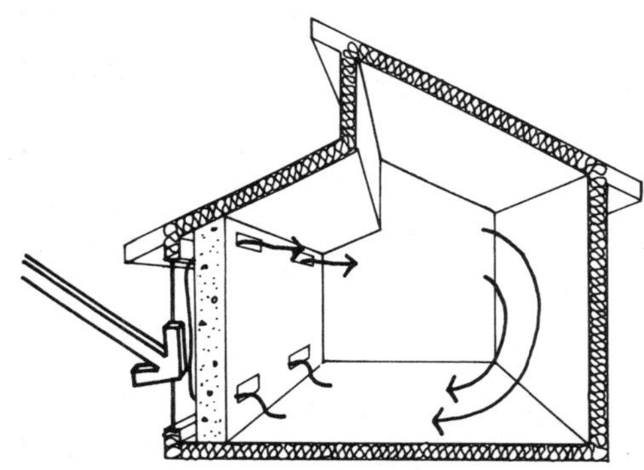

Figure 4.2

Because of the tendency of hot air to rise and cool air to drop, a natural system for providing warm air to the living space can be set up in a vented wall system. Vents placed at the top of the wall allow air, which has been heated in the airspace, to flow naturally into the adjacent living area. Vents low in the wall then pull cooler replacement air into the airspace from the living area, thus setting up a loop for circulating the heated air (thermocirculation). No mechanical means for moving the air is required. Care should be taken to ensure that the circulation pattern does not reverse itself at night, when temperatures in the airspace drop, allowing heated air from the living space to flow into the airspace, pushing cooler airspace air into the living area. The provision of simple backdraft dampers or operable louvers on the upper vents will prevent this reverse flow from occurring.

In a vented system, heated air enters the living space at roughly 90°F. The loss of this heat from the airspace consequently reduces the amount of heat available to be stored by the wall. An unvented system does not lose heat in this way and thus has the advantage of storing a greater percentage of the solar energy available to it than does a vented wall. This stored heat is not, however, readily available for immediate use, but radiates slowly into the living area.

The delay between the time when sunlight first strikes the unvented storage wall and the time when the heat has finally traveled through the wall and reached the living space is called time lag. It is characteristic of certain materials, such as masonry, that are commonly used to construct Thermal Storage Walls. Depending on the thickness and thermal properties of the selected wall materials, this lag period can last from several hours to an entire day. In a vented system, this time lag still occurs, but some of the heat is short-circuited by the vents that are providing heat to the living space throughout the day. Such a trade-off makes sense in colder climates, where daytime as well as nighttime heating requirements are high, and it is desirable to provide a certain amount of heat directly to the living space. In such situations, a vented wall should be specified. In most moderate climates, on the other hand, where daytime heating is not as important as nighttime, an unvented system will be preferable. If properly designed, such a system will be able to store heat throughout the day and release it to the living space throughout the night.

It should be noted that windows integrated into Thermal Storage Walls to provide light, view, and/or some Direct Gain daytime heating will also reduce the effective area of the storage wall.

Seasonal Operation—Cooling

The south-facing collector should be well shaded to prevent excessive heat gain during the cooling season, when heat is not required from the system (see Figure 4.3).

Shading elements, such as overhangs, should be sized so as to maximize *overall* system performance. If they are too large, they may completely shade the wall, preventing unwanted gain during the cooling season, but they will also block a significant amount of radiation during the later part of the heating season. Conversely, if they are too small, solar gain will be maximized when heat is needed, but overheating will occur during the later part of the cooling season, adding to the load on the auxiliary energy system.

It should be noted that a Thermal Storage Wall system can contain two sets of vents. In addition to the wall vents discussed earlier, vents can and should be placed on the exterior in the collector component. In the former case, the vents are placed in the storage wall itself to provide heated air to the living space. Vents placed in the collector are used to prevent overheating during the cooling season by admitting and expelling outside air to the airspace (see Figure 4.4).

These external vents can be opaque panels in the collector system, or they can simply be operable glazing elements. In situations where the collector is shaded by a fixed device, such as an overhang, an opaque vent may be preferable to glazing for use at the top of the collector. This area will generally receive little or no sun, especially if the overhang is large, but will conduct considerable amounts of heat if it is glazed. Opaque vents, if properly designed, can allow for the inclusion of adequate insulation to reduce such heat flow. Lower vents can be opaque or glazed, depending on builder preference. Glazed units have the advantage of transmitting sunlight onto an unshaded part of the wall, improving system performance.

A combination of these high and low exterior vents in the collector wall will remove much of the solar heat generated in the airspace. By opening both the high exterior vents in the collector wall and the low interior vents in the storage wall, air will be drawn across the living space and through the low vents to replace the warm air in the airspace, rising naturally out the upper vents. This method not only avoids heat buildup in the airspace but also provides some natural ventilation to the living space (see Figure 4.5).

In areas of the country with high daily temperature swings, the Thermal Stor-

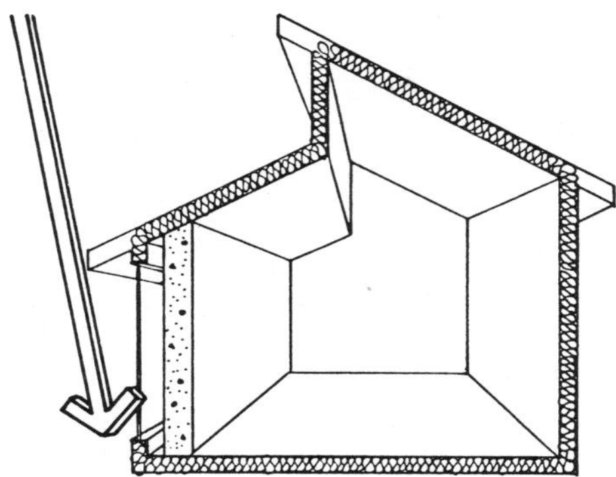
Figure 4.3

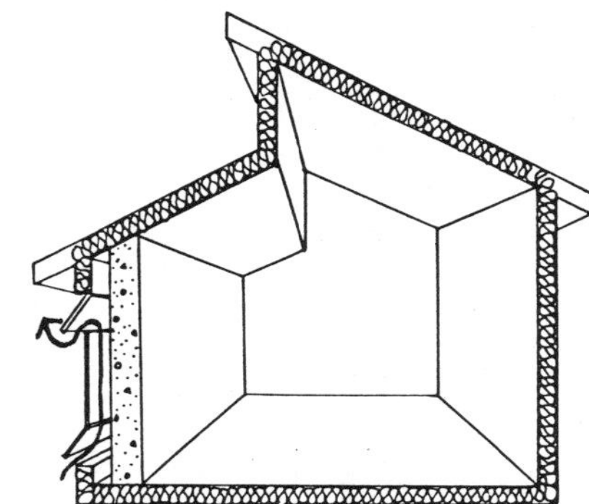
Figure 4.4

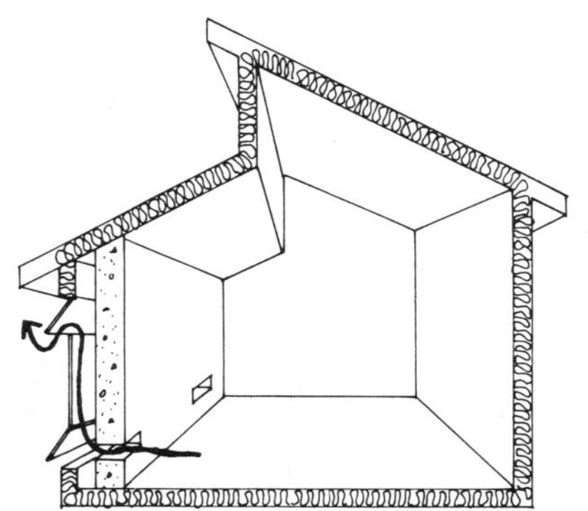
Figure 4.5

age Wall can be used to further reduce the cooling energy requirements by releasing heat to the cool night air. In the evening, the exterior vents are opened to allow night air to be drawn across the storage wall by natural convection or fans. By morning, the wall temperature will be low enough to allow the wall to absorb heat from the living space during the day.

Movable insulation should be used during this daytime period to cover the collector or the wall in order to reduce heat gain from the outside and to preserve the cool temperature of the wall for as long as possible. At the end of the day, the insulation is removed and the cy.cle is repeated (see Figure 4.6).

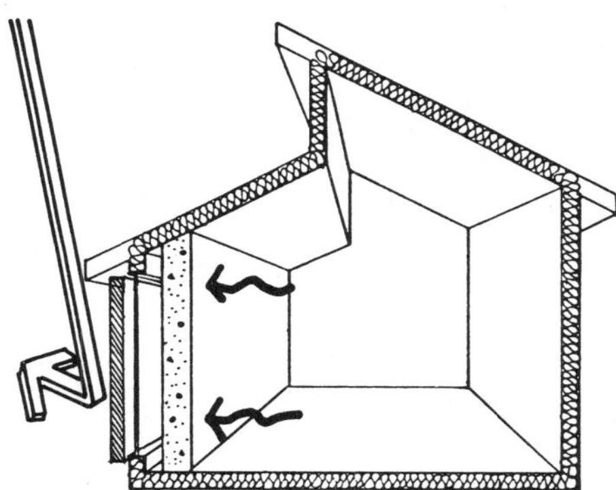
Figure 4.6

Advantages

- Glare and ultraviolet degradation of fabrics are not problems as they are in Direct Gain systems.

- The time lag between the absorption of solar energy by the exterior surface of the wall and the delivery of the resulting heat to the interior living space provides this heat in the evening, when it is most needed.

- Sufficient storage mass to achieve a desired level of thermal performance can be provided in a relatively small, concentrated area within the living space.

Disadvantages

- Two south-facing walls, one glazed and one constructed of the storage material, are required.

- Mass storage walls can be expensive and may require specially sized foundations and footings. They may also be subject to special building-code restrictions, especially in earthquake-prone areas.

- The mass storage wall can take up valuable square footage in a residence.

- Without movable insulation, the high temperature differential between outside night air and inside airspace air will cause considerable heat to be lost through the collector glazing at night during the heating season. Therefore, some form of movable insulation is highly recommended, but it can be expensive and awkward to operate.

- The storage wall can block view and daylight.

Collector

Thermal Storage Wall collectors can be constructed using off-the-shelf window units, simple patio-door replacement glass, various glazing materials, or some combination of these elements incorporated into a framing system specially designed for the specific installation.

If the collector framing system is metal, it should be separated from the storage wall, either by a space or by wood blocking, to avoid conductive heat losses from the wall, through the metal, to the outside. If the frame is wood, rough-sawn and green wood should be avoided in favor of kiln-dried members. Any paint used on the frame should be resistant to high temperatures.

Collector framing systems can be attached to, and supported by, the storage wall itself, or they can be framed conventionally and be supported on a foundation (see Figures 4.7 and 4.8).

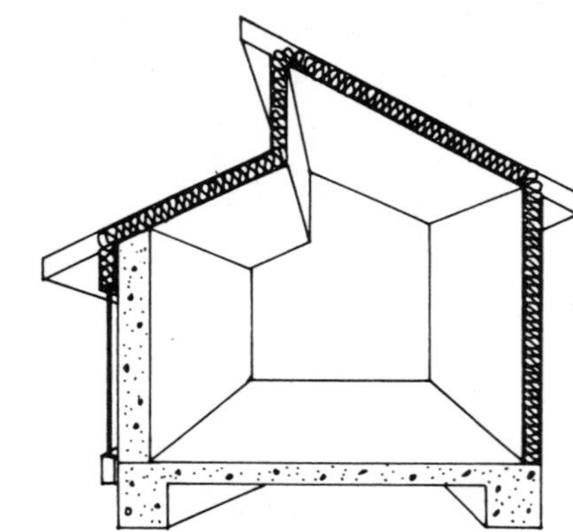

Figure 4.7

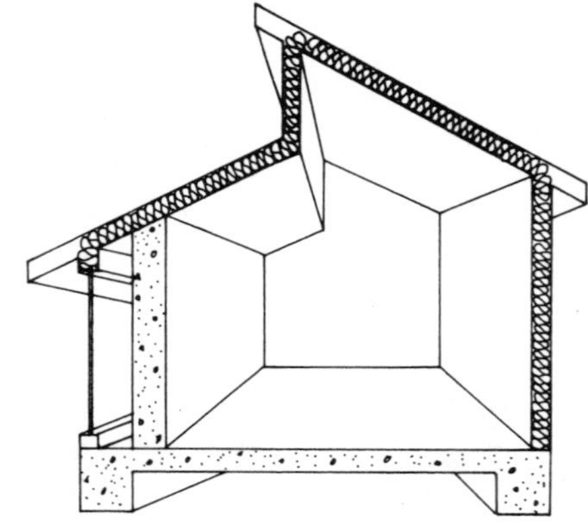

Figure 4.8

Both approaches work well, and a choice may be based on specific design considerations and local material costs. Whatever system is chosen, the dimensions of the framing members should be kept to a minimum so as to maximize the glazing area and, consequently, the amount of sunlight striking the wall.

The collector material can be glass or any one of a variety of glazing materials, including fiberglass, acrylics, polycarbonates, and certain glazing films, such as polyethylene. When reviewing the physical properties of these materials, special attention should be paid to the maximum operating temperature of the material, since temperatures within an unvented storage wall airspace can get quite high (150-180°F) and may cause some materials to degrade rapidly.

Double glazing is recommended for most installations to reduce heat losses back out through the collector. Triple glazing should be considered in severe climates if some form of movable insulation is *not* provided. Single glazing may be more cost-effective in mild climates, especially if movable insulation *is* provided.

Whatever material is chosen, special attention should be paid to expansion, contraction, and sealing. Collector frame joints should allow for significant expansion (½'' minimum) due to the high temperatures possible in the airspace, particularly in unvented walls. Caulking and sealants used in these applications must be able to accommodate such movement, and it is recommended that only high-quality products be used.

The collector should be periodically checked for possible sealant repair, as some separation may occur during the first few months of operation. It is also recommended that the collector elements be operable or removable to allow for periodic maintenance and cleaning. In systems where the collector material is used to vent summer heat to the exterior, such operability is already partially built in.

Collector Rules of Thumb

- **Orientation**:
  Orient the Thermal Storage Wall system to face south. Orientations up to 20° east or west of due south will not significantly affect performance.

- **Design**:
  The Thermal Storage Wall should be one wall of the living space(s) to be heated.

- **Area**:
  The surface area of the collector should be equal to the surface area of the storage component (see Storage Rules of Thumb in this chapter).

- **Airspace**:
  The airspace between the wall and the collector can vary from 2-6'', depending upon the space needed for movable insulation if it is to be included between the wall and collector.

- **Number of Glazings**:
  Double glazing is recommended for most installations. Triple glazing should be considered in severe climates, especially if movable insulation is *not* provided. Single glazing may be appropriate in mild climates if movable insulation *is* provided. A ½'' airspace

between glazing layers is recommended where possible.

Storage

The storage wall component of a Thermal Storage Wall system can be made of any one of a variety of materials exhibiting a high capacity to store heat (e.g., concrete, brick, phase-change materials, or water). These materials will transfer heat to the living space slowly and evenly, delivering this heat, if the system is properly designed and sized, when it is most needed.

In general, materials used for a storage wall, with the possible exception of phase-change materials, will require larger foundations and footings to accommodate the extra weight of the wall. Special reinforcing requirements must also be considered for a masonry wall if, as is typically the case, it is taken to the roof and used as a bearing wall.

The exterior face of the wall should be a dark color to increase absorption of incoming solar energy. The exposed surface can be painted, or specific colors can be specified from the material manufacturer (see Table 7.9 for the absorptivities of various materials and colors).

Mention should be made of a new type of exterior finish material that is gaining popularity in Thermal Storage Wall applications: the selective surface. The selective surface materials, usually manufactured in thin sheets that are adhered directly to the storage wall surface, absorb almost all incoming solar radiation but emit only a very small portion as heat. They thus reduce heat loss back out through the collector glazing and can increase efficiency to the point where movable night insulation may no longer be necessary.

The interior surface of the wall can be painted or left untreated. However, finishes other than paint will decrease the efficiency of the wall by restricting its ability to absorb and reradiate heat. Plaster and wallpaper will have only a minor effect in this regard. Gypsum board can be acceptable if it is attached directly to the wall and if care is taken during installation to apply a very thick coat of construction adhesive in order to increase the thermal connection between the wall and the gypsum board. Interior finishes attached on furring strips will not perform as well and should be avoided.

Water containers used as Thermal Storage Walls pose a slightly different finishing condition. Such containers are generally designed and manufactured to be freestanding and unfinished. In cases where the water containers are to be concealed for aesthetic reasons, it is not advisable to attach the finish material directly to the containers. A puncture from decorative material (e.g., wall hangings) can result in leaks from the containers. One option is to build out a stud wall with a conventional gypsum-board finish. Vents placed in the top and bottom of the wall will allow heat to be drawn off the water wall. This isolated water container configuration will not perform as well as the exposed-mass Thermal Storage Wall described earlier.

Storage Rules of Thumb

- **Area:**
  In general, the farther north the building is, the larger the Thermal Storage Wall system should be. The rough

Table 4.1

Average Winter Outdoor Temperature (Clear Day)*	Masonry Wall/Space Floor Area**			
	36°NL	40°NL	44°NL	48°NL
COLD CLIMATES				
20°F	.71	.75	.85	.98 (w/NI)
25°F	59	.63	.75	.84 (w/NI)
30°F	.50	.53	.60	.70
TEMPERATE CLIMATES				
35°F	.40	.43	.50	.55
40°F	.32	.35	.40	.44
45°F	.25	.26	.30	.33

Note: NI indicates Night Insulation.

* Temperatures listed are for December and January, usually the coldest months.

** These ratios apply to a well insulated space with a heat loss of 8 btu/degree day/square foot of floor area/°F. If space heat loss is more or less than this figure, adjust the ratios accordingly. The surface area of the wall is assumed to be the same as the glazing.

area of storage material per square foot of adjacent living space to be heated can be determined from Table 4.1 (collector area will be equal to the storage area).

For example, for Madison, Wisconsin, at 43°NL with an average January temperature of 21.8°F (data available from local and/or national weather bureau), go to the intersection of the closest latitude (44°NL) and the closest "Average Winter Outdoor Temperature" (20°F) to find that 0.85 sq ft of storage wall, and therefore glazing, is required per square foot of floor area. For a living space of 200 sq ft to be heated, 170 sq ft of Thermal Storage Wall will be required, assuming all other rules of thumb for system sizing have been followed.

- **Thickness**:
  The recommended storage component thickness ranges from 8-18". For phase-change material, consult manufacturers' literature.

- **Vent Area**:
  The area of the thermocirculation vents in the storage wall should be roughly 2 sq ft per 100 sq ft of wall, divided evenly between upper and lower vents.

- **Vent Dampers:**
  Vents should be provided with back-draft dampers to prevent reverse thermocirculation.

Control

Shading

To avoid excessive overheating in the cooling periods, it is recommended that some method of shading the Thermal Storage Wall system be provided. The range of possible methods is roughly the same as that for Direct Gain shading. Exterior devices include overhangs (see Figure 4.9), trellises, awnings, louvers (horizontal or vertical, fixed or operable), and wing walls.

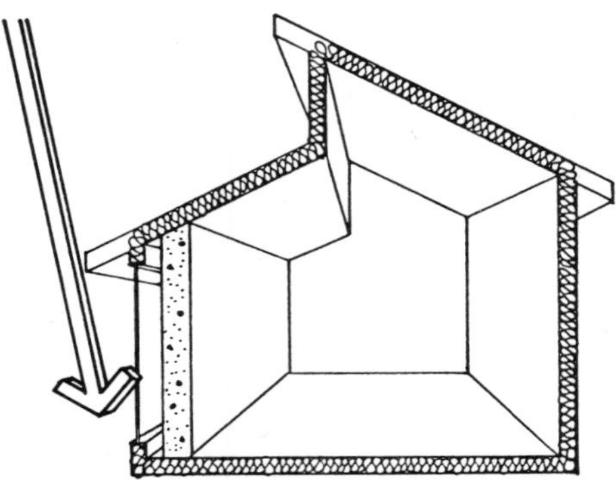

Figure 4.9

These exterior devices can form part of the building frame, or they may be attached to it. They can be stationary, or they may require daily operation. If interior devices are used, they should be placed in the airspace between the storage wall and the collector (see Figure 4.10).

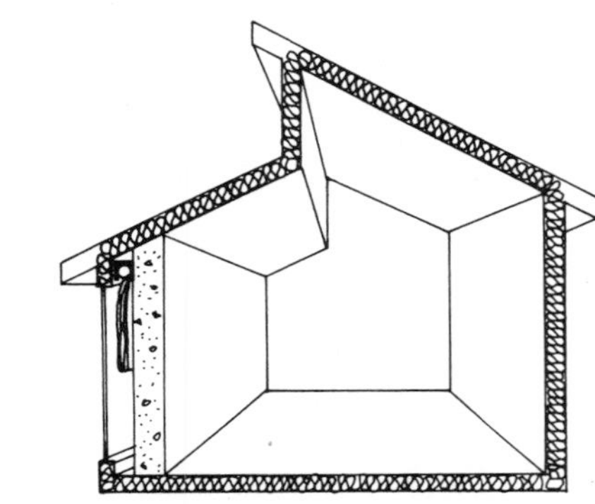

Figure 4.10

Interior shades are typically curtains, blinds, or roller shades. They must be opened and closed daily if the storage wall is used for passive cooling (see Seasonal Operation—Cooling earlier in this chapter). Thermostatically controlled, motor-driven interior shades are generally most acceptable to the homeowner, but manually operated interior shades are generally less expensive, with fewer maintenance problems. In both cases, special attention must be paid to the method of operation of a specific shade to ensure safe and maintenance-free operation from the living space. For optimum performance, these interior shading elements should also be designed to provide insulation during the day in the cooling season and at night in the heating season (see Insulation—Movable Insulation later in this chapter).

Reflecting

Reflectors placed horizontally above or below the Thermal Storage Wall collector can increase overall system performance by increasing the amount of sunlight reaching the collector (see Figure 4.11).

In certain cases where physical obstructions (e.g., trees or buildings) on or around the building site shade the collector, the provision of reflectors can increase solar collection 30-40%.

Reflectors are usually panels, coated on one side with a material of high reflectance, which are placed directly in front of the south-facing wall. If placed on the ground, they should slope slightly downward (5° is recommended), away from the wall, to increase the amount of reflected sunlight and to facilitate drainage. If possible, they should be designed so that they can also be used as movable insulation panels (see Figure 4.12).

Light-colored exterior landscape elements, such as patios or terraces, can also serve as reflectors, although with less thermal efficiency than reflective panels.

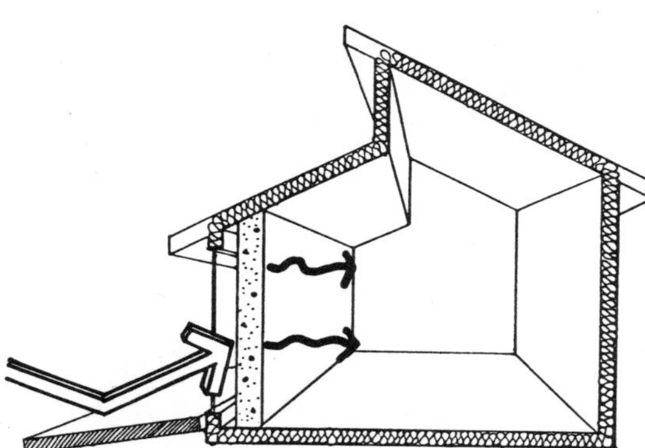

Figure 4.11

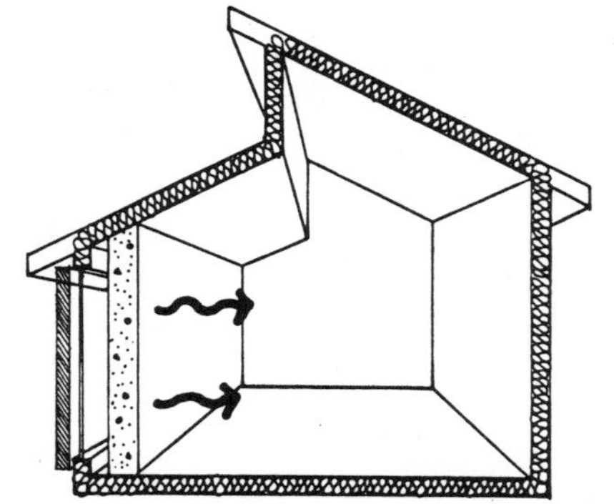

Figure 4.12

Insulating—Movable Insulation

Wherever possible, movable insulation covering the entire Thermal Storage Wall should be provided during the heating season to prevent heat gained during the day from being lost at night or during prolonged overcast periods. This insulation can also be used to shield the wall from unwanted heat gain during the cooling season.

There are two basic applications of movable insulation: (1) insulation that is applied to the exterior of the collector; and (2) insulation that is placed in the airspace between the collector and the storage wall. The former are usually rigid panels and can be difficult to store, maintain, and operate. If designed properly, however, they can be used as both reflecting panels during the day and insulation at night (see Reflecting above). They will also be more effective at reducing unwanted heat gain during the cooling season because they serve as shades intercepting the sunlight before it can enter the collector and be converted to heat. Care must be taken to ensure that there is a tight, well-sealed fit between the exterior insulation and the collector to avoid heat losses at the edges of the insulation (see Figure 4.12). Placing the insulation in the airspace generally solves the problems of storage and reduces those of maintenance (see Figure 4.13).

Providing for easy operation of airspace insulation, however, can pose significant problems. It is recommended that an adequate operating mechanism, whether manual or automatic, be developed very

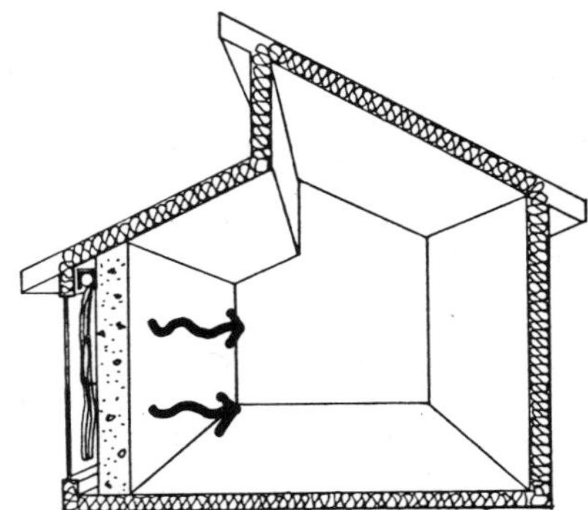

Figure 4.13

Projection (L) = $\frac{\text{window opening, i.e., collector (H)}}{\text{F (see table below)}}$

North Latitude	F Factor*
28°	5.6-11.1
32°	4.0- 6.3
36°	3.0- 4.5
40°	2.5- 3.4
44°	2.0- 2.7
48°	1.7- 2.2
52°	1.5- 1.8
56°	1.3- 1.5

*Select a factor according to your latitude. The higher values will provide 100% shading at noon on June 21, the lower values on August 1.

early in the design process to avoid potential future problems. Care should also be taken that this airspace insulation not interfere with the proper operation of the interior vents (if provided) or the exterior summer cooling vents.

Control Rules of Thumb

Shading

- **Overhang:**

  Generally it is not cost-effective to construct a south-facing overhang with a projection that exceeds 2′. Overhangs of this dimension and less can be built using standard truss construction. However, an overhang sized to provide a specified level of shading has the advantage of further reducing both the home's cooling energy requirements and potential glare in the cooling season.

  The projection of the overhang that will be adequate (provide 100% shading at noon on June 21) at particular latitudes can be quickly calculated by using the following formula:

  A slightly longer overhang may be desirable at latitudes where this formula does not provide enough shade during the later part of the cooling season (see Figure 4.14).

  For example, for Madison, Wisconsin, at 43°NL, select an F factor between 2.0 and 2.7 (values for 44°NL). Assuming shade is desired on August 1, choose the lower value of 2.0. If the window opening (H) is 8′, the projection (L) of the overhang placed at the top of the collector will be: $\frac{8}{2.0} = 4'$.

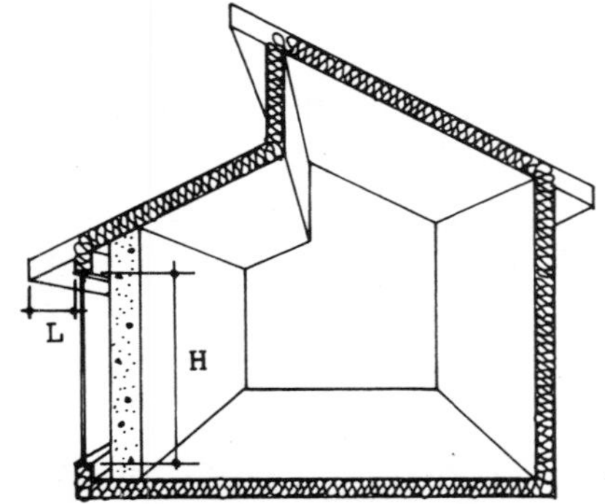

Figure 4.14

Reflecting

- **Slope:**
  A downward slope (away from collector) of roughly 5° is recommended for reflector (see Figure 4.15).

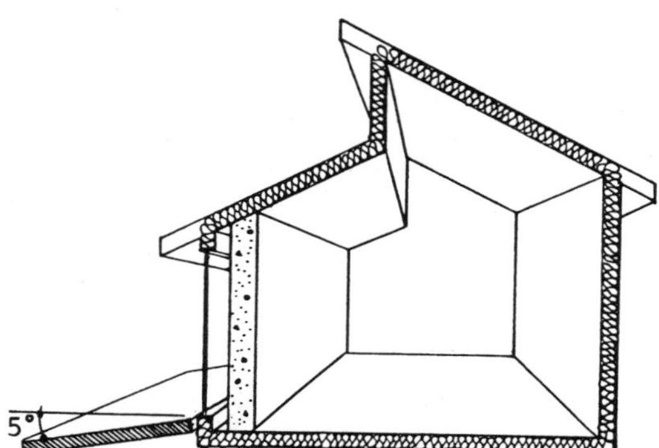

Figure 4.15

- **Sizing:**
  The reflector should be the same length as the collector and roughly as wide as the collector is high.

Insulating

- **Movable Insulation:**
  Movable insulation of R-4 through R-9 is recommended.

Thermal Storage Wall Details

The drawings on the following pages illustrate common construction detailing for Thermal Storage Wall systems. The details are divided into three sections denoting the major material and construction options available in the design of a Thermal Storage Wall system.

Each section focuses on one of the primary storage materials: masonry, water, and phase-change materials. The first page in each section consists of two sets of details illustrating the major construction type options. Variations that highlight alternate construction methods and the use of materials other than those shown in the basic details are found on the subsequent page(s). Following the variations are the construction notes referenced from the details. The notes provide guidelines, troubleshooting tips, and other information useful in building a Thermal Storage Wall.

In the first section, masonry Thermal Storage Wall details, six pages are added after the notes suggesting masonry material substitutions. These material alternates can be substituted for the storage material shown on the first page of the section.

THERMAL STORAGE WALL

(Masonry)

DETACHED GLAZING

ATTACHED GLAZING

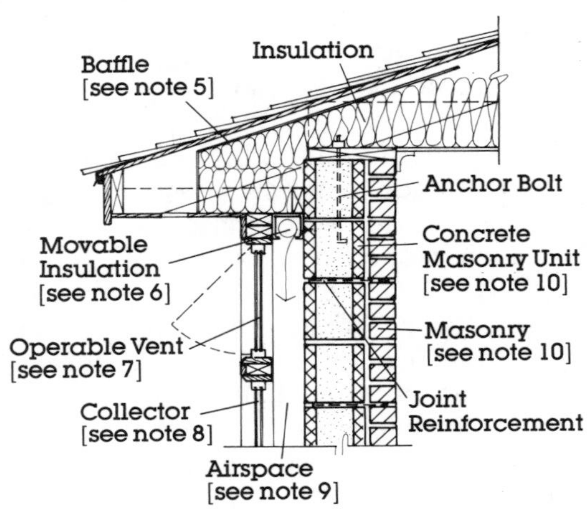

Collector at Roof

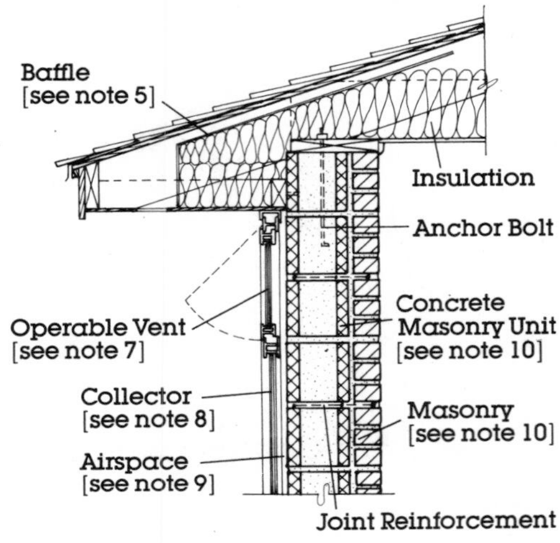

Collector at Roof

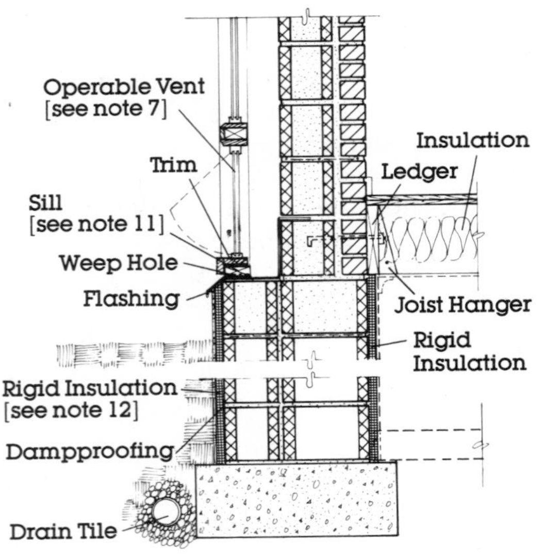

Collector at First Floor

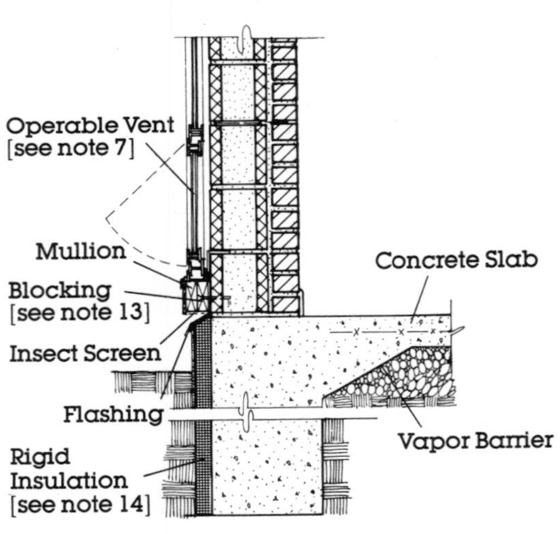

Collector at Slab-on-Grade

Variations: Detached Glazing
(Masonry)

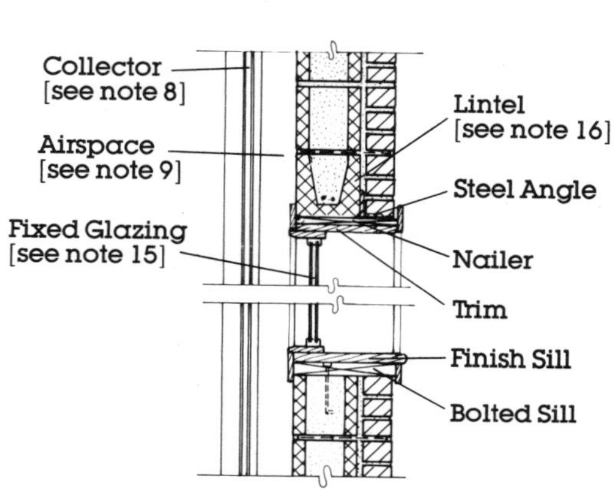

Collector at Window

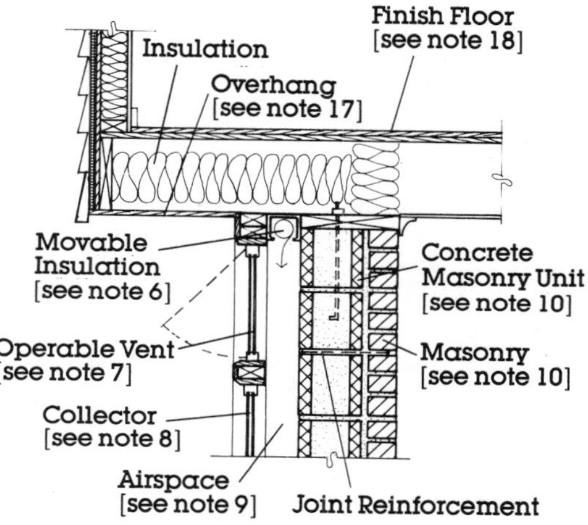

Collector at Floor Overhang

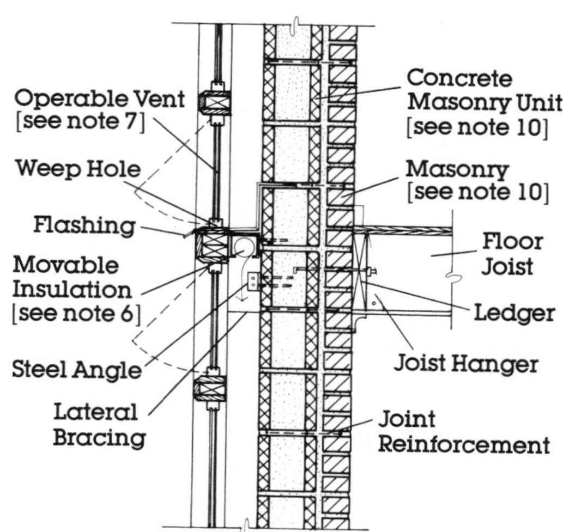

Collector at Second Floor

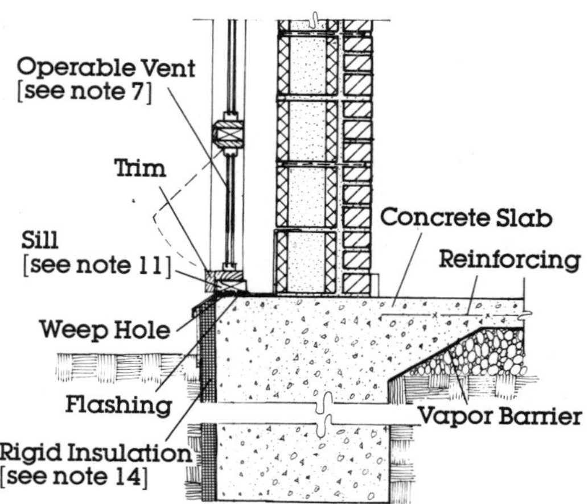

Collector at Slab-on-Grade

Variations: Attached Glazing

(Masonry)

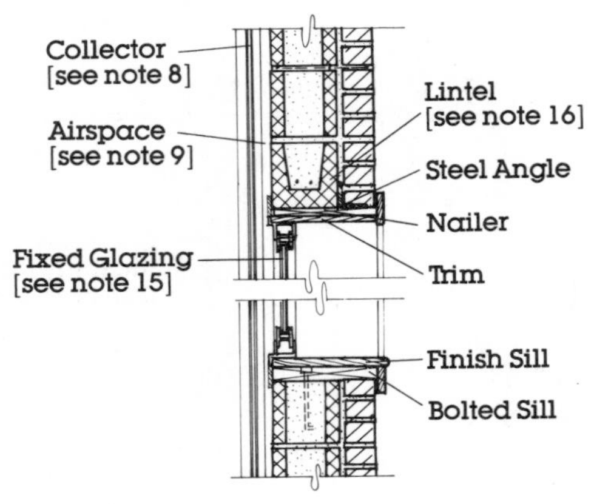

Collector at Window

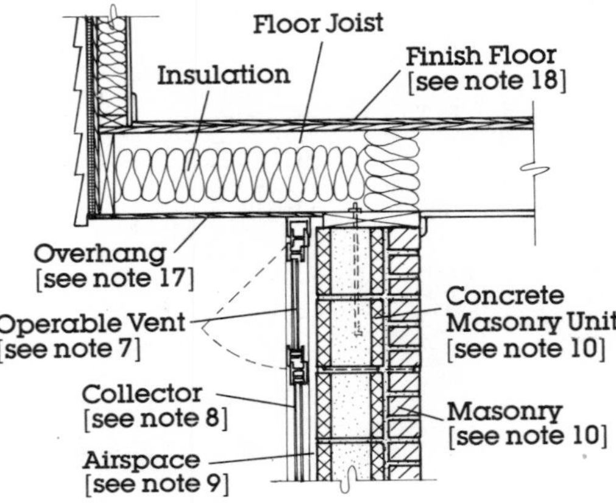

Collector at Floor Overhang

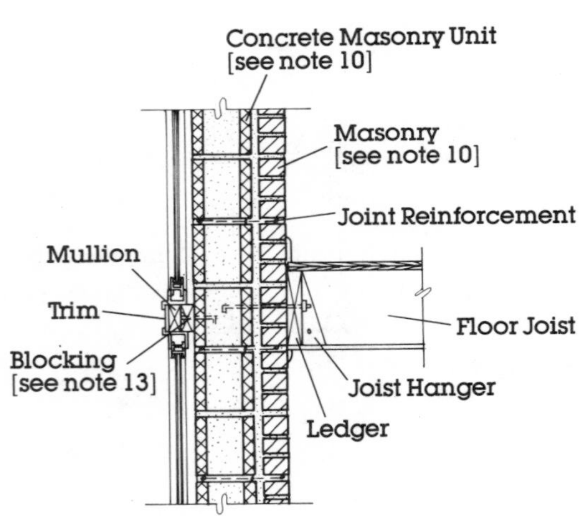

Collector at Second Floor

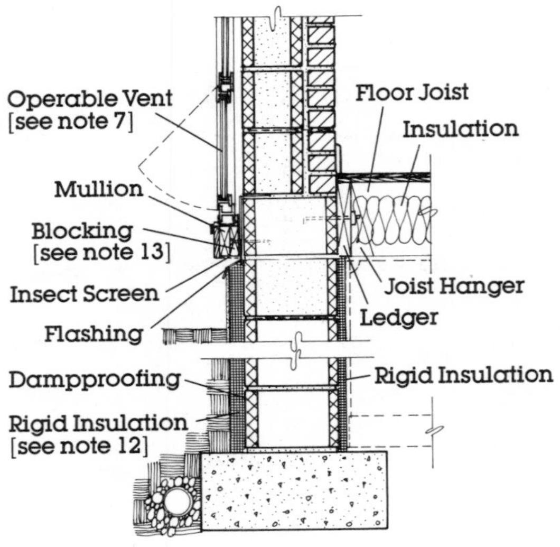

Collector at First Floor

Notes

(Masonry)

GENERAL NOTES:

1. All footings must bear on undisturbed soil. Adjust footing size, reinforcing, and depth below grade as required by site conditions and/or local code. Where groundwater problems exist, provide sufficient granular fill to prevent water penetration at slab-on-grade.

2. Placement of reinforcing, metal ties, anchor bolts, flashing, and weep holes is to be determined by local building codes and/or accepted practice. Reinforcing in concrete slab is not recommended except where required by local conditions and/or code.

3. Insulation levels at foundation/slab and/or floor, and roof/ceiling must meet or exceed requirements of local building/energy codes. The storage mass should not be insulated. In crawl-space or basement construction, it is recommended that the Thermal Storage Wall foundation be insulated on both sides to prevent losses to the outside and to the ground except where basement or crawl space is heated. In these cases, insulation is not required on the inside.

4. Other roof configurations may be used at builder's option. Consult local building codes for restrictions and special requirements concerning roof/wall connections, especially in seismic and/or high-wind areas. Roof/attic section should be vented to ensure proper performance of insulation. Vents may be located at sides of roof construction or at soffit and/or ridge vents. Provide gutters and downspouts as required. Overhang at collector should be sized to optimize solar gains during heating season and prevent overheating during cooling season (see Control Rules of Thumb in this chapter and Control Components in Chapter 7).

DETAIL NOTES:

5. Provide baffles where necessary to maintain 1'' minimum airspace for venting attic.

6. Movable insulation should be provided to control nighttime and cloudy-day heat loss and to prevent overheating during the cooling season. Insulation may be installed within the airspace or placed on the exterior of the collector. In the latter case, insulating shutters or panels can be used. Care must be taken to ensure a tight fit between the insulation and the collector to reduce heat loss at the shutter or panel edges. When provided within the airspace, care must be taken to ensure adequate space for proper installation and operation. Where upper and lower portions of the Thermal Storage Wall are separated, movable insulation must be installed in each section. For further information, see Control Components in Chapter 7.

7. Portions of the collector should be operable to allow venting of the Thermal Storage Wall during periods of excessive heat gain. Awning-type or other operable windows may be speci-

fied and should be equipped with demountable insect screens that can be removed during the heating season.

8. Double glazing is generally recommended to ensure optimum thermal performance. A ½'' airspace between glazing layers is recommended where feasible. The glazing material can be glass or plastic (see Collector Components in Chapter 7). The collector unit should be demountable or externally operable to aid in maintenance and cleaning. The collector frame may be detached from the masonry wall and designed to support the glazing, glazing frame (if preframed), roof, and other related loads, or it may be attached to and supported by the masonry wall. Framing should be kept to a minimum, since it reduces the effective collector area and may cast shadows, both of which will affect the thermal performance of the wall. All structural framing and connections should be designed per local building-code restrictions and/or accepted practice. Frames should be continuous at edges. 2x framing members bolted to the masonry wall can serve as spacers, ensuring adequate airspace. They should be discontinuous, allowing for free circulation of air behind the collector.

9. It is generally recommended that the airspace between the glazing and the storage mass be a minimum of ¾'' and a maximum of 6'' to avoid excessive heat loss.

10. Brick or concrete brick, or architectural concrete masonry units, such as split, slump, fluted, scored, or recessed blocks, may be used for facing. Concrete masonry backup unit must be solid, or hollow with grout-filled cores. For information on finishes, material variations, and thermal performance characteristics, see Storage in this chapter and Storage Components in Chapter 7.

11. Sill can be wood or masonry at builder's option, although use of wood is generally recommended for the collector frame. Where applicable, lumber should be pressure-treated.

12. It is recommended that the Thermal Storage Wall foundation be insulated on both sides to prevent heat losses to the outside. In heated basements and heated crawl spaces, insulation is not required on the inside. Insulation levels on both sides should be equal and should meet code requirements for foundation insulation. Insulation must be protected where exposed outside above grade. Cement plaster over wire lath or other methods may be employed. Where dampproofing is used, allow for complete curing before applying insulation. Consult insulation manufacturer for recommended installation methods.

13. Blocking can be discontinuous to provide weep holes where required. It is recommended that insect screens be provided at weep holes.

14. Insulation must be protected where exposed outside above grade. Cement plaster over wire lath or other methods may be employed. Where dampproofing is used, allow for complete curing

before applying insulation. Consult insulation manufacturer for recommended installation methods.

15. Conventional window for light, view, and/or Direct Gain may be incorporated into the Thermal Storage Wall design. Glazing may be single or double (see Collector Rules of Thumb in Chapter 3).

16. Lintel may be bond beam, lintel, or standard concrete masonry units with depressed, cut-out, or grooved webs, or precast concrete. Provide 6'' minimum bearing at each end. Reinforcing is to be designed for loading and clear span per local building codes and/or accepted practice. Facing unit is supported by steel angle.

17. Overhang at collector should be sized to optimize solar gains during heating season and prevent overheating during cooling season. For further information, see Control Rules of Thumb in this chapter and Control Components in Chapter 7.

18. Direct Gain system may be incorporated into upper-level design. For further information on Direct Gain systems, see Chapter 3.

THERMAL STORAGE WALL

Alternates

(Masonry)

OPAQUE INSULATED EXTERIOR VENT

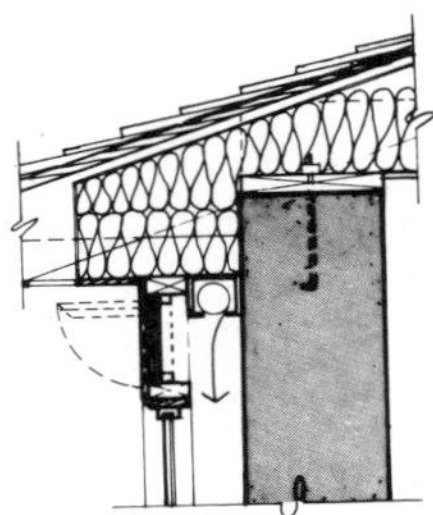

Detached Glazing at Roof

- The exterior vent of a Thermal Storage Wall is frequently constructed of an opaque, rigid, insulating material. This is justified by the fact that the collector directly underneath the roof overlay receives little sunlight.

STEEL ANGLE GLAZING SUPPORT

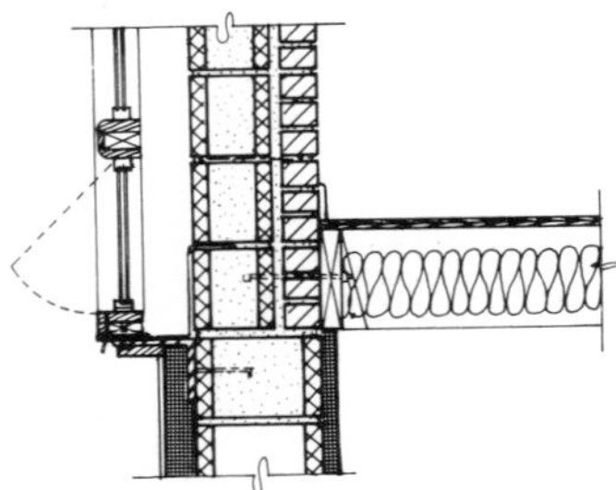

Detached Glazing at Floor

- For detached glazing applications, the collector can be supported by a metal angle fastened directly into the masonry wall.

INTERIOR OPAQUE WALL VENT

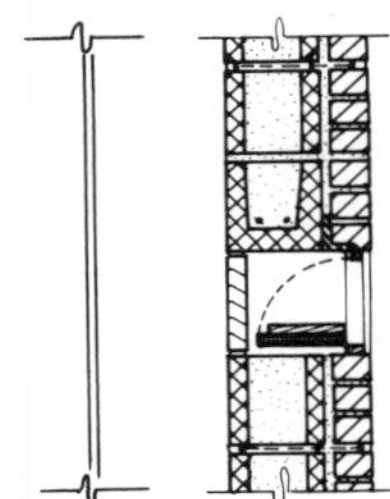

Trombe Wall Vent

- When the Thermal Storage Wall is vented into the living space, it is referred to as a vented or Trombe wall. This detail illustrates one possible vent construction. The damper shown here prevents "reverse thermocirculation." A hinged insulating panel isolates the airspace from the living space during the cooling season.

THERMAL STORAGE WALL

Wall Types

(Masonry)

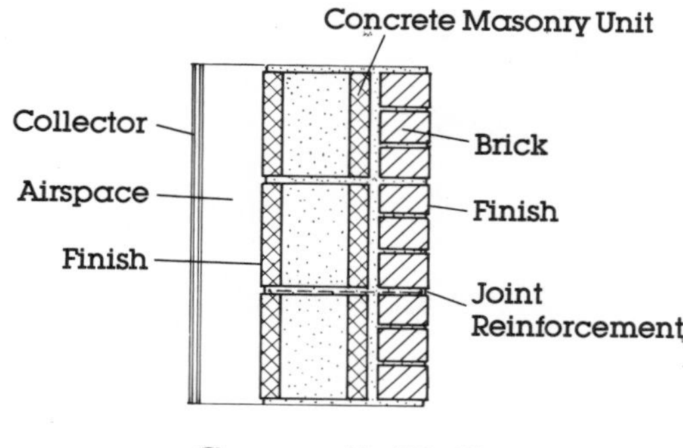

Composite Wall

GENERAL NOTES

- The surfaces of the wall should be covered with clear plastic for one week to allow the wall to outgas properly prior to glazing.

- Thicknesses of individual wythes can vary. For information on recommended composite wall thickness, see Storage Rules of Thumb in this chapter.

- It is recommended that wythes be laid simultaneously. The first wythe laid is parged (backplastered) with mortar not less than ⅜'' thick before the adjacent wythe is laid.

- Openings in masonry walls may be spanned by arches, steel angle lintels, or reinforced masonry lintels (constructed using standard and/or specially shaped masonry units, mortar, grout, and reinforcing steel). Specifications for all three should conform to local building-code requirements, depending on load, span, and type of wall construction.

AIRSPACE

- It is generally recommended that the airspace between the storage and collector components be a minimum of ¾'' and a maximum of 6'' to ensure optimum performance of the Thermal Storage Wall system.

- If movable insulation is installed within the airspace, sufficient space should be provided to allow for proper installation, operation, and maintenance.

FINISH

- For optimum thermal performance, masonry exposed to solar radiation should be a dark color and should be left exposed wherever feasible. A slurry mix of portland cement and sand can be used as a finish and will not affect the thermal performance of the wall. Particular colors can be specified from the masonry manufacturer, or the masonry/slurry can be stained or painted. For further information on finishes and their effects on thermal performance, see Storage in this chapter and Storage Components in Chapter 7.

- For optimum thermal performance, interior masonry finish should be left exposed wherever feasible. Certain finishes (e.g., plaster and gypsum board) can be acceptable if properly applied. Particular colors can also be specified from the manufacturer, or the masonry can be stained or painted. For further information on finishes and their effect on thermal performance, see Storage in this chapter and Storage Components in Chapter 7.

MASONRY

- Concrete block, concrete brick, or architectural facing units, such as split, slump, fluted, scored, or recessed block, may be used for thermal storage. To ensure optimum thermal performance, units must be solid, or hollow with grout-filled cores. For further material information, see Storage Components in Chapter 7. For recommended thicknesses, see Storage Rules of Thumb in this chapter.

- Any of the wide variety of standard bricks are suitable for thermal storage. To ensure proper thermal performance, bricks must be solid, or hollow with grout-filled cores. For further material information, see Storage Components in Chapter 7. For recommended thicknesses, see Storage Rules of Thumb in this chapter.

JOINT REINFORCEMENT

- Joint reinforcement should be provided to control cracking caused by thermal movement and shrinkage. Truss, ladder, or x-tie configurations are suitable and can be embedded in the wall at vertical intervals per local code requirements or industry recommendations. Splices in the reinforcing should be lapped a minimum of 6'', and the width of the reinforcement should be 2'' less than the nominal width of the storage wall.

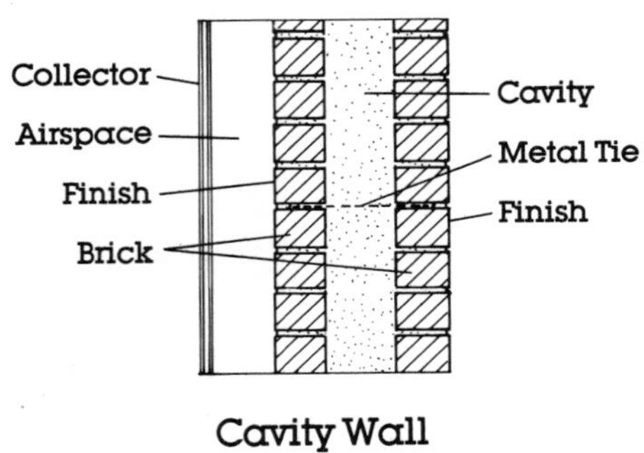

Cavity Wall

GENERAL NOTES

- The surfaces of the wall should be covered with plastic for one week to allow the wall to outgas properly prior to glazing.

AIRSPACE

- It is generally recommended that the airspace between the storage and collector components be a minimum of ¾'' and a maximum of 6'' to ensure optimum performance of the Thermal Storage Wall system.

- If movable insulation is installed within the airspace, sufficient space should be provided to allow for proper installation, operation, and maintenance.

FINISH

- For optimum thermal performance, masonry exposed to solar radiation should be a dark color and should be left exposed wherever feasible. Particular colors can be specified from the manufacturer, or the masonry can be stained or painted. For further information on finishes and their effect on thermal performance, see Storage in this chapter and Storage Components in Chapter 7.

- For optimum thermal performance, interior masonry should be left exposed wherever feasible. Certain finishes (e.g., plaster and gypsum board) can be acceptable if properly applied. Particular colors can also be specified from the manufacturer, or the masonry can be painted. For further information on finishes and their effect on thermal performance, see Storage in this chapter and Storage Components in Chapter 7.

MASONRY

- Any of the wide variety of standard bricks, concrete bricks, or architectural facing units, such as split, slump, fluted, scored, or recessed blocks, are suitable for thermal storage. To ensure proper thermal performance, masonry must be solid, or hollow with grout-filled cores. For further material information, see Storage Components in Chapter 7. For recommended thicknesses, see Storage Rules of Thumb in this chapter.

CAVITY

- It is generally recommended that the cavity be 2-4½'' in width. The cavity must be filled solid in order to ensure proper thermal performance. The cavity may be filled with grout or other suitable materials. A curing period of not less than three days is recommended before filling the cavity with grout. This is necessary to prevent "blow-outs" caused by fluid pressure from the grout.

METAL TIE

- Adjacent wythes must be tied together. Use of 3/16'' corrosion-resistant steel ties or metal tie wire of equivalent stiffness, embedded in the horizontal mortar joints, is generally recommended. Use one tie for each 4½ sq ft of wall area. The maximum vertical distance between ties should not exceed 36''. Individual ties in alternate courses should be staggered.

- Continuous horizontal wall ties (joint reinforcement) of either truss or ladder type may be used instead of individual metal ties. Individual lengths of reinforcement should be lapped at least 6'' to provide continuity. Vertical spacing of joint reinforcement at 8'', 16'', or 24'' is recommended, depending on length and height of wall and the number of openings.

- Metal ties or joint reinforcement should be of sufficient strength to resist "blow-out" due to fluid pressure of grout when filling cavity.

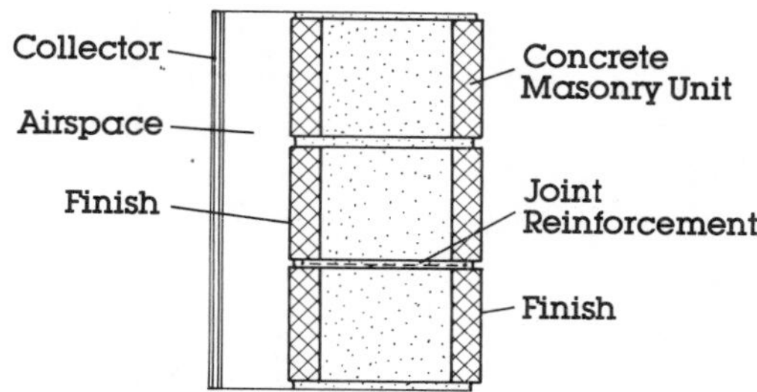

Solid Concrete Masonry Unit

GENERAL NOTES

- Single-wythe concrete masonry units may be built with conventional mortar joints or may be surface bonded. For further information on surface bonding, see Storage Components in Chapter 7.

AIRSPACE

- It is generally recommended that the airspace between the storage and collector components be a minimum of ¾'' and a maximum of 6'' to ensure optimum performance of the Thermal Storage Wall system.

- If movable insulation is installed within the airspace, sufficient space should be provided to allow for proper installation, operation, and maintenance.

FINISH

- For optimum thermal performance, masonry exposed to solar radiation should be a dark color and should be left exposed wherever feasible. A slurry mix of portland cement and sand can be used as a finish and will not affect the thermal performance of the wall. Particular colors can be specified from the masonry manufacturer, or the masonry/slurry can be stained or painted. For further information on finishes and their effects on thermal performance, see Storage in this chapter and Storage Components in Chapter 7.

CONCRETE MASONRY UNIT

- Concrete block, concrete brick, or architectural facing units, such as split, slump, fluted, scored, or recessed block, may be used for thermal storage. To ensure optimum thermal performance, units must be solid, or hollow with grout-filled cores. For further material information, see Storage Components in Chapter 7. For recommended thicknesses, see Storage Rules of Thumb in this chapter.

JOINT REINFORCEMENT

- Joint reinforcement should be provided to control cracking caused by thermal movement and shrinkage. Truss, ladder, or x-tie configurations are suitable and can be embedded in the wall at vertical intervals per local code requirements or industry recommendations. Splices in the reinforcing should be lapped a minimum of 6'', and the width of the reinforcement should be 2'' less than the nominal width of the storage wall.

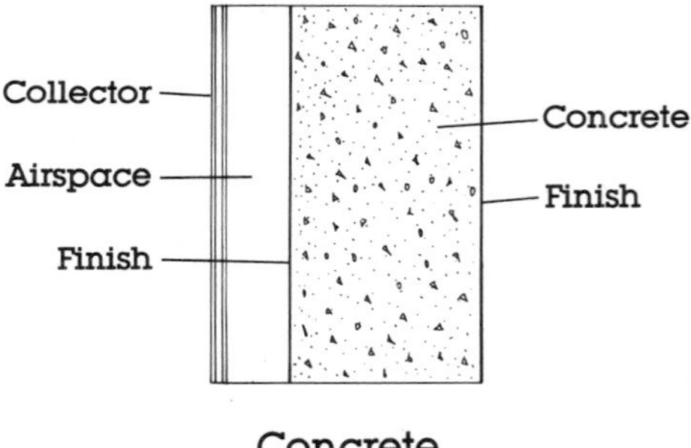

Concrete

CONCRETE

- The use of normal-weight concrete (140-150 lb) is generally recommended. Placement of formwork and pouring of concrete should follow standard practice and/or local building codes. For further information, see Storage Components in Chapter 7. For recommended material thicknesses, see Storage Rules of Thumb in this chapter.

FINISH

- For optimum performance, concrete exposed to solar radiation should be a dark color wherever feasible. Particular colors can be specified from the supplier, or the concrete can be stained or painted. For further information on finishes and their effect on thermal performance, see Storage in this chapter and Storage Components in Chapter 7.

- For optimum thermal performance, masonry exposed to solar radiation should be a dark color and left exposed wherever feasible. A slurry mix of portland cement and sand can be used as a finish and will not affect the thermal performance of the wall. Particular colors can be specified from the masonry manufacturer, or the masonry/slurry can be stained or painted. For further information on finishes and their effects on thermal performance, see Storage in this chapter and Storage Components in Chapter 7.

AIRSPACE

- It is generally recommended that the airspace between the storage and collector components be a minimum of ¾'' and a maximum of 6'' to ensure optimum performance of the Thermal Storage Wall system.

- If movable insulation is installed within the airspace, sufficient space should be provided to allow for proper installation, operation, and maintenance.

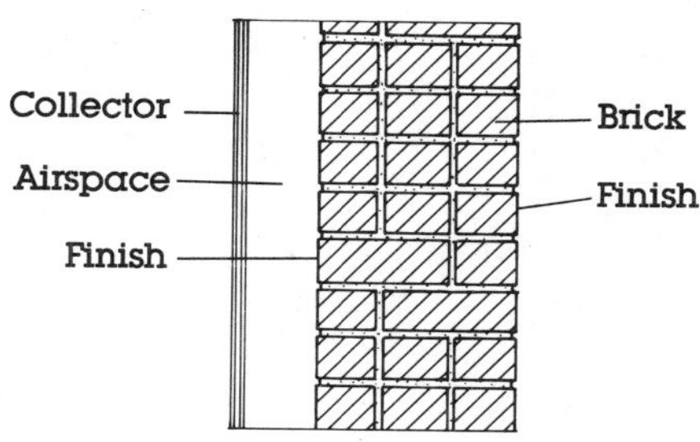

Solid Brick

GENERAL NOTES

- Structural bonding of the wall may be accomplished by overlapping (interlocking) masonry units. The maximum vertical and horizontal distance between headers is generally recommended to be 24'', subject to local code restrictions.

- Openings in brick walls may be spanned by arches, steel angle lintels, or reinforced brick lintels (constructed using standard and/or specially shaped brick units, mortar, grout, and reinforcing steel). Specifications for all three should conform to the Brick Institute of America recommendations and local building-code requirements, depending on load, span, and type of wall construction.

AIRSPACE

- It is generally recommended that the airspace between the storage and collector components be a minimum of ¾'' and a maximum of 6'' to ensure optimum performance of the Thermal Storage Wall system.

- If movable insulation is installed within the airspace, sufficient space should be provided to allow for proper installation, operation, and maintenance.

FINISH

- For optimum thermal performance, masonry exposed to solar radiation should be a dark color and should be left exposed wherever feasible. A slurry mix of portland cement and sand can be used as a finish and will not affect the thermal performance of the wall. Particular colors can be specified from the masonry manufacturer, or the masonry/slurry can be stained or painted. For further information on finishes and their effects on thermal performance, see Storage in this chapter and Storage Components in Chapter 7.

BRICK

- Any of the wide variety of standard bricks are suitable for thermal storage. To ensure proper thermal performance, bricks must be solid, or hollow with grout-filled cores. For further material information, see Storage Components in Chapter 7. For recommended thicknesses, see Storage Rules of Thumb in this chapter.

- Through-the-wall units may also be specified for thermal storage. Such units are generally cored and must be grout-filled to ensure proper performance.

THERMAL STORAGE WALL

(Water)

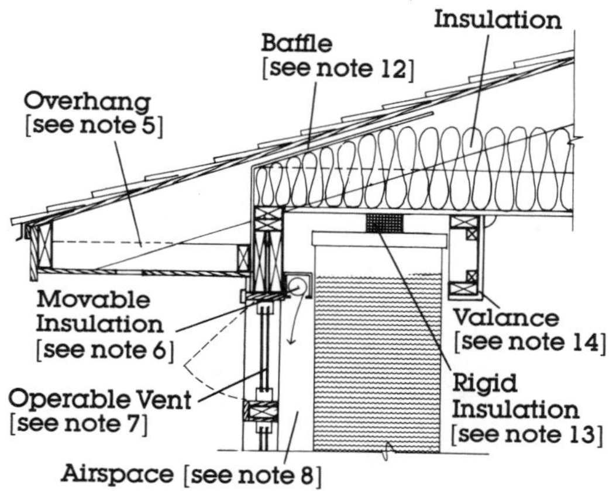

Water Tubes at Roof

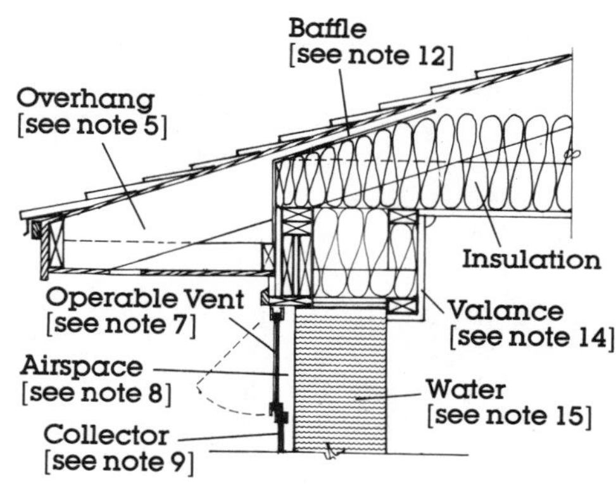

Water Containers at Roof

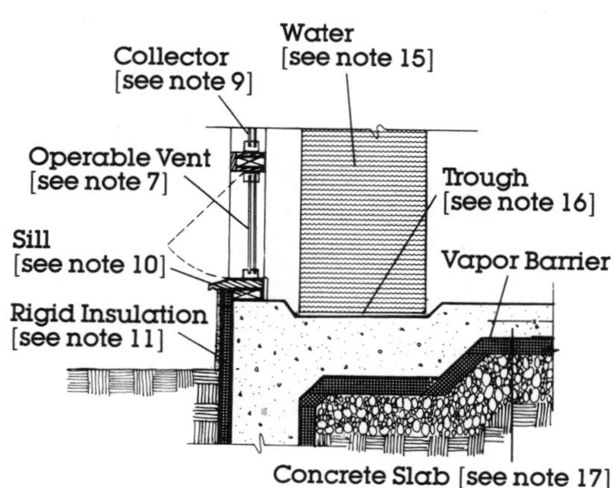

Water Tubes at Foundation

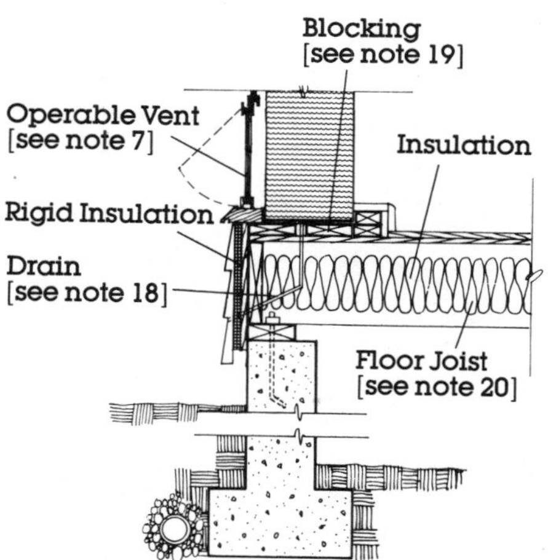

Water Containers at First Floor

Notes

(Water)

GENERAL NOTES:

1. All footings must bear on undisturbed soil. Adjust footing size, reinforcing, and depth below grade as required by site conditions and/or local code. Where groundwater problems exist, provide sufficient granular fill to prevent water penetration at slab-on-grade.

2. Placement of reinforcing, metal ties, anchor bolts, flashing, and weep holes is to be determined by local building codes and/or accepted practice. Reinforcing in concrete slab is not recommended except where required by local conditions and/or code.

3. Insulation levels at foundation/slab and/or floor, and roof/ceiling must meet or exceed requirements of local building/energy codes. The storage mass should not be insulated. In crawl-space or basement construction, it is recommended that the Thermal Storage Wall foundation be insulated on both sides to prevent losses to the outside and to the ground except where basement or crawl space is heated. In these cases, insulation is not required on the inside.

4. Other roof configurations may be used at builder's option. Consult local building codes for restrictions and special requirements concerning roof/wall connections, especially in seismic and/or high-wind areas. Roof/attic section should be vented to ensure proper performance of insulation. Vents may be located at sides of roof construction or at soffit and/or ridge vents. Provide gutters and downspouts as required.

DETAIL NOTES:

5. Overhang at collector should be sized to optimize solar gains during heating season and prevent overheating during heating season. For further information, see Control Rules of Thumb in this chapter and Control Components in Chapter 7.

6. Movable insulation to control nighttime and cloudy-day heat loss and to prevent overheating during the cooling season should be provided. Insulation may be installed within the airspace or placed on the exterior of the collector. In the latter case, insulating shutters or panels can be used. Care must be taken to ensure a tight fit between the insulation and the collector to reduce heat loss at the shutter or panel edges. When provided within the airspace, care must be taken to ensure adequate space for proper installation and operation. Where upper and lower portions of the Thermal Storage Wall are separated, movable insulation must be installed in each section. For further information, see Control Components in Chapter 7.

7. Portions of the collector should be operable to allow venting of the Thermal Storage Wall during periods of excessive heat gain. Awning-type or

other operable windows may be specified and should be equipped with demountable insect screens that can be removed during the heating season.

8. It is generally recommended that the airspace between the glazing and the storage mass be a minimum of ¾″ and a maximum of 6″ to avoid excessive heat loss.

9. Double glazing is generally recommended to ensure optimum thermal performance. A ½″ airspace between glazing layers is recommended where feasible. The glazing material can be glass or plastic (see Collector Components in Chapter 7). The collector unit should be demountable or externally operable to aid in maintenance and cleaning. The collector frame may be detached from the wall and designed to support the glazing, glazing frame (if preframed), roof, and other related loads, or it may be attached to and supported by the wall system. Framing should be kept to a minimum, since it reduces the effective collector area and may cast shadows, both of which will affect the thermal performance of the wall. All structural framing and connections shall be designed per local building-code restrictions and/or accepted practice. Frames should be continuous at edges. 2x framing members bolted to the wall assembly can serve as spacers, ensuring adequate airspace, and should be discontinuous, allowing for free circulation of air behind the collector.

10. Continuous sill sealer is recommended to provide protection against infiltration.

11. It is recommended that the Thermal Storage Wall foundation be insulated on both sides to prevent heat losses to the outside. In heated basements and heated crawl spaces, insulation is not required on the inside. Insulation levels on both sides should be equal and should meet code requirements for foundation insulation. Insulation must be protected where exposed outside above grade. Cement plaster over wire lath or other methods may be employed. Where dampproofing is used, allow for complete curing before applying insulation. Consult insulation manufacturer for recommended installation methods.

12. Provide baffles where necessary to maintain 1″ minimum airspace for venting attic.

13. The airspace in a Thermal Storage Wall system must be thermally isolated from the adjacent living spaces for optimum performance. Blocking made of rigid insulation serves to provide such isolation and also may provide some lateral bracing for water storage tubes.

14. It is recommended that freestanding water storage containers be braced against lateral movement. Bracing may be in the form of a framed valance.

15. Water, stored in appropriate containers, is an effective material for

use in Thermal Storage Wall systems. Suitable containers include fiberglass tubes, polyethylene or steel drums, and metal culverts. Prefabricated tank systems designed to be set within or on structural wall framing are also available. Care should be taken to ensure that steel and metal containers are lined with corrosion-resistant materials. In all cases, water should be treated with algae-retardant chemicals. For further information on the use of water as thermal storage, including descriptions of appropriate containers, see Storage Components in Chapter 7.

16. Troughs should be provided as a precaution against leakage of water from containers or from condensation. Troughs should be formed of corrosion-resistant materials attached or set into the supporting floor system, so as to provide added lateral bracing at the base of the container. Troughs may be designed for a single container or a row of containers. It is generally recommended that each trough be provided with a slight slope of the base to a central drain that is tied to the main plumbing system of the building or other drainage system.

17. Concrete slab should be thickened to take the additional weight of the water containers. Consult local codes for requirements.

18. PVC drains should be provided to eliminate water from drainage trough as required.

19. Blocking is provided beneath the trough to allow space for the drain.

20. Consideration must be given to the additional weight of the water containers. Consult structural design loading tables and local codes to determine suitable grade and spacing of joists, and grade and thickness of subfloor.

THERMAL STORAGE WALL

(PCM)

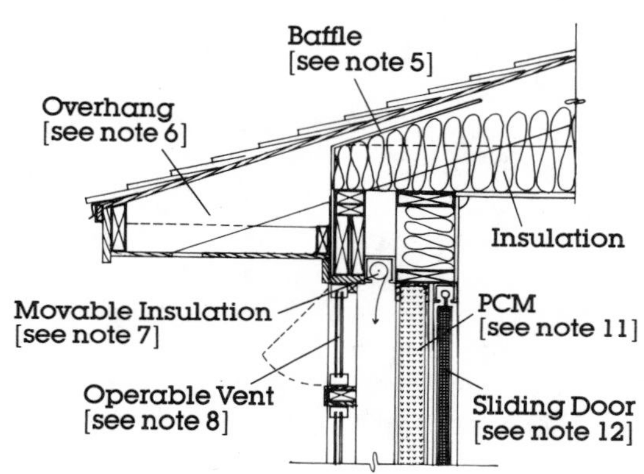

Vertical Rods at Roof

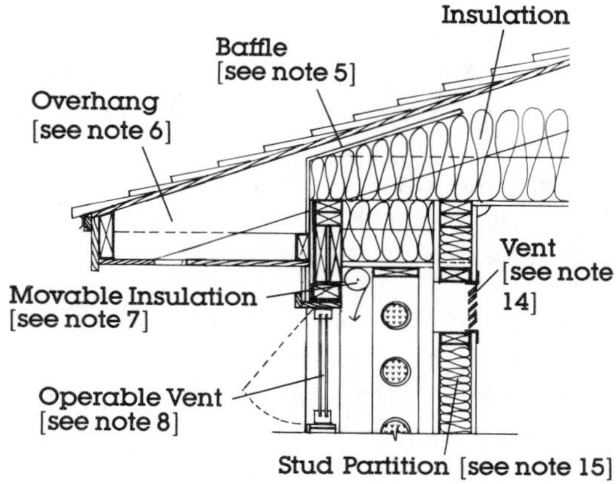

Horizontal Rods at Roof

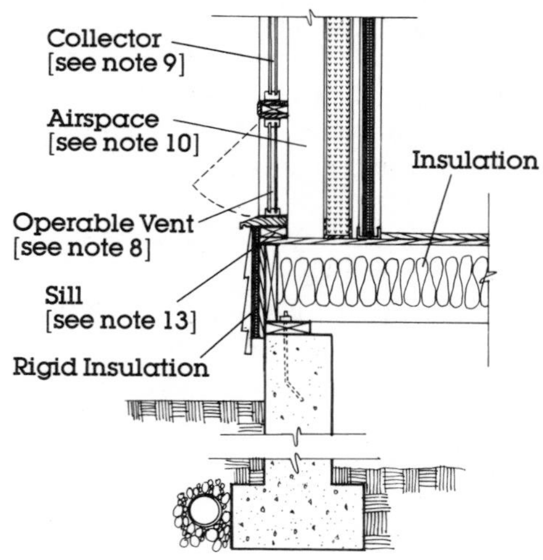

Vertical Rods at First Floor

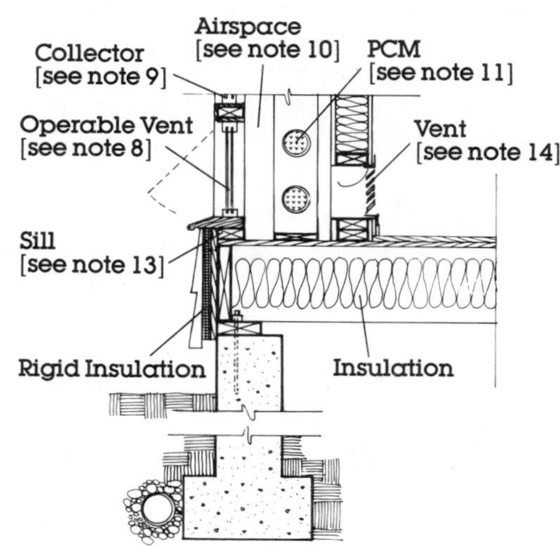

Horizontal Rods at First Floor

Notes
(PCM)

GENERAL NOTES:

1. All footings must bear on undisturbed soil. Adjust footing size, reinforcing, and depth below grade as required by site conditions and/or local code. Where groundwater problems exist, provide sufficient granular fill to prevent water penetration at slab-on-grade.

2. Placement of reinforcing, metal ties, anchor bolts, flashing, and weep holes to be determined by local building codes and/or accepted practice. Reinforcing in concrete slab is not recommended except where required by local conditions and/or code.

3. Insulation levels at foundation/slab and/or floor, and roof/ceiling must meet or exceed requirements of local building/energy codes. The storage mass should not be insulated. In crawl-space or basement construction, it is recommended that the Thermal Storage Wall foundation be insulated on both sides to prevent losses to the outside and to the ground except where basement or crawl space is heated. In these cases, insulation is not required on the inside.

4. Other roof configurations may be used at builder's option. Consult local building codes for restrictions and special requirements concerning roof/wall connections, especially in seismic and/or high-wind areas. Roof/attic section should be vented to ensure proper performance of insulation. Vents may be located at sides of roof construction or at soffit and/or ridge vents. Provide gutters and downspouts as required.

DETAIL NOTES:

5. Provide baffles where necessary to maintain 1'' minimum airspace for venting attic.

6. Overhang at collector should be sized to optimize solar gains during heating season and prevent overheating during cooling season. For further information, see Control Rules of Thumb in this chapter and Control Components in Chapter 7.

7. Movable insulation to control nighttime and cloudy-day heat loss and to prevent overheating during the cooling season should be provided. Insulation may be installed within the airspace or placed on the exterior of the collector. In the latter case, insulating shutters or panels can be used. Care must be taken to ensure a tight fit between the insulation and the collector to reduce heat loss at the shutter or panel edges. When provided within the airspace, care must be taken to ensure adequate space for proper installation and operation. Where upper and lower portions of the Thermal Storage Wall are separated, movable insulation must be installed in each section. For further information, see Control Components in Chapter 7.

8. Portions of the collector should be operable to allow venting of the Thermal Storage Wall during periods of

excessive heat gain. Awning-type or other operable windows may be specified and should be equipped with demountable insect screens that can be removed during the heating season.

9. Double glazing is generally recommended to ensure optimum thermal performance. A ½'' airspace between glazing layers is recommended where feasible. The glazing material can be glass or plastic (see Collector Components in Chapter 7). The collector unit should be demountable or externally operable to aid in maintenance and cleaning. The collector frame may be detached from the PCM wall assembly and designed to support the glazing, glazing frame (if preframed), roof, and other related loads, or it may be attached to and supported by the PCM wall assembly. Framing should be kept to a minimum, since it reduces the effective collector area and may cast shadows, both of which will affect the thermal performance of the wall. All structural framing and connections shall be designed per local building-code restrictions and/or accepted practice. Frames should be continuous at edges. 2x framing members bolted to the PCM wall assembly can serve as spacers, ensuring adequate airspace, and should be discontinuous, allowing for free circulation of air behind the collector.

10. It is generally recommended that the airspace between the glazing and the storage mass be a minimum of ¾'' and a maximum of 6'' to avoid excessive heat loss.

11. Phase-change materials contained in rods may be specified for use in Thermal Storage Wall systems. Rods may be installed vertically or horizontally. If rods are set so as to be in contact with wood framing members (or other nonfireproof material), gaskets should be provided to eliminate direct contact between the rods and the wood. Particular PCM products should be installed per manufacturers' recommendations. For further information on PCMs, including availability, performance characteristics, and applications, see Storage Components in Chapter 7.

12. An insulated sliding door provides thermal isolation of the rods during the day, allowing for optimum performance. At night, with doors opened, the rods heat adjacent living spaces.

13. Continuous sill sealer is recommended to provide protection against infiltration.

14. Vents are provided at top and bottom of the adjacent stud wall to allow for air heated by the rods to circulate into the living spaces. A backdraft damper should be provided in the lower vent to prevent reverse thermocirculation. Vents should be closed completely during the cooling season to prevent heated air from entering into the living spaces.

15. The partition is insulated to provide thermal protection at the wall to reduce heat losses. Insulation level should meet or exceed that specified by local building/energy codes.

TVA Solar House #2
Thermal Storage Wall

Designed to be built into a south-facing slope, the lower level of this 2,000-sq-ft house is nearly hidden from view on three sides. In addition to berming, the home is buffered on the north by a variety of circulation and service functions, including the air-lock front entry, the kitchen, and a bathroom. The house opens up to the south to reveal a two-story Thermal Storage Wall punctured with operable windows for light and view, and an integrated second-floor sun porch, which functions as an Attached Sunspace.

The Thermal Storage Wall, constructed of grout-filled 12" concrete blocks, is coated at its exterior with a dark-stained cement plaster to increase solar absorption. A light-colored plaster interior finish retains the thermal integrity of the wall. Operable insulated vents at the top and bottom of the collector frame provide ventilation of the airspace in the summer. Roof and first-floor south wall overhangs are dimensioned to allow the sun to strike the wall in winter and shade it in summer. Wing walls offer added protection against overheating.

The sun porch, with access to the kitchen and dining areas, occupies the southeast corner of the building. The interior concrete wall is directly heated by the winter sun in the morning and early afternoon hours.

Above the sun porch, a thermostatically controlled fan vents heated air through an insulated ceiling duct to remote interior spaces. To cool the sun porch in the summer, an insulated ceiling damper is opened. The attic space in turn is ventilated by a thermostatically controlled exhaust fan in the roof.

The home is well insulated, including 2" of rigid insulation placed at the upper edge of the bermed foundation wall; 1" minimum R-8 rigid waterproof insulation down to the footings; and R-30 batt insulation in the roof.

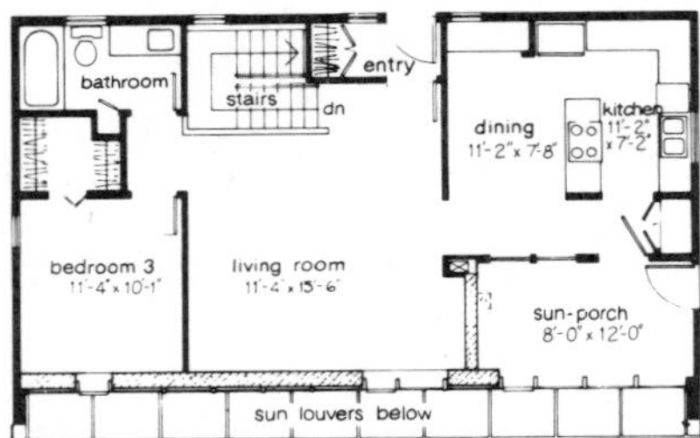

Upper Level

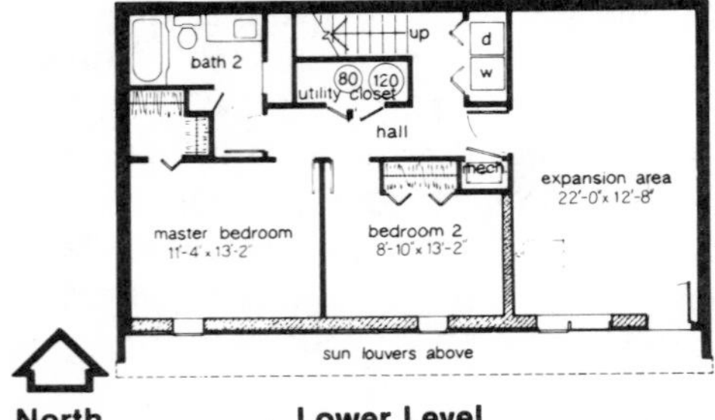

North

Lower Level

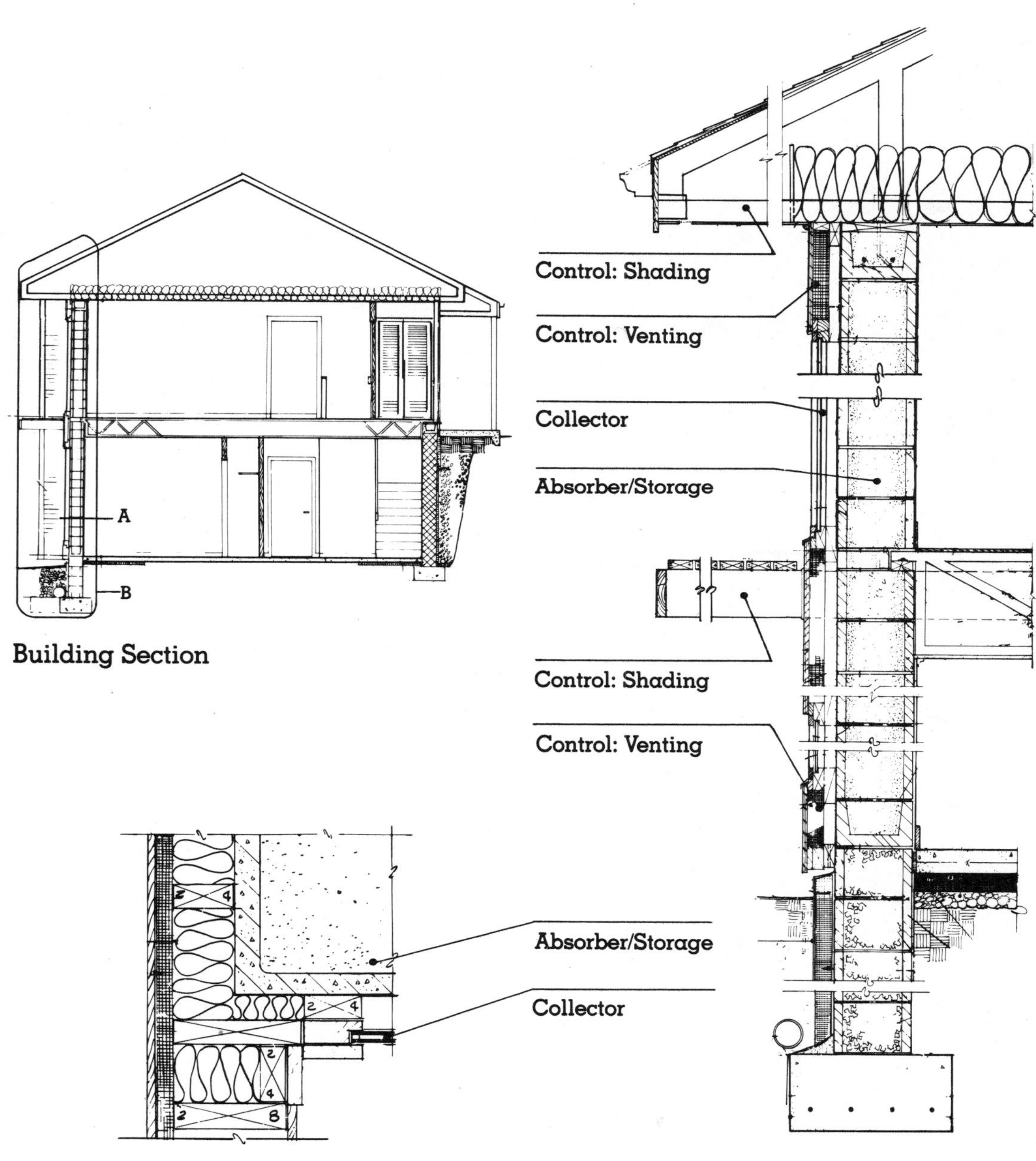

Building Section

A Corner Detail: Plan

B Thermal Storage Wall Detail: Section

SSEC Louisville House

Thermal Storage Wall

This 2,300-sq-ft two-story home designed for Louisville, Kentucky, is protected on the north by an air-lock entry and garage. The south facade integrates Thermal Storage Wall, Attached Sunspace, and Direct Gain systems into a coherent and effective design that substantially reduces energy costs.

The two-story Thermal Storage Wall is composed of 10″ concrete masonry blocks filled solid with grout, with the exterior painted black to increase the solar absorption. The double-glazed collectors are wood framed and heavily caulked with a silicone sealant. A ¾″ plaster interior surface creates an attractive and thermally effective wall finish.

The Thermal Storage Wall is fully vented, providing three modes of operation. During the heating season, the interior vents at the top and bottom of each floor can be opened during the day to circulate heated air through the home, or closed to capture and store the heat in the thermal mass of the wall. In the cooling season, exterior vent panels in the collector serve to remove heat by circulating outside air in the airspace, cooling the wall. For ease of use, a pulley system enables remote operation of these panels from below.

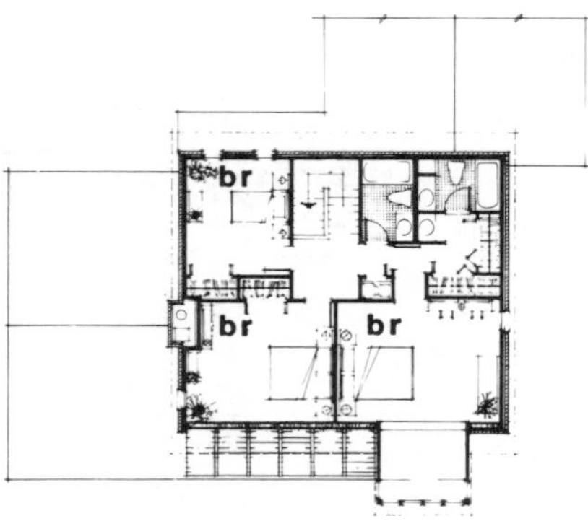

Upper Floor

A wooden trellis with angled slats and second-story overhang are both designed as sunscreens to permit the winter sun to strike the wall while preventing direct summer gain.

Flanking the Thermal Storage Wall, two Attached Sunspaces collect and store heat while providing additional living space. Sliding glass doors allow heat exchange between the single-height solarium and both the living and dining rooms. Brick-veneered west wall and brick paver floor provide heat storage.

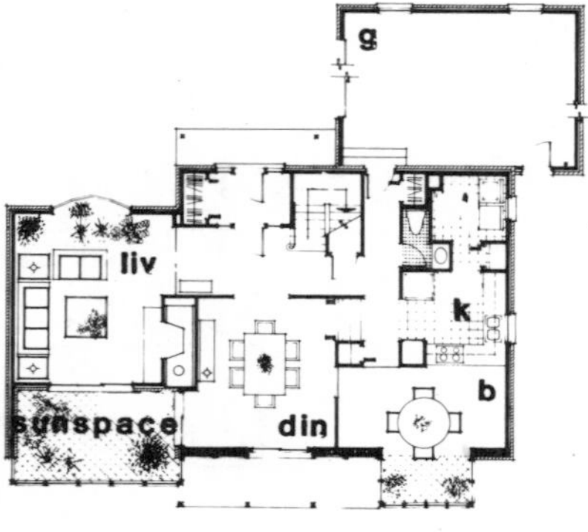

Lower Floor

A double-height open Sunspace creates a dramatic breakfast area and provides additional solar gain. Heat storage is provided by the interior masonry-veneered walls and tile floor. A roof overhang protects this space from overheating.

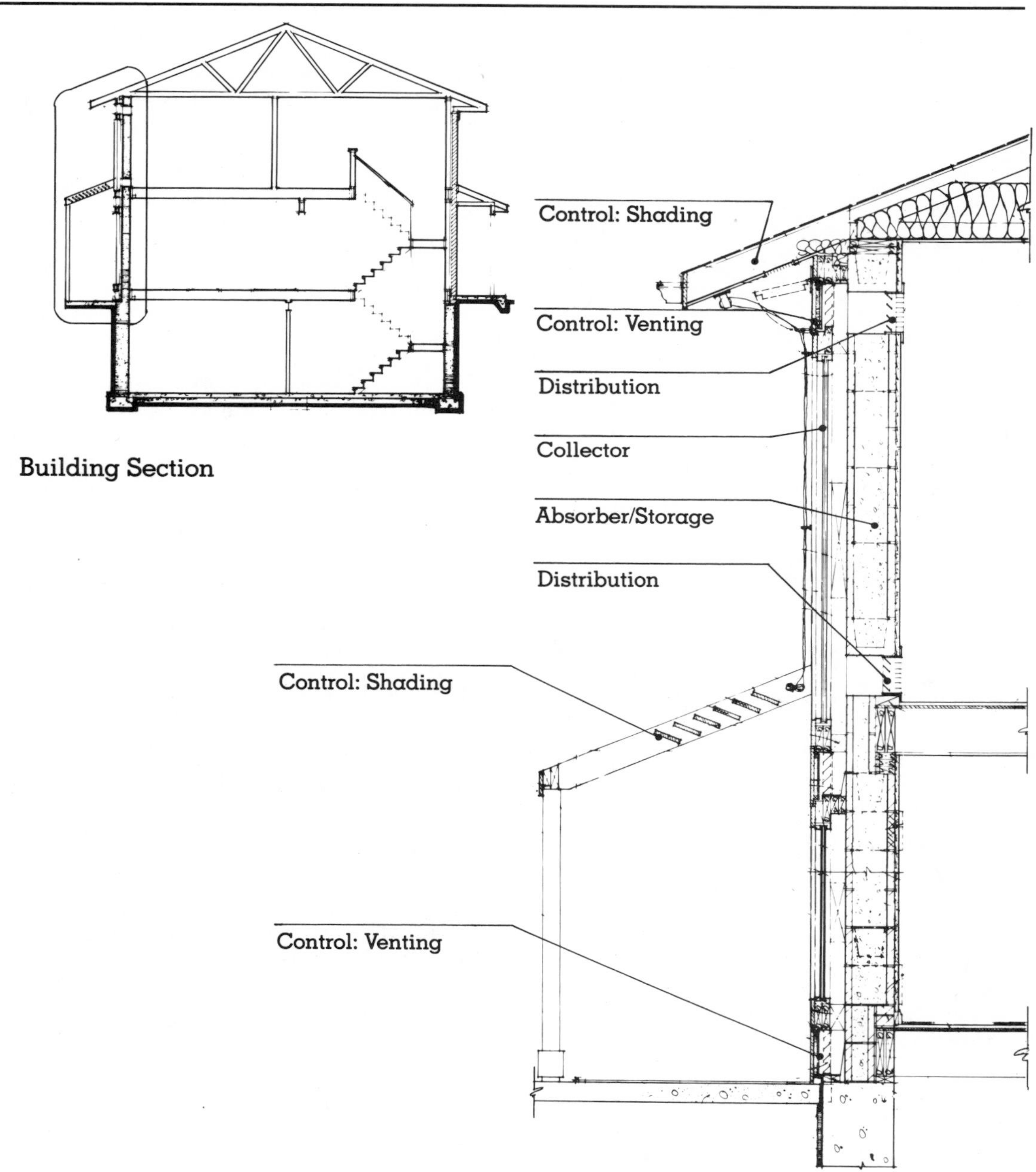

Building Section

Thermal Storage Wall Detail: Section

Pine Hall House

Thermal Storage Wall

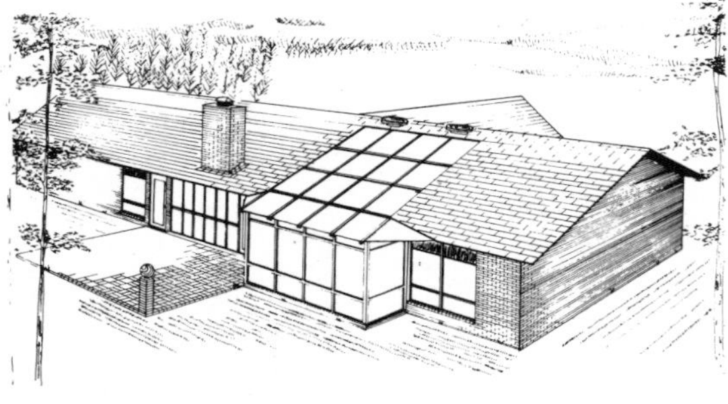

The 2,000-sq-ft single-story Winston-Salem, North Carolina, house employs Thermal Storage Wall and Attached Sunspace systems for passive solar heating. The floor plan is laid out with the garage protruding as a windbreak on the north. The recessed entry and service spaces buffer much of the north facade, allowing the living areas to be located on the south.

The Thermal Storage Wall, which is constructed of three wythes of brick, is located at the center of the living room with glazed doors at each end for daylighting. The wall is vented, with openings at its base to draw cool room air into the airspace and at the top of the wall to deliver warm, solar-heated air into the living areas. Hinged exterior vents in the collector frame are opened in the cooling season to prevent overheating of the space and wall. An overhang protects the Thermal Storage Wall from the direct summer sun.

North

Main Floor Plan

The Thermal Storage Wall doubles as a fireplace to take advantage of the heat-retaining qualities of the masonry. A brick-veneered interior wall between the kitchen and living room and the brick paver floor provide additional heat storage.

Adjacent to both the living room and main bedroom is a Sunspace, or solar atrium. The roof and south wall of the Sunspace are glazed, flooding the interior with light and solar heat. Interior walls of the Sunspace are also glazed, isolating it from the other living spaces while preserving the openness of the plan. Its interior doors and windows can be opened to release collected heat to the rest of the home. Exterior shades placed over the glazing and two exhaust fans help keep the Sunspace cool in the summer.

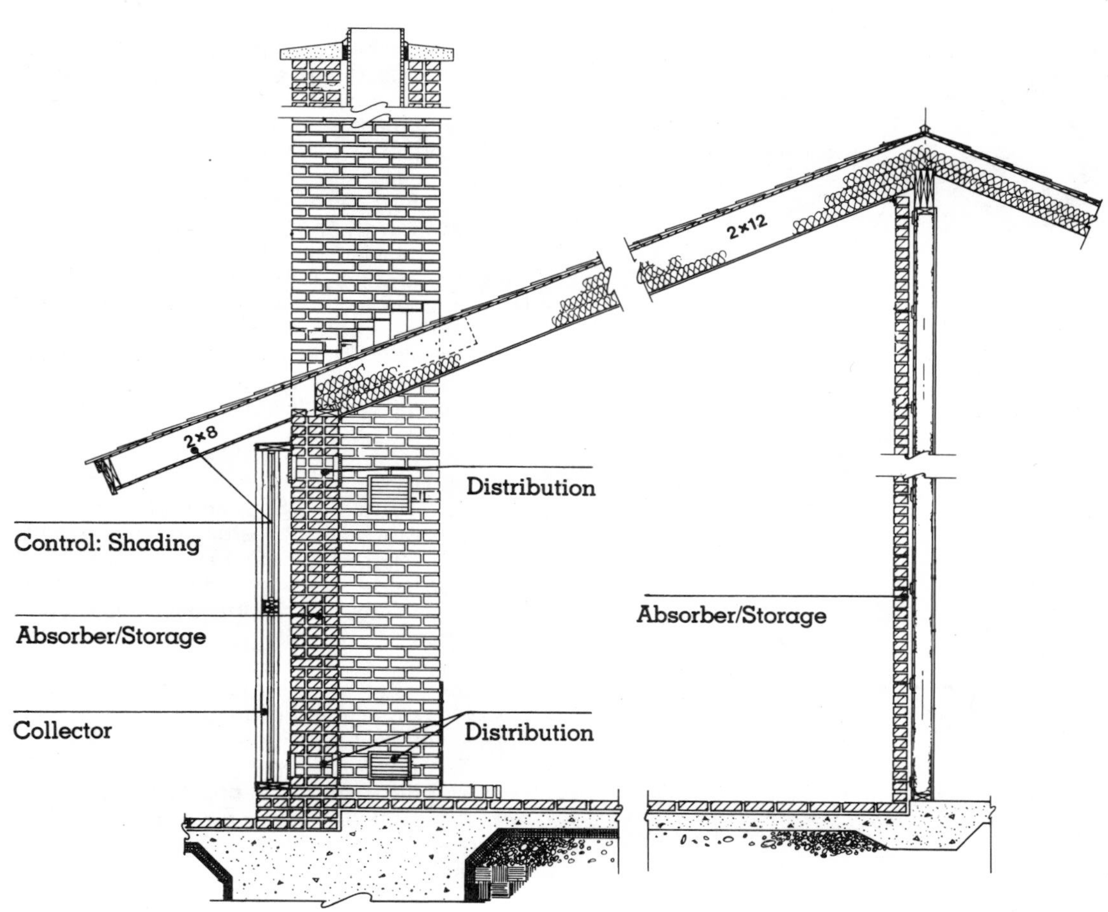

Building Section

PCA Split-Level
Thermal Storage Wall

This 2,100-sq-ft split-level passive home is designed for a sloping site and benefits from being earth-sheltered on its north and east sides. On the south, concrete masonry unit panels alternate with expanses of operable windows, combining the advantages of the Thermal Storage Wall and Direct Gain systems.

The partially bermed north entry facade is insulated with R-10 exterior foam board, reducing slab edge heat losses. The kitchen, a heat-generating space, is ideally placed with the garage, air-lock entry, and other circulation and service areas along the north, buffering the home's major living spaces. Living spaces are concentrated on the southern side of the house with direct access to passive solar gain.

The entry-level concrete slab floor is finished with quarry tile for heat storage. Clerestory windows above the 1½-story living and family rooms permit winter sun to strike most of the floor surface area in these rooms. An energy-conserving fireplace aids in reducing the home's fuel costs.

The south-facing, two-story Thermal Storage Wall is constructed of fully grouted 8″ concrete blocks with 4″ concrete bricks as the interior finish. For greater absorption, the exterior face of these composite panels is painted black. Generous overhangs and wing walls effectively shade the south wall from summer sun and overheating.

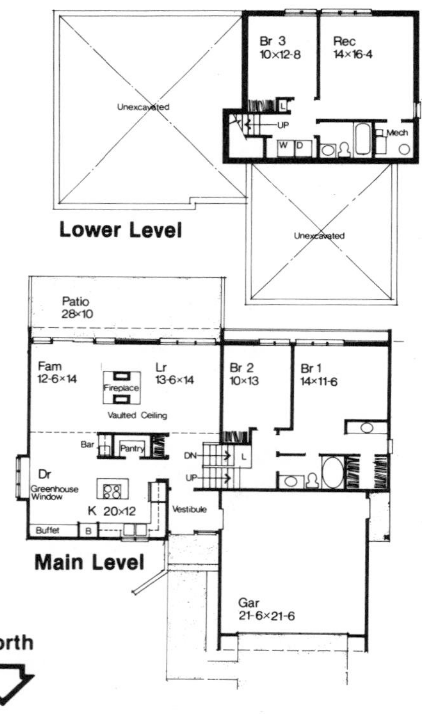

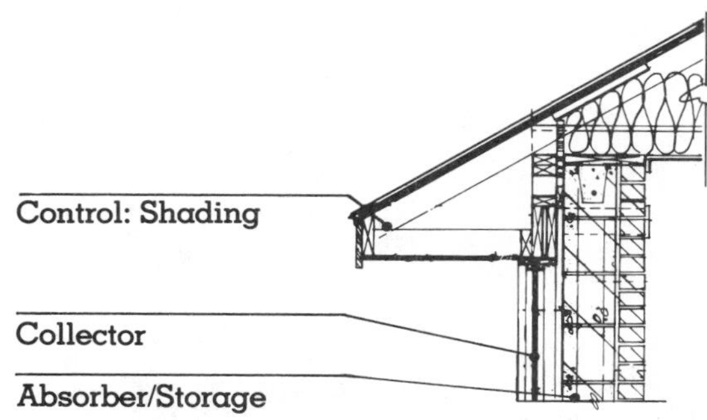

A Thermal Storage Wall: Roof Detail

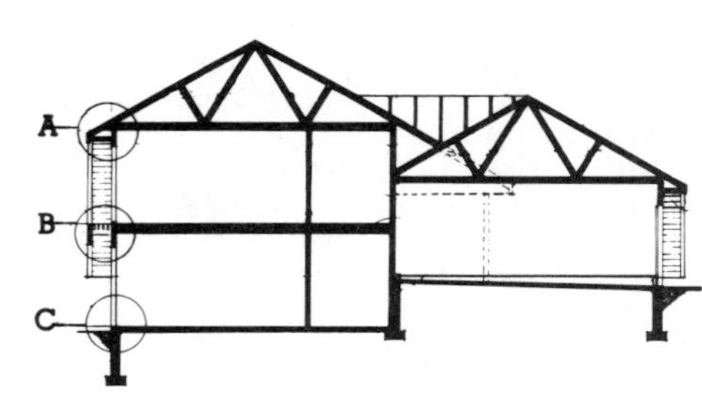

Building Section

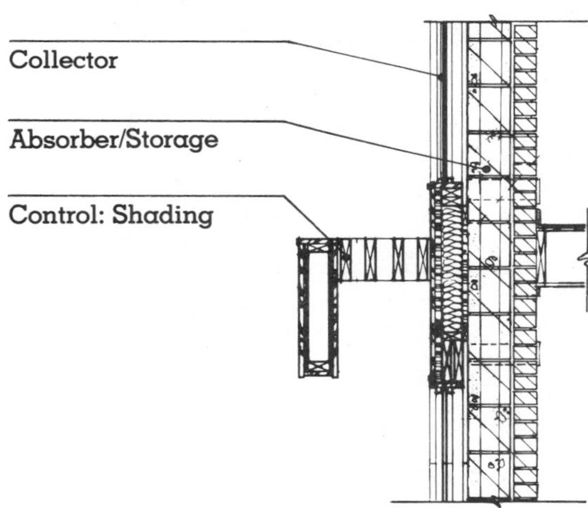

B Thermal Storage Wall: Upper Floor Detail

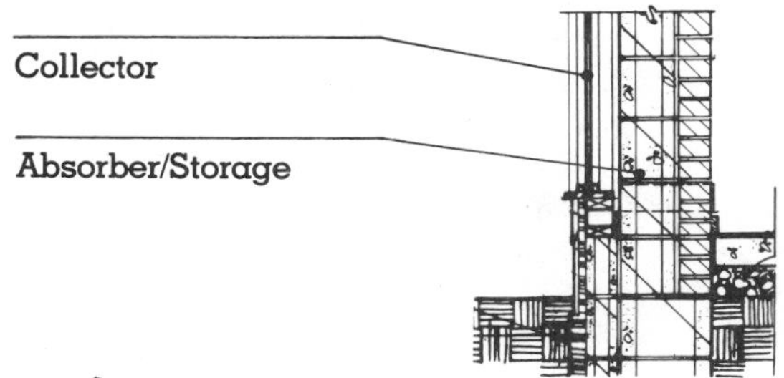

C Thermal Storage Wall: Foundation Detail

TVA Solar House #9
Thermal Storage Wall

A two-story Thermal Storage Wall provides passive solar heating for this 1,200-sq-ft traditional farm-style house. Designed for a flat or slightly sloping site, kitchen and other service and circulation spaces buffer the north. An optional garage can be constructed to the north, further protecting the home.

The south-facing Thermal Storage Wall alternates with operable windows. The wall is designed to be constructed of 12" heavy-weight grout-filled, hollow-core concrete blocks for effective heat storage.

During the summer, operable exterior insulated vents at the top and bottom of the collector frames allow outside air to enter the airspace, producing potential heat gain to the wall. Roof and intermediate-level overhangs screen the Thermal Storage Wall from heat gain in summer, further reducing the home's cooling requirements. Wing walls offer additional protection from the west.

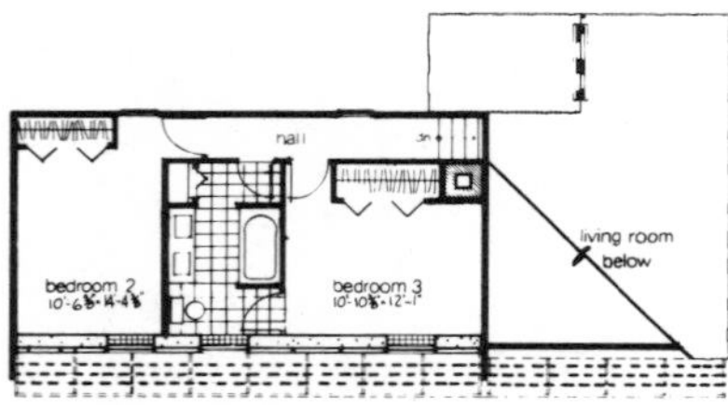

Upper Level 549sf

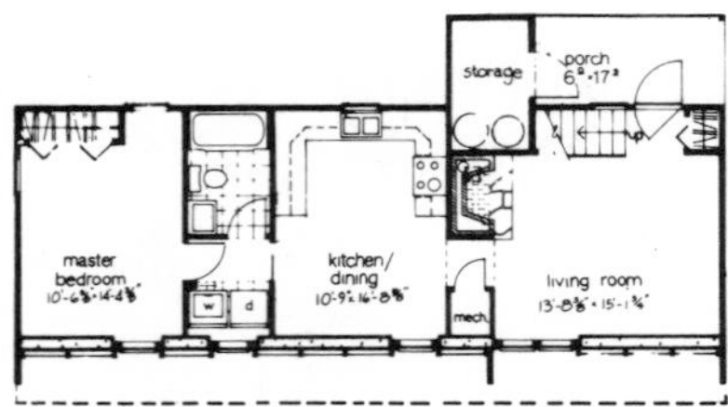

Lower Level 822 sf
+549 sf upper level
1,371 sf gross house

North

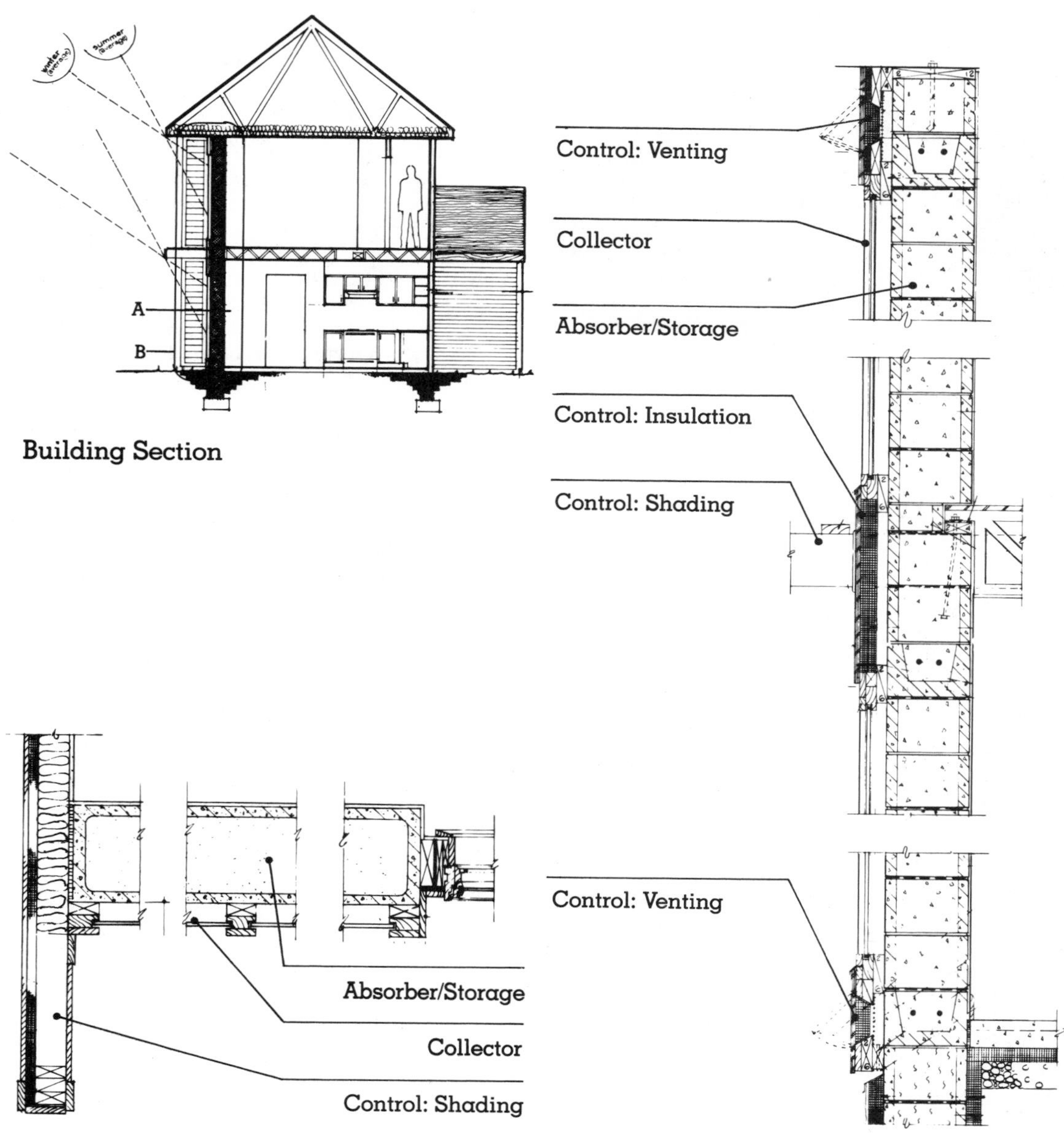

Building Section

A Corner Detail: Plan

B Thermal Storage Wall Detail: Section

General Shale House
Thermal Storage Wall

The 2,350-sq-ft house is designed for a site that slopes to the south. Partially earth-sheltered, the home is entered on the upper level from the north, with bedrooms beneath the entry level. Circulation and service spaces, including an air-lock entry vestibule, provide buffers for the north side of the home. Passive solar heating is provided by Thermal Storage Wall and Direct Gain systems, covering most of the south facade.

Windows incorporated into the brick and concrete masonry Thermal Storage Wall provide Direct Gain heating and daylighting. Sunscreens overhang at both floors to protect the south glass from excessive summer heat gain.

An Attached Sunspace family room that opens onto a second-story deck connects the garage and house proper. The brick pavers laid on a concrete slab floor store solar heat in the winter, tempering the space throughout the day and night. R-5 insulating drapes effectively control nighttime heat losses, and a roof overhang controls summer heat gain.

R-30 ceiling and R-19 wall insulation levels are specified, completing a thermally tight building envelope. Supplementary heating is provided by a woodburning fireplace.

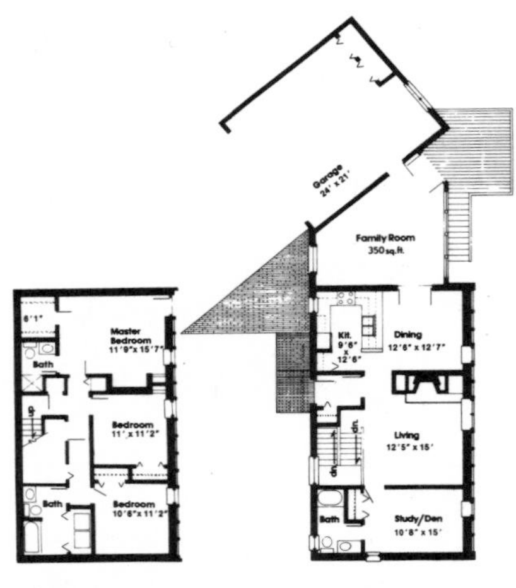

Upper Floor Plan

Main Floor Plan

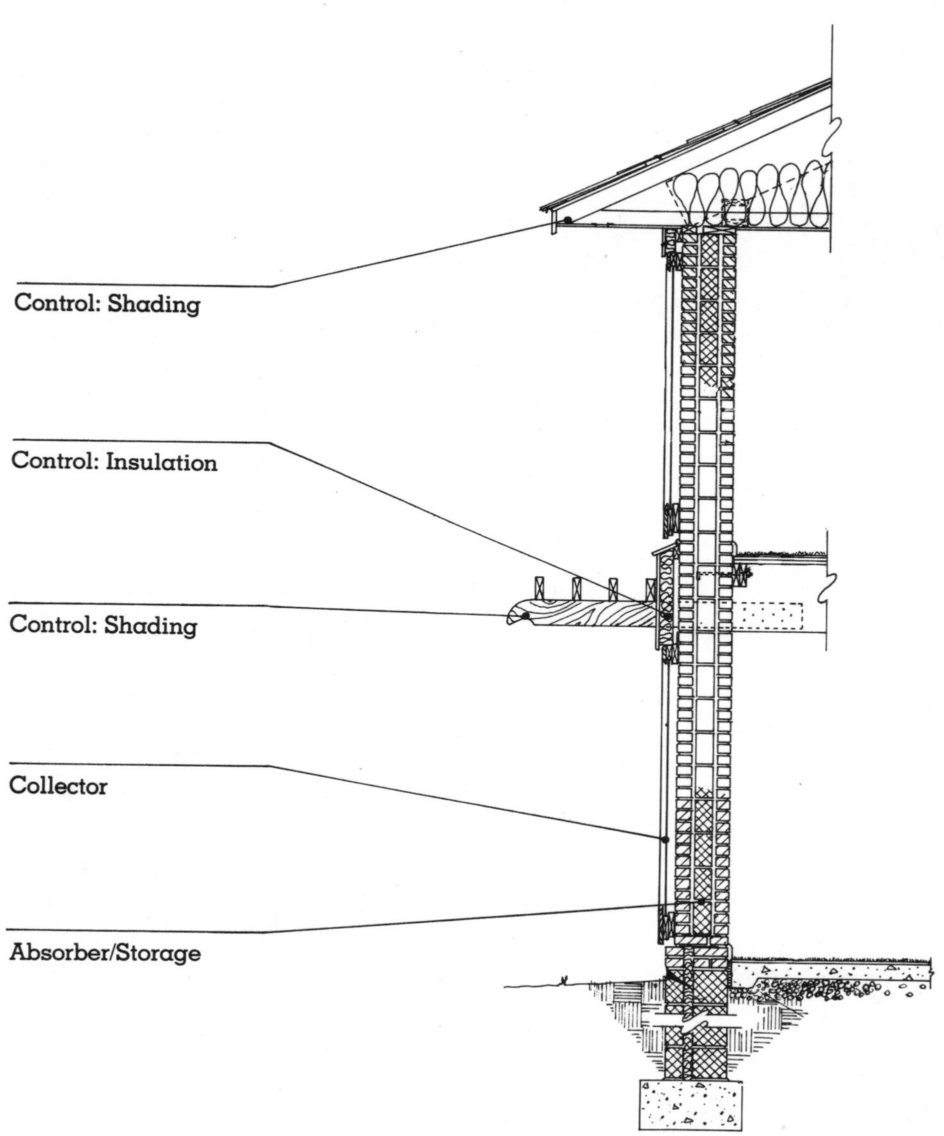

Thermal Storage Wall Detail: Section

PEG House #2
Thermal Storage Wall

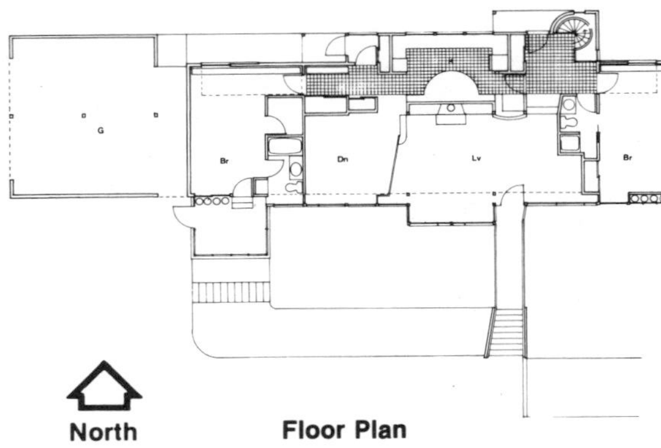

Floor Plan

The Geisel house combines Direct Gain with earth-sheltered construction to significantly reduce the energy needed to heat its 2,900 sq ft of living space. The living spaces on the north side of the home receive direct sunlight through a band of clerestory windows. A small attached greenhouse provides additional solar gain.

The 260-sq-ft Thermal Storage Wall is constructed using floor-to-ceiling, water-filled fiberglass cylinders behind Teflon-coated translucent glazing panels. Vents at the top and bottom of the wall allow heated air to circulate into the adjacent living spaces. Circulation is controlled by flaps constructed of polyethylene plastic sheets.

Outside, hinged insulated panels with a reflective surface increase incident light on the water wall when they are opened during the day. At night, when closed, their insulation reduces heat loss. The panels can be closed all summer, reducing unwanted heat gain. On the second floor, roll-up canvas shades afford similar protection.

A small greenhouse at the southwest corner of the home contains water storage and serves as a sunroom for the main bedroom. All 224 sq ft of the home's Direct Gain insulative glazing have been fitted with movable R-9 insulating shades to reduce losses. One layer of 6″ batt insulation in the walls and two layers in the roof, as well as 2″ of rigid polystyrene sheathing at the perimeter foundation wall, help create a thermally tight envelope.

Light entering the clerestory windows on the roof must also pass through the interior horizontal skylights before illuminating the living spaces. Movable reflective insulating shutters between the clerestories and skylights increase the solar gain. Along the hallway beneath the skylights, water-filled steel drums and a masonry floor provide needed thermal storage.

Project: Passive Solar Residence

Designer
Harrison Fraker, Architects
Harrison Fraker, Michael Guerin
Princeton, New Jersey

Sponsor
Private Individual
Princeton, New Jersey

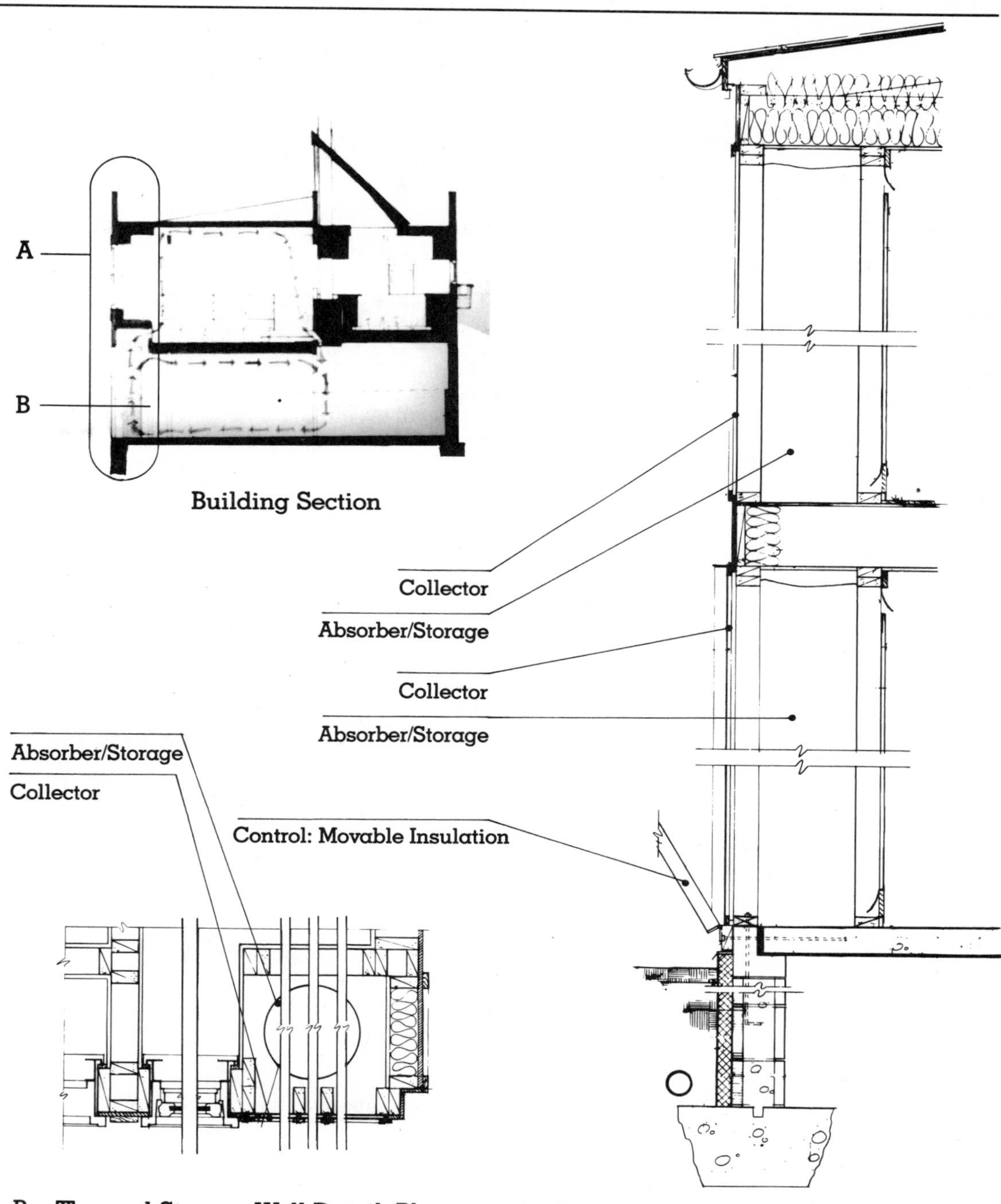

B Thermal Storage Wall Detail: Plan

A Thermal Storage Wall Detail: Section

SERI House #7

Thermal Storage Wall

Second Floor Plan

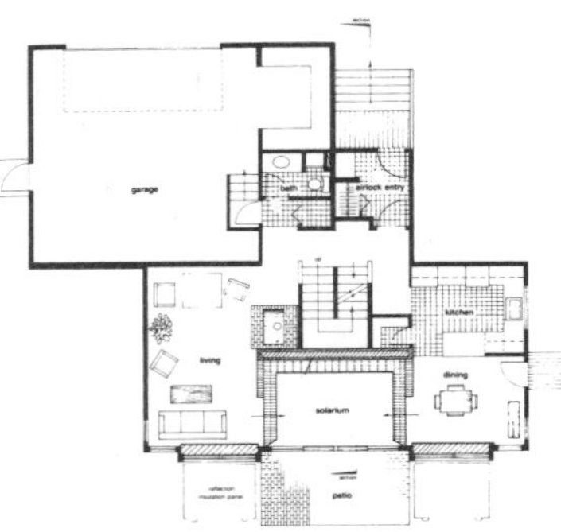

Main Floor Plan

North

Two Thermal Storage Walls flank an enclosed two-story Sunspace with a mass storage wall beyond, providing passive solar heating to this 1,900-sq-ft home in Colorado. Coupled with a solar domestic home water system mounted on the roof, this solar design substantially reduces its dependence on fossil fuels for heating.

Located at each end of the south wall, the thermal storage panels consist of 12" heavy-weight concrete blocks filled with grout and coated with a heat-resistant black exterior paint and an interior stucco finish. Exterior insulating panels with reflective interiors are raised by hand winches to reduce night heat losses. When lowered, their exposed reflective surfaces serve to increase effective collector gain.

The double-height Sunspace, onto which all the living spaces face, is double-glazed, with storage provided in its brick and concrete wall construction and brick paver over 4" concrete slab floor. Above the Sunspace, an operable insulated ceiling turbine vents the Sunspace to provide cooling during the summer.

A south-facing clerestory window provides solar gain directly to the north side of the second floor. Its cable-operated insulating shutter panel is coated with an aluminum surface that, when opened, reflects additional sunlight into the space.

Trellis, balcony, and roof overhangs screen the south wall from excessive solar gain in the summer, while insulating shutters, drapes, and shades provide extra thermal protection at night in the winter. Auxiliary heating needs are met by a woodburning stove and a gas-fired forced-air furnace. Insulation levels of R-30 ceiling, R-19 wall, and R-8 foundation help keep the building envelope heat losses to a minimum.

Project: SERI #7

Builder
Kurowski Development Company
John Kurowski
Westminster, Colorado

Designer
Sunflower Architects/Environmental Designers
David Barrett, AIA
Boulder, Colorado

Sponsor
Solar Energy Research Institute
Denver Metro Home Builders Assoc.
Denver, Colorado

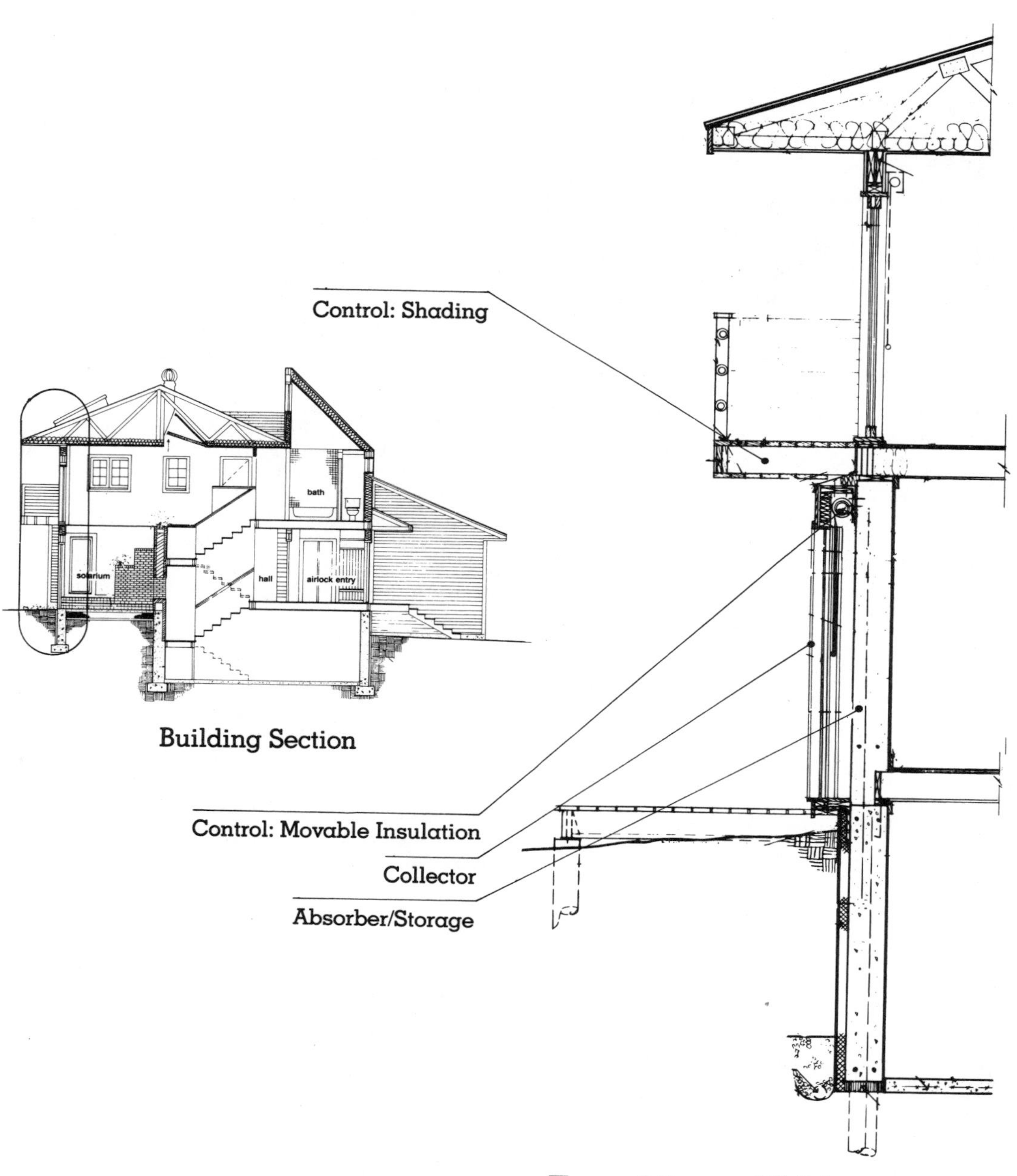

Building Section

Thermal Storage Wall Detail: Section

Optimum Technology House

Thermal Storage Wall

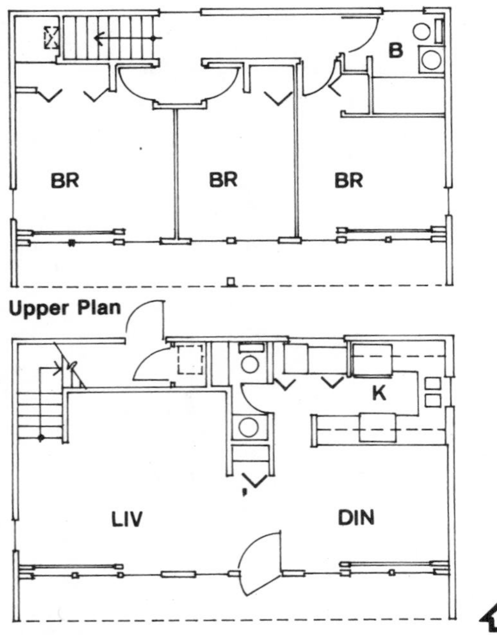

This 1,000-sq-ft expandable two-story home has been designed to be both affordable and energy efficient. Adaptable to a flat or gently sloping site, it can be built as shown, or expanded with an attached garage and optional Sunspace. Service and circulation functions completely buffer the north exposure, while evergreens provide an effective windbreak. The open-plan living spaces are free to face south, with both Direct Gain and Thermal Storage Wall systems integrated into the design.

Two sections of floor-to-ceiling phase-change storage rods provide thermal mass for the storage wall. Insulated folding doors on the interior side of the rods are closed during the day to control heat emission from the wall. At night, the doors are folded back to reveal a black-painted wire-mesh interior finish over the rods through which radiant and convective heat is released into the living spaces. Rolling insulated shades in the airspace between the rods and the glazing ensure that little heat is lost to the outside. In the cooling season, the shades and doors are closed to minimize heat gain.

Wing walls and overhangs on the roof and intermediate levels permit the entry of solar heat in winter but effectively shade the glazing from summer sun. The area framed by these overhangs can be enclosed to provide either a greenhouse or porch Sunspace addition. Insulated drapes have been supplied for all south-facing fixed and operable glazing.

Double and triple glazing, insulated metal exterior doors, R-30 ceiling and R-19 wall batt insulation, and 2″ of R-5 closed-cell polystyrene perimeter insulation beneath the 4″ concrete slab all contribute to the good thermal performance of the building. Auxiliary heat is provided by a heat pump, and a whole-house ventilating fan in the attic aids in cooling. Gable vents encourage natural cross-circulation through the open interior.

Project: Optimum Technology House

Builder
Itawamba Junior College
Building Trades Class
Nancy J. White
Fulton, Mississippi

Sponsor
Appalachian
Regional
Commission
Washington, D.C.

Designer
Architectural Branch
Tennessee Valley Authority
Lee J. Ingram
Knoxville, Tennessee

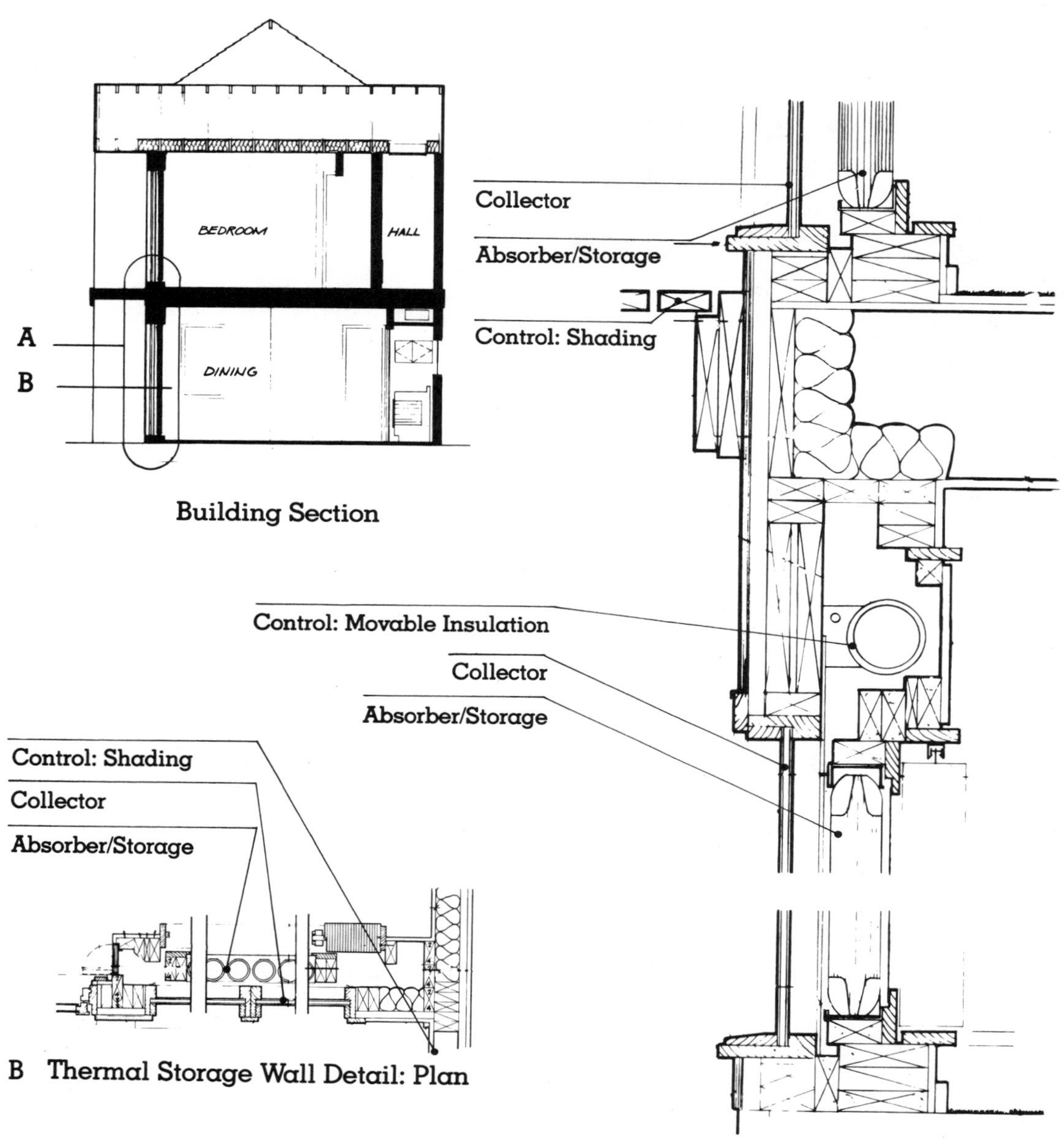

Building Section

B Thermal Storage Wall Detail: Plan

A Thermal Storage Wall Detail: Section

Star Tannery—One Design House

Thermal Storage Wall

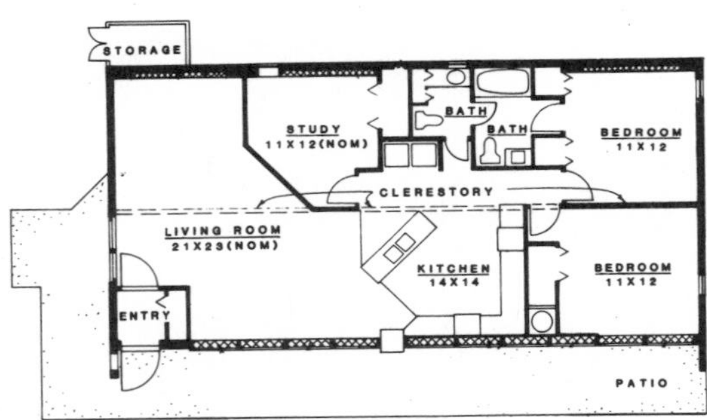

Floor Plan

North

Situated atop a knoll near Winchester, Virginia, the single-story 1,250-sq-ft Star Tannery house employs water Thermal Storage Walls and Direct Gain for passive solar heating.

An air-lock vestibule at the southwest corner of the house provides entry into the living room through insulated metal doors. Virtually the entire south wall is constructed of the prefabricated water-wall units. A 2′ × 4′ partition wall, finished with gypsum board on the interior, screens the water wall from view. Above, a row of unobstructed transom windows provides Direct Gain into the living spaces.

Sunlight strikes the prefabricated water containers, which in turn heats the air in the space between the containers and gypsum board. This heat is vented into the living space through vents at the top of the partition wall. Manually operated exterior vents dump collected heat to the outdoors during the cooling season.

A light-colored reflective concrete patio slab increases light incident on the water wall. An overhang fitted with movable louvers, landscapings, and wing walls provide shade during the summer.

At mid-roof, a continuous expanse of south-facing clerestory windows brings sunlight into the north side of the home. Another overhang above shades these windows from Direct Gain in the summer. Light passing through the clerestories strikes the dark-painted 4″ brick placed on the inside of the north wall.

Building envelope heat losses are reduced by R-19 wall and R-30 ceiling insulation levels and 2″ of closed-cell insulation wrapping the foundation wall. A propane-fueled, forced-air furnace supplies backup heating as needed.

Project: Star Tannery House

Builder
Holt Brothers Construction
Wayne Holt
Strasburg, Virginia

Designer
One Design, Inc.
Tim Maloney
Winchester, Virginia

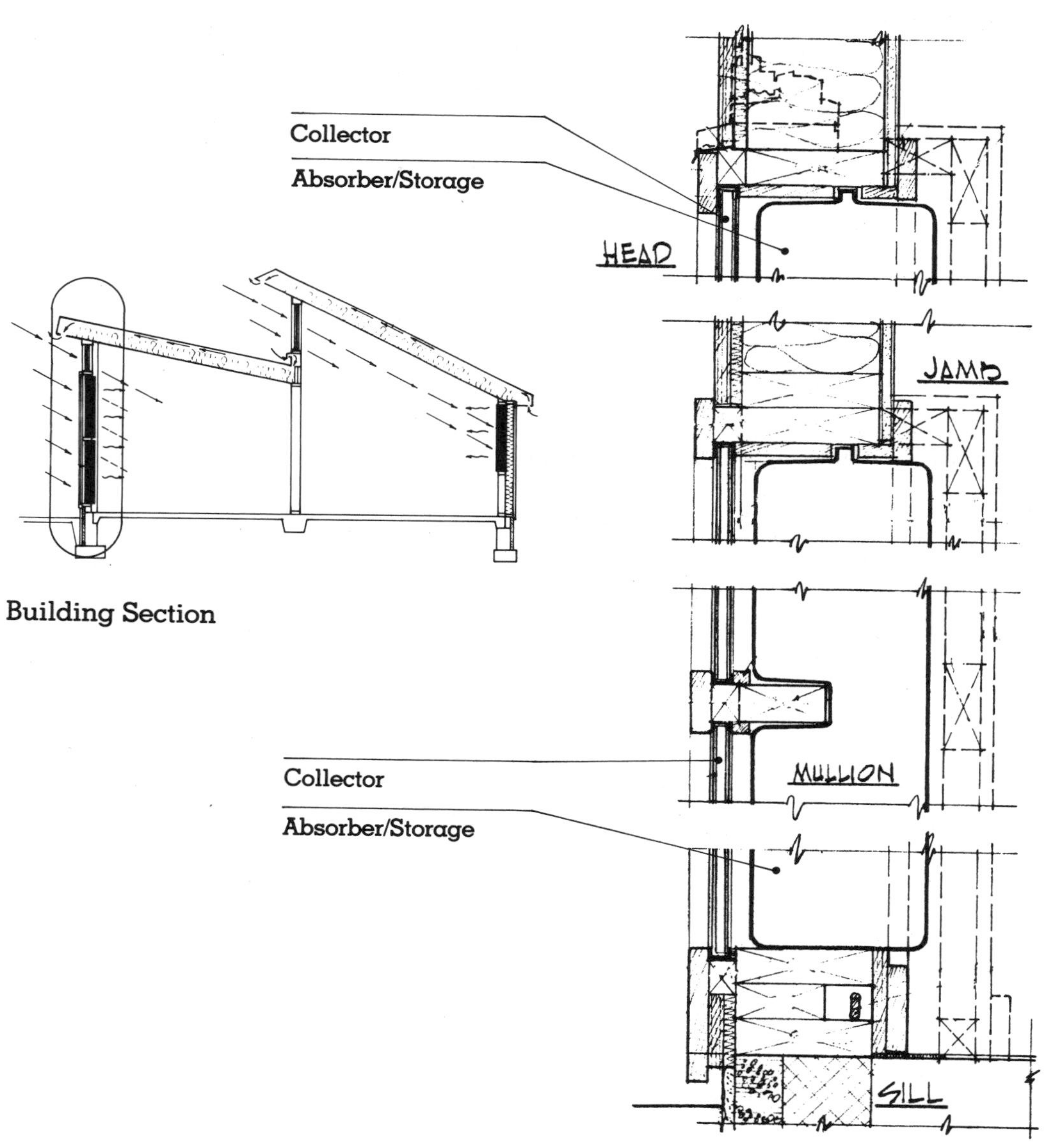

Building Section

Thermal Storage Wall Detail: Section/Plan

Chapter 5

ATTACHED SUNSPACE

Design Overview

Attached Sunspaces are designed to be used primarily for heating and are effective in areas with moderate to severe winter climates. As the name implies, they are spaces designed for passive solar gain that are "attached" to the south side of a building. Such spaces often serve as working greenhouses. In these cases, they are referred to as attached greenhouses or solar greenhouses, to distinguish them from the standard nonsolar varieties that are generally not as well constructed and detailed (see Figure 5.1).

A Sunspace can project from the house, or the house can be designed to "wrap around" the Sunspace and partially enclose it. The latter configuration has the advantages of: (1) reducing heat losses from the Sunspace; (2) transferring heat easily to a greater area of the surrounding house; and (3) allowing for the inclusion of a larger amount of thermal storage material than is possible in a projecting Sunspace.

Whether they project from the house or are enclosed by it, Attached Sunspaces can have two basic floor configurations. In the first and most common, the Sunspace floor is level with, or only slightly below or above, the adjacent living area. In the second, the so-called pit Sunspace, the floor level is located below the frost line and thus significantly below the floor of the adjacent living spaces. The pit variety has the advantage of increased headroom and of reduced floor and perimeter heat losses due to the relatively constant temperatures of the surrounding earth. A pit Sunspace can actually *gain* heat from the ground in winter if, as is often the case, the temperature in the Sunspace is allowed to drop below the surrounding subsurface ground temperature. In these instances, the Sunspace floor slab need not be insulated (see Figure 5.2).

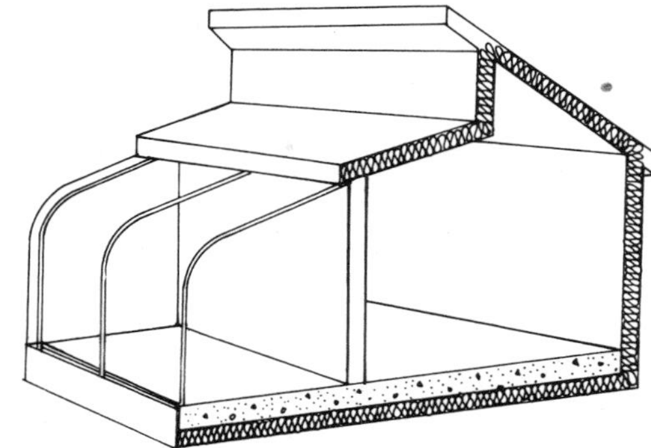

Figure 5.1

The solar energy collected and absorbed by an Attached Sunspace is used to heat both the Sunspace itself and the adjacent living areas. The ways in which this energy is stored and distributed form the basis for distinguishing five separate Attached Sunspace subsystems, which differ significantly from each other in terms of operation and construction: open wall, direct gain, air exchange, thermal storage wall, and remote storage. Where appropriate in this chapter, each of these subsystems is treated separately, with the construction details organized to reflect this subsystem orientation.

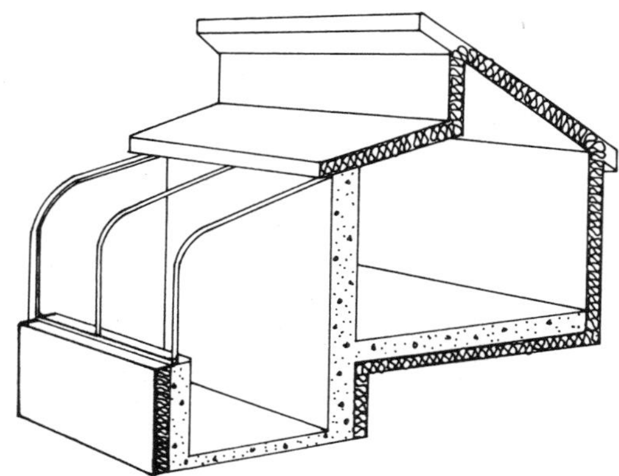

Figure 5.2

There are two basic modes of operation for any Attached Sunspace subsystem. In the first, the Sunspace serves as an extension of the living area. It is *not* thermally isolated from the living space, and its temperature is not allowed to fluctuate outside the comfort range. In most cases, this type of Sunspace will require some form of auxiliary energy system to maintain comfortable temperatures (see Figure 5.3).

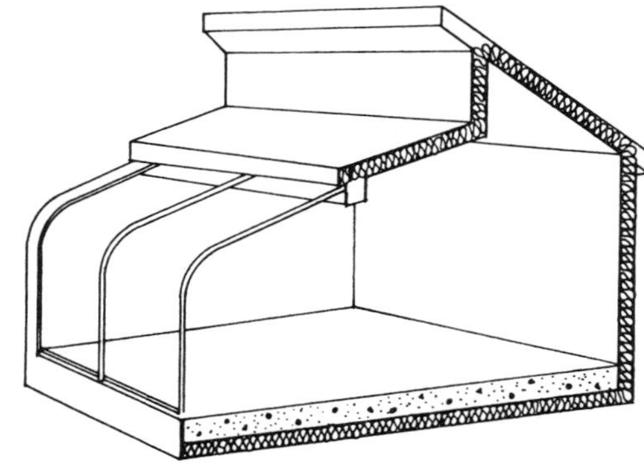

Figure 5.3

In the second mode, the Sunspace is a separate, distinct area that *is* thermally isolated from the living space. It is not a direct extension of the living space, and its temperature can be allowed to fluctuate outside the comfort range. In this case, an auxiliary energy system is optional and, if present, is used only to heat the Sunspace (see Figure 5.4).

Each of the five subsystems, in its pure form, operates in one or the other of these two modes. It should be noted, however, that very often two or more systems are combined to function together. In such hybrid configurations, the actual mode of operation will depend on the particular subsystems involved and the way in which the space is intended to be used. This will, in turn, determine the amount and location of insulation (fixed and/or movable) employed and whether or not an auxiliary energy system is necessary.

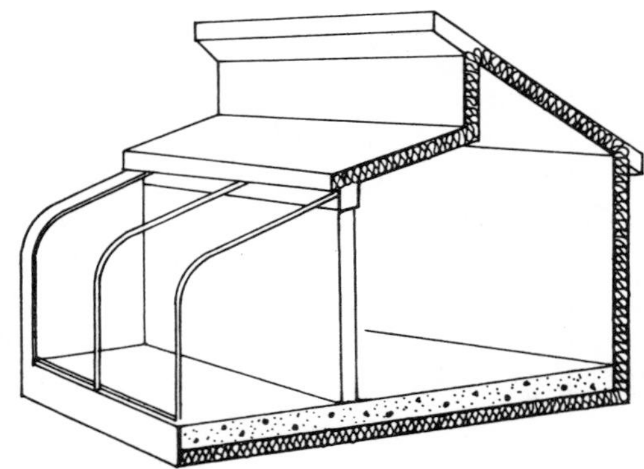

Figure 5.4

Seasonal Operation—Heating

During the heating season, sunlight enters an Attached Sunspace through the south-facing collector (an expanse of glass or plastic), is absorbed by elements within the Sunspace, and is converted to heat. This basic process is the same for all the Attached Sunspace subsystems. It is only in the next stages of operation (storage, distribution, and control) that these subsystems differ significantly from each other. The following descriptions summarize the operating characteristics for each of the subsystems during the heating season.

Open Wall Subsystem

In this subsystem, the Sunspace opens directly to the living spaces and is basically an extension of these areas. Because of this openness, there is a direct and unimpeded transfer of warm air between the two spaces. While the sun is shining, heat is generated in the Sunspace, and the resulting warm air moves freely to the adjacent living spaces. Some of this heat is also stored in storage components (e.g., concrete slab floor) located in the Sunspace and/or living spaces (see Figure 5.5).

It is important to note that this subsystem, unlike the other four that follow, is essentially an extension of the living area, and its temperature is maintained at the same level as the other living spaces. The provision of storage material in the open wall sunspace will help dampen temperature fluctuations, but the Sunspace in this subsystem will still need to be heated by conventional heating equipment at night and during prolonged periods of cloudy weather. At these times, the direction of heat transfer is reversed, and the Sunspace will receive heat from the adjacent living area as well as from the storage components located in the Sunspace itself (see figure 5.6).

In order to avoid excessive loss of this heat, some form of movable insulation that can be used during these periods should be provided at the Sunspace collector. The Sunspace floor will also need to be insulated to the same level as the standard living-space floor.

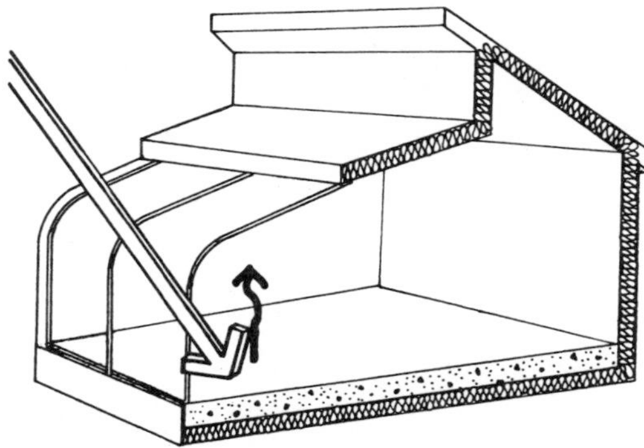
Figure 5.5

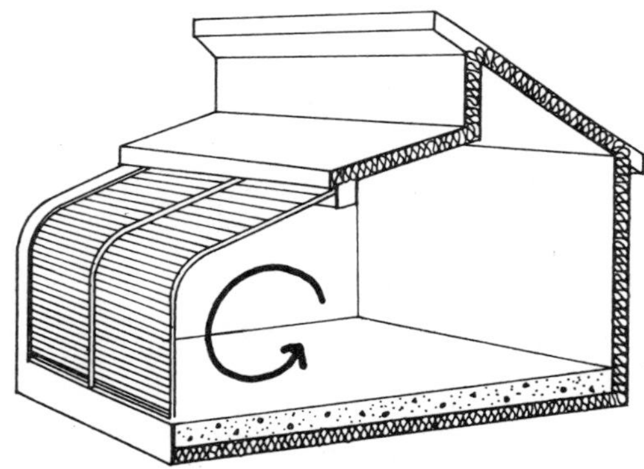
Figure 5.6

Direct Gain Subsystem

Although this subsystem is similar to the open wall system in construction, it is generally operated as an isolated space and is typically not provided with backup heating from an auxiliary energy system. The open connection between the Sunspace and the living areas is replaced by a glazed, shared wall. Because of the thermal separation that this wall provides, temperatures in the Sunspace can be allowed to fluctuate without seriously affecting the amount of heat lost from the living space. However, it may still be advisable to double-glaze the shared wall in severe climates (see Figure 5.7).

The use of movable insulation located at the shared-wall glazing will further reduce heat losses from the living areas. This may not be cost-effective, however, because the Sunspace, if properly designed, will act as a buffer between the outside air and the living areas, reducing the rate of heat loss. Movable insulation will be effective, however, if placed on the Sunspace collector. Such insulation, if correctly operated, will both enhance the Sunspace's effectiveness as a heat generator and prolong the period over which comfortable temperatures are maintained in the Sunspace.

Heat generated in the Sunspace can be used both to warm the Sunspace itself and to supply heat to the adjacent living areas. When providing heat only for itself, the Sunspace is acting simply as a climatic buffer for the adjacent living spaces. Some Direct Gain radiation does reach the living spaces through the glazed shared wall. However, if the glazed shared wall can be opened (e.g., patio doors), the

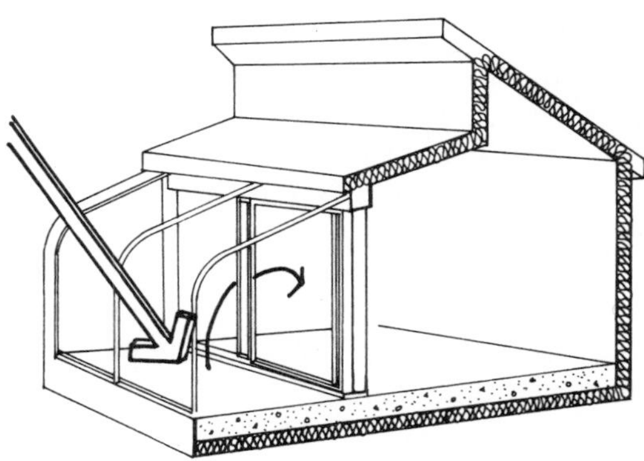
Figure 5.7

Sunspace can provide heated air, in addition to solar radiation, to the living spaces when it is required.

When this heat is no longer desired or when the Sunspace is no longer receiving solar radiation, the homeowner can simply close the glazed wall (see Figure 5.8).

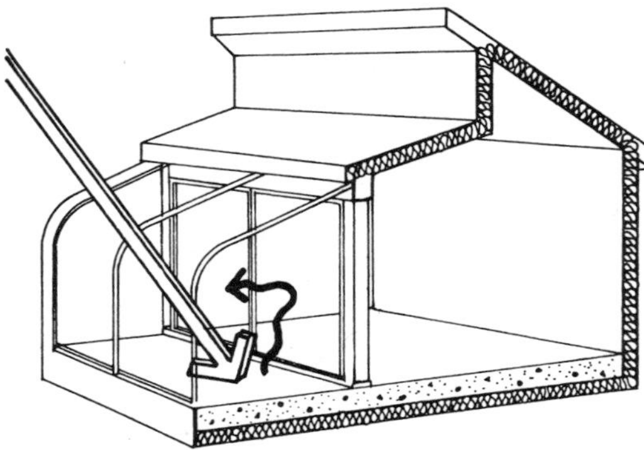

Figure 5.8

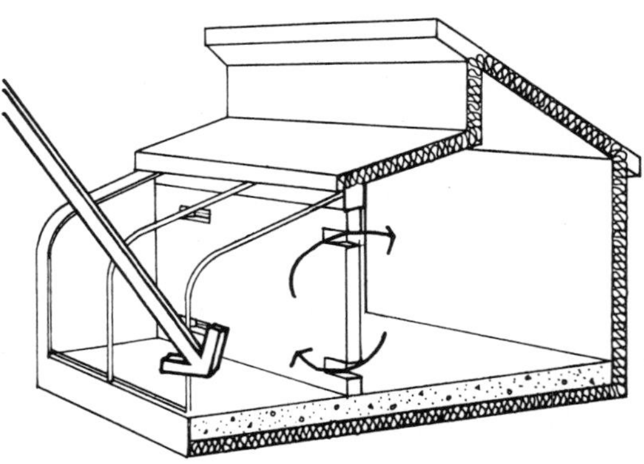

Figure 5.9

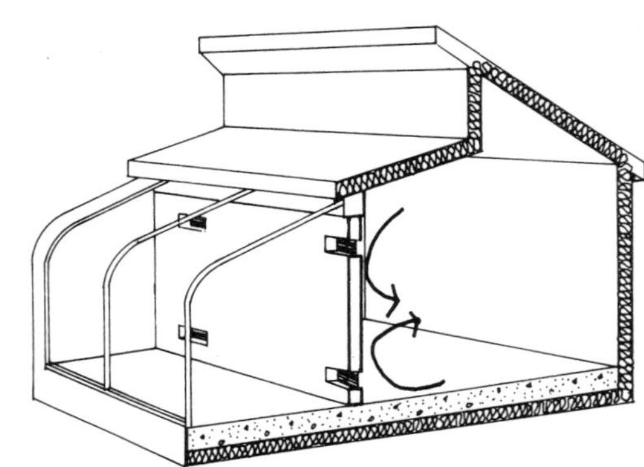

Figure 5.10

Air Exchange Subsystem

In this subsystem, the living areas and the Sunspace are separated by an opaque frame wall containing operable vents. These vents allow for the transfer of heated air from the Sunspace directly to the living areas during periods of heat gain. The vents, which are either manually or automatically operated, can be closed off when heat is no longer wanted or when the Sunspace temperature drops below that of the living space (see Figure 5.9).

Closing the vents effectively turns the Sunspace into a buffer zone, reducing the overall heat loss from the living spaces (see Figure 5.10).

Because the temperature difference is reduced, it may not be necessary to insulate this opaque shared wall. One advantage such insulation might have, however, is to reduce heat gain through the wall during the cooling season, especially if the wall receives direct solar radiation during this period.

Thermal Storage Wall Subsystem

In this subsystem, the shared wall between the living areas and the Sunspace is a mass wall that functions much like a standard Thermal Storage Wall. In fact, it may be easiest to visualize this subsystem as simply a Thermal Storage Wall with an enlarged airspace between the collector and storage components.

The wall absorbs incoming solar radiation and stores it. The resulting heat then slowly migrates through the wall to the adjacent living area. The wall reradiates the heat it has stored both to the living space and back into the Sunspace itself. In doing so, it not only heats the living area but also helps to maintain a relatively constant temperature in the Sunspace (see Figure 5.11).

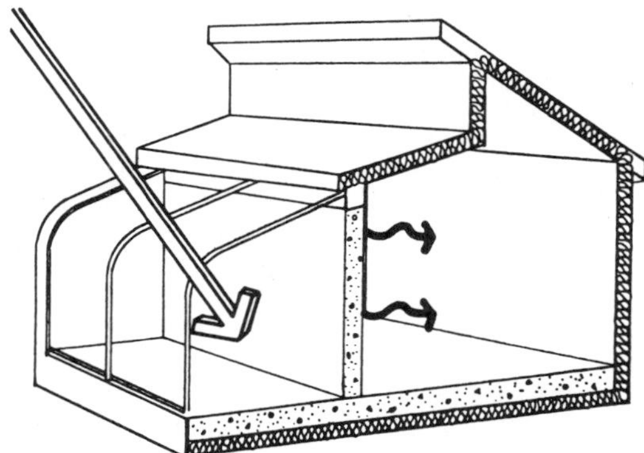

Figure 5.11

Because this Sunspace temperature is higher than the surrounding outside air, the Sunspace acts as a thermal buffer zone, reducing the amount of heat lost from the living space at night and during prolonged overcast periods. Movable insulation placed at the collector will significantly improve the overall efficiency of the system. By reducing the amount of stored heat that escapes, this insulation will maintain higher temperatures in the Sunspace during the heating season. Due to these higher temperatures, the shared mass wall will have less of a tendency to transfer its heat into the Sunspace and will thus provide a greater portion of its heat to the adjacent living areas (see Figure 5.12).

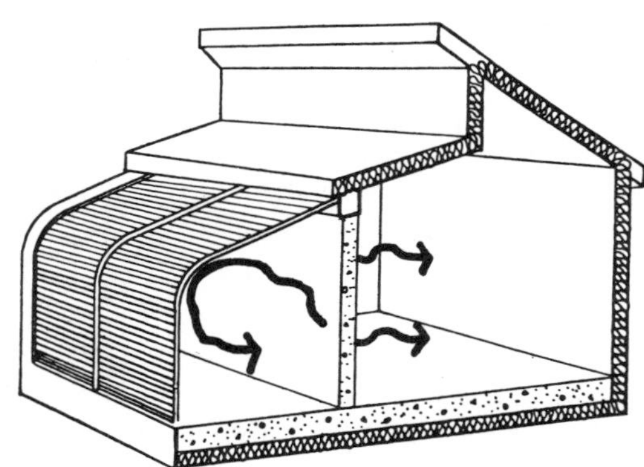

Figure 5.12

Such movable insulation should be located at the collector, but placement on the outside face of the shared storage wall is also possible.

It should be noted that the storage wall can also function as a vented Thermal Storage Wall. In this instance, the Sunspace is actually a hybrid combination of the thermal storage wall and the air exchange subsystems (see Figure 5.13).

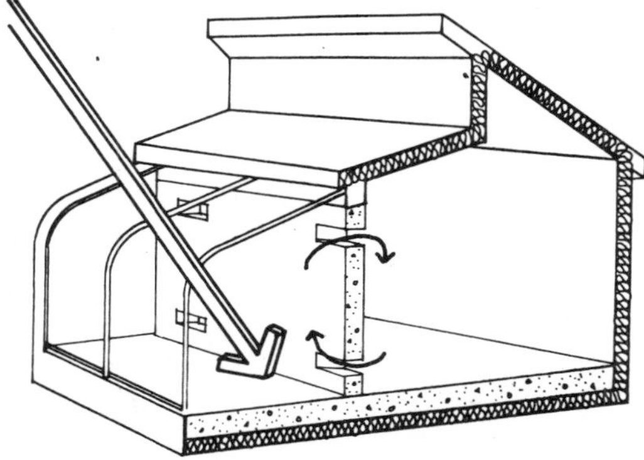

Figure 5.13

Remote Storage Subsystem

In the remote storage subsystem, the common wall between the living spaces and the Sunspace is opaque but is not vented, or is constructed of mass materials. Its primary purpose is to thermally separate the Sunspace from the living spaces. Heat, which is generated in the Sunspace, is removed (usually by some form of fan system) for storage in some remote location away from the Sunspace and the living areas (see Figure 5.14).

The remote storage usually takes the form of a rock or gravel bed located under the uninsulated floor slab of the living space. Heat from the hot air blown through this gravel is stored in the rocks and then radiated up through the slab into the room to be heated. The temperature of this air is usually low (90°F or less), and the resulting radiant heat reaching the living space will not cause the room to overheat. In this configuration, the underside of the rock bed must be insulated to prevent heat losses to the ground. An alternate design that, however, is not as thermally effective, isolates the rock bed from the living space by insulating it on all sides. When required, air warmed in the rock bed is distributed to the living spaces through ducts.

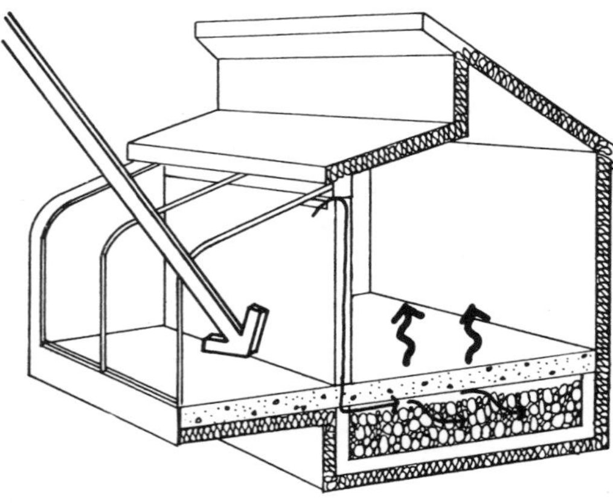

Figure 5.14

Seasonal Operation—Cooling

During the cooling season, the Sunspace can easily overheat, adding to the cooling load on the auxiliary system. Therefore, the collector in any of the five subsystems should be well shaded to prevent unwanted heat gain. Deciduous trees and other types of vegetation can be used, as can building elements such as overhangs, louvers, and shades. If movable insulation has already been provided, it can serve a dual purpose and also act as a shading device.

The Sunspace should also be properly vented. If possible, low vents, to admit cool air, should be located in the shade on the windward side of the Sunspace, and high vents, to exhaust hot air, on the leeward side. These vents can either be operable glazing elements or they can form part of the frame. Gable vents on the east and west end walls are also possible. In order to reduce infiltration losses at the edges of these vents, it is preferable to have a few large vents serve the space rather than a series of smaller ones with a larger aggregate edge area (see Figure 5.15).

If properly managed, the storage mass within the Sunspace can also help cool it. Placing vents so as to allow cool night air to pass over the storage mass will cause the mass to become cooler. If properly shaded from the sun, this cooler mass will absorb heat generated in the Sunspace during the day, effectively cooling it and,

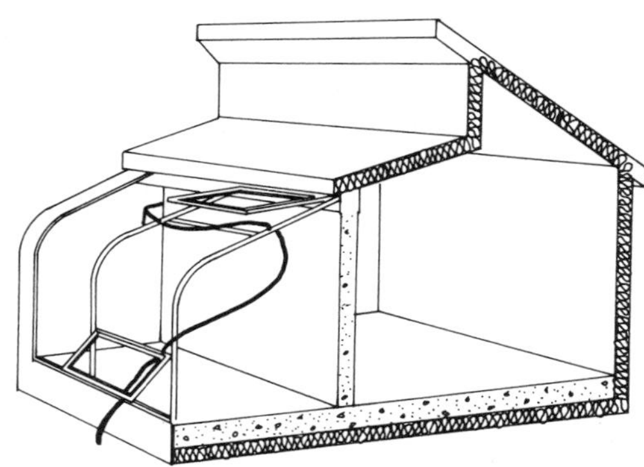

Figure 5.15

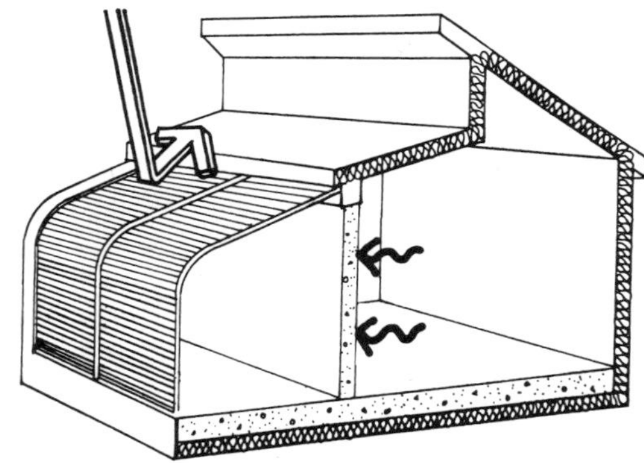

Figure 5.16

depending on the subsystem involved, the adjacent living areas as well (see Figure 5.16). The mass will then dissipate this collected heat to the night air, completing the diurnal cycle. Openings placed between the Sunspace and the adjacent living spaces will enhance the cooling effect of the mass by inducing cooled air to circulate through the rest of the home.

It must be reiterated that the majority of Sunspace configurations built consist of hybrid combinations of the subsystems described above. It is not uncommon for a thermal storage wall to include operable vents to allow for direct transfer of heated air, nor for the opaque shared wall in a remote storage subsystem to include windows that allow for some direct gain heating in the living areas. As noted earlier, the thermal performance of such hybrids is, in general, very difficult to quantify. It should be assumed, however, that combining subsystems will not result in significant decreases in overall thermal performance if the general rules of thumb for Attached Sunspaces are followed.

Advantages

- The Sunspace can be used for growing houseplants and/or food.

- The Sunspace acts as a buffer zone to reduce heat loss from the adjacent living spaces.

- The Sunspace can serve more than one function, including being an extension of the living space.

- The Sunspace can become an aesthetic asset to a house.

Disadvantages

- The thermal performance can vary significantly from one design to another, and predicting performance accurately is difficult.

- Costs can be kept low by using less expensive materials, but these tend to degrade fairly quickly, and the installation as a whole will be less durable. Quality construction, on the other hand, can be expensive.

- If used as a working greenhouse, certain characteristics such as excess humidity, odors, control of insects, provision for running water, and drainage must be accommodated in the design.

Collector

As in all passive solar applications, the purpose of the Attached Sunspace collector is to admit and trap solar energy. Although often perceived as a net energy loser, the large expanse of south-facing glazing in a well-designed Sunspace collector actually gains much more heat than it loses. The collector component functions in the same way in each of the Attached Sunspace subsystems, and the design and materials information presented below applies to all five subsystems.

There are a wide variety of possible collector configurations. A large number of Sunspaces are site-fabricated to meet specific and often idiosyncratic design specifications. The collectors for the majority of these site-built installations are framed in wood, whereas prefabricated, off-the-shelf collector units are usually framed in metal. Both systems have advantages and disadvantages.

The prefabricated systems are generally simple to erect, are relatively impermeable to the effects of weather, and require little maintenance. They can be expensive, and their dimensional modularity may inhibit design and system sizing flexibility. They may also be difficult to insulate in precisely the manner necessary to ensure maximum thermal performance. The metal-frame collector system should be equipped with thermal breaks in the frame to reduce heat loss.

Site-built systems can be adapted to any particular design situation and can be insulated and shaded to meet specific conditions. Unless carefully detailed, however, such systems are notorious for failing to maintain a tight seal against the weather. Special attention must be paid to caulking and sealing the glazing material, and periodic inspections should be made to ensure that the seals are still holding. Rough-sawn or green wood should be avoided in such applications in favor of kiln-dried lumber. Wood should also be treated to resist moisture damage. Copper naphthenate should be used in greenhouse applications rather than creosote or pentachlorophenol, which give off fumes that are potentially harmful to vegetation. Regardless of the material, the collector frame should be painted a light color in order to reflect more light into the Sunspace.

At present, the most common Sunspace collector material is glass, and it is recommended that double glazing be used in applications where glass alone, not in combination with any other collector material, is used. For many prefabricated greenhouse manufacturers, double glazing is currently standard. In site-built applications, the simplest and most common glazing type is patio-door replacement glass, which can usually be specified double glazed.

Materials other than glass are, however, gaining in popularity. These include various forms of fiberglass, acrylics, polycarbonates, and certain glazing films, such as polyethylene. Each of these has its own advantages and disadvantages. Each can also be used in combination with any of the other materials to maximize the best qualities of each. For example, the roof of the Sunspace might be constructed of corrugated fiberglass because of its ease of handling during construction, while the vertical portion might be standard double glass to provide a view to the outside. Combinations and variations such as this should be considered for each individual application. The advantages of using a light-diffusing collector material to promote a balanced distribution of incoming solar energy should also be considered if storage mass is evenly distributed throughout the Sunspace.

In general, the most thermally effective Sunspace is one in which little or no glazing is placed on the roof. Sunspaces with completely opaque and insulated roofs, however, limit the penetration of sun into the Sunspace (often inhibiting plant growth) and are difficult to find in prefabricated models. A Sunspace with a partially glazed roof is an excellent compromise between utility and thermal effectiveness.

In general, collector material on the east and west walls is thermally inefficient, losing more heat than it gains. It is recommended that such walls be opaque and well insulated where possible, or that the house itself "wrap around" the Sunspace on the east and west.

Collector Rules of Thumb

- **Orientation**:
  Extend the Sunspace along the south wall of the building for the most efficient performance. Orientation can be up to 20° east or west of true south without significant loss in performance. East is generally preferred to allow morning sun to "wake up" the Sunspace.

- **Sloped Glazing**:
  The angle of tilt for any sloped surface of the Sunspace collector should be 45-60°. The tilt should be slightly flatter (nearer 45°) in overcast or foggy areas, and slightly steeper (nearer 60°) in mostly clear and sunny areas.

- **End Walls**:
  East and west walls of the Sunspace should be opaque and well insulated. "Wrapping" the main building around the Sunspace is a good approach.

- **Sizing**:
  Table 5.1 can be used to determine the collector size.

For example, for Madison, Wisconsin, at 43°NL with an average January temperature of 21.8°F (data available from local and/or national weather bureau), select the closest "Average Winter Outdoor Temperature" (20°F). Reading across the table, it is recommended that 0.9-1.5 sq ft of glazing be used per square foot of floor area. Because Madison is at a relatively high latitude, the upper end of this range should be employed, and a figure of 1.3 will be adequate. Thus, to heat a living space of 200 sq ft, 260 sq ft of south-facing glazing will be required.

Storage

The five Attached Sunspace subsystems embody unique approaches to the problem of thermal storage, and each is treated separately below.

Open Wall Subsystem

In this subsystem, solar radiation is absorbed by and stored in the Sunspace floor, in water containers placed within the Sunspace, and, where applicable, in the opaque end walls if they are constructed of suitable materials (see Figure 5.17).

Storage elements can also be provided in the living areas and will prove very effective if sunlight can penetrate deeply enough

Table 5.1

Average Winter Outdoor Temperature (Clear Day)*	Square Feet of Sunspace Glazing Needed for Each One Square Foot of Floor Area**
COLD CLIMATES	
20°F	0.90 - 1.50
25°F	0.78 - 1.30
30°F	0.65 - 1.17
TEMPERATE CLIMATES	
35°F	0.53 - 0.90
40°F	0.42 - 0.69
45°F	0.33 - 0.53

* Temperatures listed are for December and January, usually the coldest months.

** Within each range choose a ratio according to your latitude. For southern latitudes (i.e., 35°NL) use the lower glazing to floor area ratios; for northern latitudes (i.e., 48°NL), use the higher ratios. For a poorly insulated sunspace or building, always use slightly more glazing.

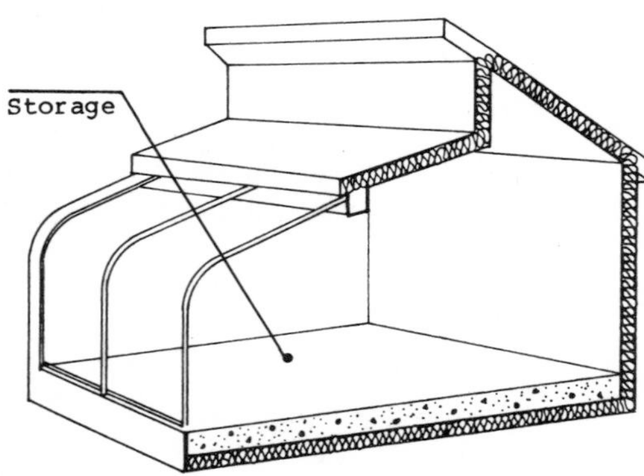

Figure 5.17

into the living spaces to strike them directly. The majority of the heat reaching the living areas, however, will be in the form of warm air generated in the Sunspace during the day.

The storage components in the Sunspace and the adjacent living spaces can be made from any of a variety of high thermal storage capacity materials and should be as dark a color as possible to increase absorption of solar radiation.

Direct Gain Subsystem

In this subsystem, the storage components are the same as in the open wall subsystem and function in a similar manner (see Figure 5.18).

The chief difference is that the living area storage components, while still receiving direct solar radiation through the glazed shared wall, do not normally receive convected warm air from the Sunspace. The heat collected and stored in the Sunspace is available to the living areas only when the shared-wall glazing is opened. At other times, when the glazing is closed, this heat is used to maintain warm temperatures in the otherwise unheated Sunspace. In this mode of operation, the Sunspace serves as a buffer zone, reducing heat losses from the adjacent living areas, thereby reducing overall fuel consumption.

In this subsystem, as in the open wall subsystem, the storage elements can be

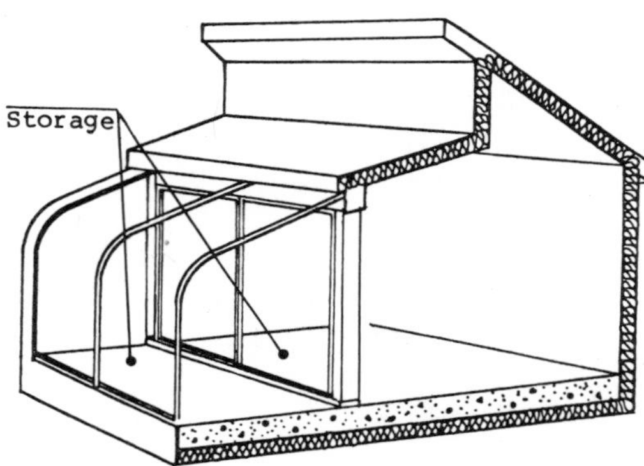

Figure 5.18

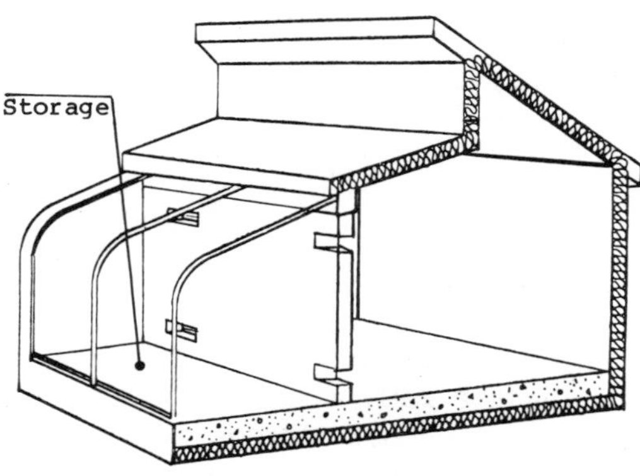

Figure 5.19

of the same materials and construction as those found in a Direct Gain system.

Air Exchange Subsystem

In this subsystem, the living areas are separated from the Sunspace by an opaque shared wall and receive no direct solar radiation. Vents in this wall allow warm air generated in the Sunspace to be convected into the living areas (see Figure 5.19).

As in the other subsystems, storage mass located in the Sunspace will absorb and store heat. When heat is required in the living spaces, the wall vents are opened and warm air is allowed to flow from the Sunspace into the living areas. When the vents are closed, the stored heat is used to heat the Sunspace itself, helping it to act as a thermal buffer, reducing heat losses from the adjacent living areas. Once again, the materials and construction techniques used in Direct Gain applications are appropriate for the air exchange sunspace storage components.

Thermal Storage Wall Subsystem

This subsystem acts very much like a Thermal Storage Wall with a much larger airspace between the storage wall and the collector. The mass wall separating the living areas from the Sunspace is used to absorb and store solar radiation, which is in turn used to heat both the living areas and the Sunspace itself. The wall can be built of any of the Thermal Storage Wall materials discussed in Chapter 4 and will be constructed in much the same fashion (see Figure 5.20).

Whatever material is chosen, the Sunspace side of the wall should be painted a dark color to increase absorption of incoming solar energy (see Table 7.9 for the absorptivities of various materials and

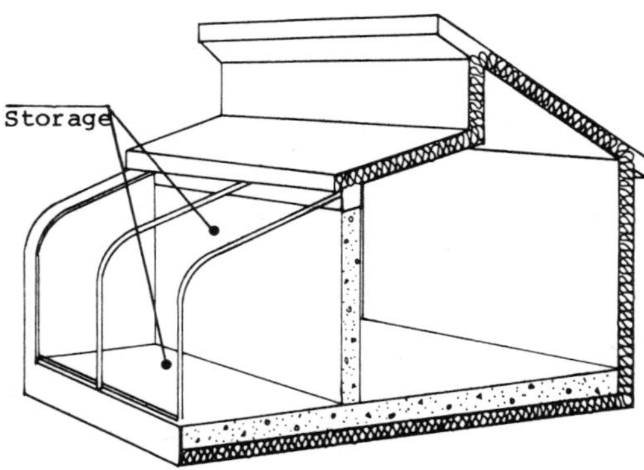

Figure 5.20

colors). The interior surface can be painted or left unfinished. Finishes other than paint, however, will decrease the efficiency of the wall by restricting its ability to reradiate heat. Plaster and wallpaper will have little impact on thermal performance. Gypsum board can also be acceptable over masonry walls if it is attached directly to the wall (not on furring strips), and if care is taken during installation to apply a very thick coat of construction adhesive in order to increase the thermal connection between the wall and the gypsum board. Finishes that are attached to the wall with furring strips will create an insulating airspace, significantly reducing the performance of the system.

The heat radiating from the storage wall back into the Sunspace, along with heat stored in elements within the Sunspace itself, will help maintain higher-than-ambient temperatures in this space. As in the other subsystems, these higher temperatures will help reduce heat losses from the adjacent living spaces.

Remote Storage System

In this subsystem, the storage is typically a rock bed, literally a bed of rocks usually located under the floor of the major living spaces. Heat is delivered to this bed in the form of heated air from the Sunspace (see Figure 5.21).

Allowing the delivered hot air to convect up directly through openings in a frame floor or to be ducted from a well-insulated rock bed can work, but it is generally better to simply allow the heat from the bed to radiate through the floor of the home. For this reason, the bed should lie under as large a portion of the living-space floor as possible. In this configuration, the floor should be made of a masonry material.

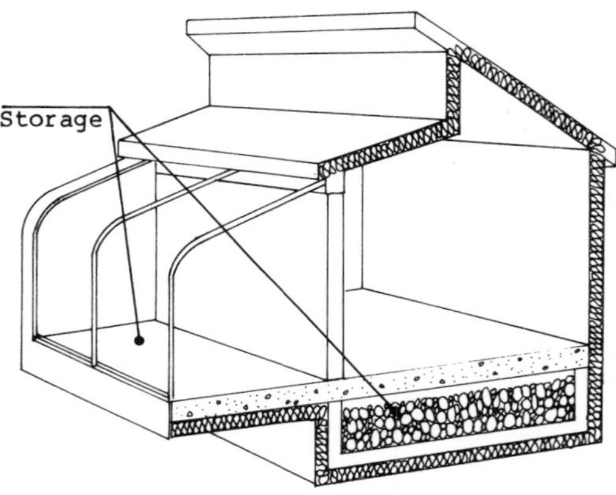

Figure 5.21

The masonry floor can be constructed on top of a wood floor over a crawl space, in which case the rocks should come up right under the wood subfloor, or it can simply be a slab laid directly on top of the bed. In any rock storage system, care should be taken to ensure that the heated Sunspace air is distributed evenly throughout the rock bed.

Storage Rules of Thumb

Floors (All Subsystems)

- **Thickness:**
  Standard slab thicknesses are usually appropriate. Consult Storage Rules of Thumb in Chapter 3.

Walls (Thermal Storage Wall Subsystem)

- **Thickness:**
  The recommended Attached Sunspace storage wall thickness is 8-18". For dimensional requirements using phase-change materials, consult manufacturers' literature.

Rock Bed (Remote Storage Subsystems)

- **Area:**
  1. In cold climates, the bed area should be 75-100% of the floor area of the living space to be heated.

  2. In moderate climates, the bed area should be 50-75% of the floor area of the living space to be heated.

- **Volume of Rock Bed:**
  1. In severe climates, the bed should contain ¾-1½ cu ft of fist-sized (2-3" diameter) rocks* per sq ft of Sunspace glazing.

  2. In moderate climates, the bed should contain 1½-3 cu ft of fist-sized (2-3" diameter) rocks* per sq ft of Sunspace glazing.

*The smaller rock or gravel sizes can be more efficient due to the greater amount of surface area available for heat transfer. General rule: bed depth in the direction of airflow = 20 × diameter of rocks.

Vents (Air Exchange Subsystem)

- **Size:**
  Interior vents should be a minimum of 3% of the area of the shared wall between the Sunspace and the adjacent living spaces. To achieve a more balanced airflow, upper vents should *not* be located directly over lower vents, but should be moved to one side if possible. Backdraft dampers or operable louvers should be provided for upper vents to prevent reverse thermocirculation.

Control

Shading

In each of the Attached Sunspace subsystems, it is recommended that some method for shading the collector be provided to help avoid unwanted summer heat gain. Most of the devices used in Direct Gain are applicable for at least the vertical or near vertical wall section of the collector, including vegetation (trees), trellises, vertical and/or horizontal louvers, and attached awnings.

The glazed portion of the collector roof, however, represents a slightly different shading condition. The most efficient way to shade this area is to use operable wood, aluminum, or plastic roller shades attached directly to the collector, preferably on the outside. Prefabricated collectors are often equipped with these shades integrated into the framing system. In site-built installations, provision should be made for the inclusion of such elements as part of the site-built frame.

Wherever possible, it is recommended that the roof portion of the collector be opaque at its connection to the home. This opaque roof will serve as an overhang for shading the Sunspace. In working greenhouse applications, care should be taken to balance the amount of shading with the daylight requirements of the plants (see Figure 5.22).

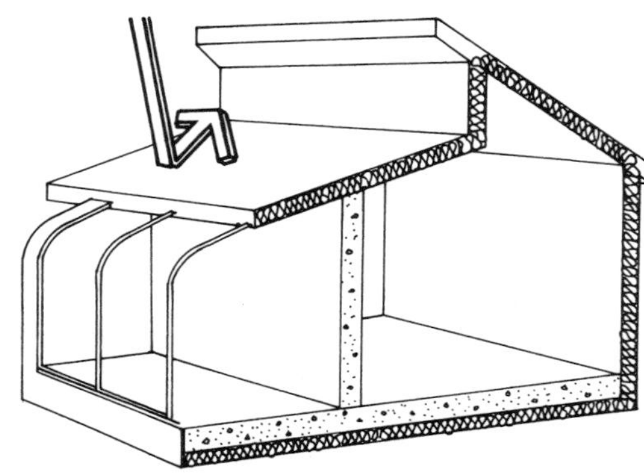

Figure 5.22

Reflecting

Reflecting panels can be used in any of the Attached Sunspace subsystems to enhance the amount of solar energy reaching the collector. They should be located so as to reflect sunlight onto the mass within the space without directing excessive radiation directly onto the plants.

Insulating—Fixed Exterior Insulation

In general, all exposed opaque surfaces in the Sunspace, including roof and end walls, should be well insulated to reduce heat flow to the exterior. Standard insulating techniques are appropriate except in the case where a particular construction is intended to function as a storage component. For example, if an end wall in the Sunspace is to function as a Direct Gain storage component, the construction procedures and insulation techniques should follow the discussions and details in Chapter 3.

Foundation insulation will, of course, reduce heat losses. These reductions can be increased if the floor slab is also insulated at its perimeter. However, such perimeter insulation may or may not be advisable, depending upon the particular mode in which the Sunspace is operated.

If the Sunspace is thermally isolated from the living area and its temperatures are allowed to fluctuate freely, under-slab insulation will not be necessary if perimeter insulation is already provided. If the Sunspace is not thermally isolated from the living space and/or if an auxiliary energy system is used to maintain temperatures in the Sunspace within the comfort range, perimeter insulation and under-slab insulation should be considered. In general, the floor in such Sunspaces should be insulated to the same extent as a Direct Gain floor would be in the same location (see Insulating—Fixed Exterior Insulation in Chapter 3).

In general, it is not necessary to provide fixed insulation for the shared wall between the Sunspace and the adjacent living spaces in any of the five subsystems. In the open wall and direct gain subsystems, such considerations do not apply. In the air exchange, thermal storage wall, and remote storage subsystems, the shared wall need not be insulated because movable insulation is already provided at the collector. In the thermal storage wall subsystem, insulating the shared storage wall will effectively nullify any passive benefits that the wall might provide.

Insulating—Movable Insulation

The provision of movable insulation at the Sunspace collector is recommended for all five subsystems. Such insulation will reduce heat losses during prolonged overcast periods and at night, as well as prevent unwanted heat gain during the cooling season. This will generally result in significant increases in the thermal performance of all of the subsystems. In fact, in more severe climates, certain Attached Sunspace configurations may be net energy losers if such insulation is *not* provided.

A limited selection of movable insulation products is available for off-the-shelf Attached Sunspace collectors. It is possible, however, to adapt existing insulating products commonly used in Direct Gain or Thermal Storage Wall applications. It is advisable to select or design an insulating system that, when closed, provides a tight air seal between the collector and Sunspace. Sunspace insulating materials are typically suspended from guide wires or run in a track that may be attached to or integrated with the collector frame.

These insulating materials may be bulky and difficult to store if not designed into the collector framing system. Manual operation of movable insulation is often very simple but requires an active commitment on the part of the homeowners. It is therefore recommended that automatic systems, either switch or thermostatically controlled, be considered.

Local codes should also be consulted to assess the eligibility of certain materials for use in exposed situations. For example, some insulating materials such as polystyrene and urethane may be prohibited due to potential fire hazards.

Control Rules of Thumb

Shading

- **Overhangs**:
  If the roof contains additional collector area that is shaded by a movable screen (see Shading, earlier in this chapter), the Sunspace can be designed to allow the screen to extend over the vertical glass, shading the entire collector area.

  If a separate shade is required for the vertical collector, it is generally not cost-effective to construct an overhang with a projection that exceeds 2′. Overhangs of this dimension and less can be simply achieved by extending the Sunspace roof construction. However, an overhang sized to provide a specified level of shading has the advantage of further reducing both the home's cooling energy requirements and potential glare in the cooling season.

  The projection of the overhang that will be adequate (that will provide 100% shading at noon on June 21) at particular latitudes can be quickly calculated by using the following formula:

Projection (L) = $\frac{\text{window opening, i.e., collector (H)}}{\text{F (see table below)}}$

North Latitude	F Factor*
28°	5.6-11.1
32°	4.0- 6.3
36°	3.0- 4.5
40°	2.5- 3.4
44°	2.0- 2.7
48°	1.7- 2.2
52°	1.5- 1.8
56°	1.3- 1.5

*Select a factor according to your latitude. The higher values will provide 100% shading at noon on June 21, the lower values on August 1.

A slightly longer overhang may be desirable at latitudes where this formula does not provide enough shade during the later part of the cooling season (see Figure 3.15).

For example, for Madison, Wisconsin, at 43°NL, select an F factor between 2.0 and 2.7 (values for 44°NL). Assuming shade is desired on August 1, choose the lower value of 2.0. If the window opening (H) is 8′, the projection (L) of the overhang placed at the top of the collector will be: $\frac{8}{2.0} \doteq 4'$.

Reflecting

- **Sizing:**
  The reflector should be the same length as the collector area served and roughly as wide as the collector area is high.

- **Slope:**
  A downward slope (away from collector) of roughly 5° is recommended.

Insulating

- **Fixed Insulation:**
  Fixed exterior insulation R-values should meet or exceed the levels that would be maintained by the same element in a nonsolar building, which should in turn meet or exceed local building-code requirements.

- **Movable Insulation:**
  Movable insulation should have an R-value between R-4 and R-9, subject to cost-benefit considerations.

- **Seal:**
  All insulation should provide a tight seal to avoid heat loss at the edges.

Vents (Exterior)

- **Area:**
  The exterior vent area should be one-sixth of the area of the wall shared between living space and Sunspace. The top vents should be one-third larger than bottom vents. The vertical distance between vents should be as great as possible.

Attached Sunspace Details

The drawings on the following pages illustrate common construction detailing for Attached Sunspace systems. The details are divided into five one-page sections corresponding to the subsystem types discussed above: open wall, direct gain, air exchange, thermal storage wall, and remote storage.

The five section pages suggest one design and material option for each subsystem. Components and materials shown for one subsystem can generally be substituted for another, provided the basic functional characteristics of the system are unchanged. For example, the various collector components illustrated can be used with any of the subsystem types.

Three pages of construction notes follow the details. The notes, referenced from the details, provide guidelines, troubleshooting tips, and other information useful in constructing an Attached Sunspace. Note that details for remote storage systems appear in Chapter 6, along with Convective Loop details. These storage systems can be incorporated, where appropriate, in the remote storage subsystem.

ATTACHED SUNSPACE
Open Wall Subsystem

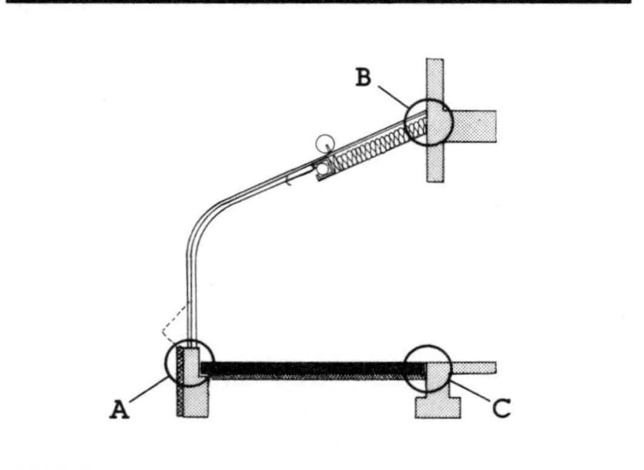

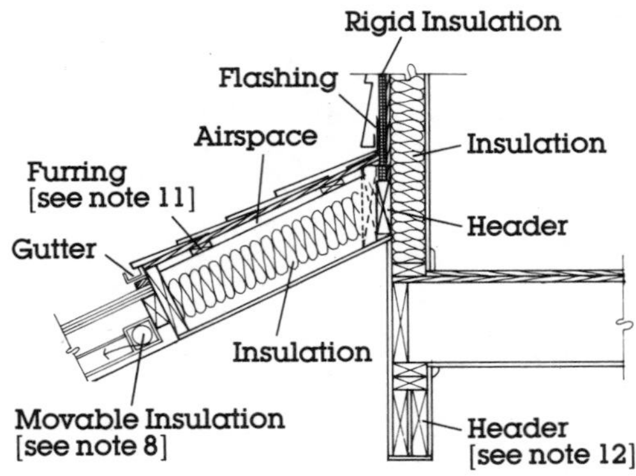

B. Collector at Framed Opening

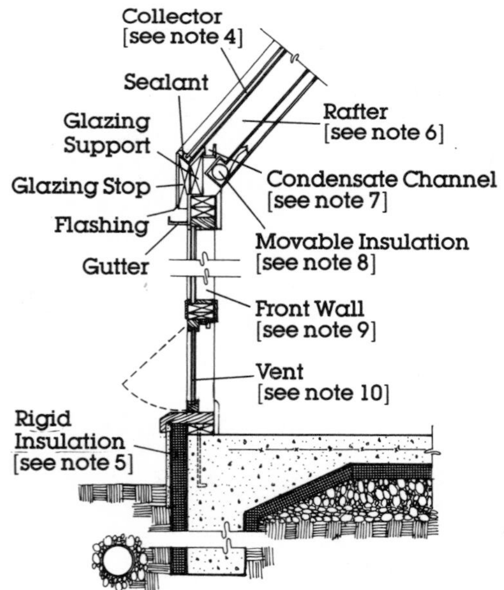

A. Front Wall

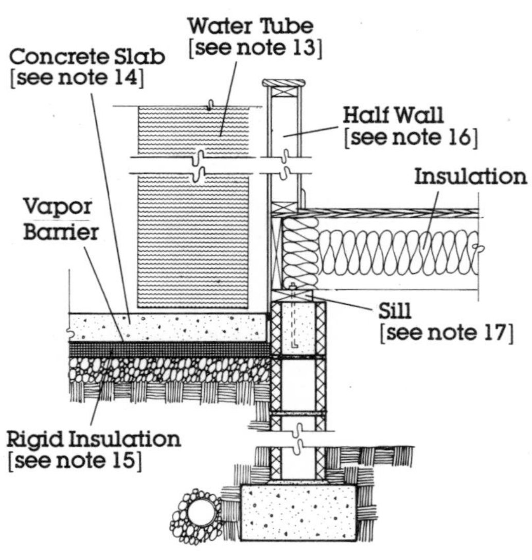

C. Low Wall at Framed Opening

Direct Gain Subsystem

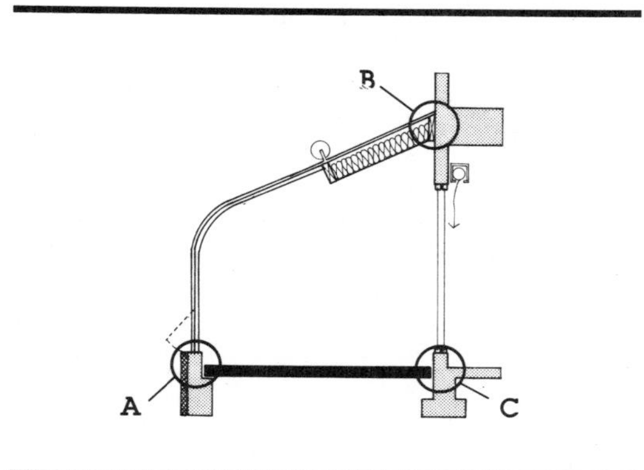

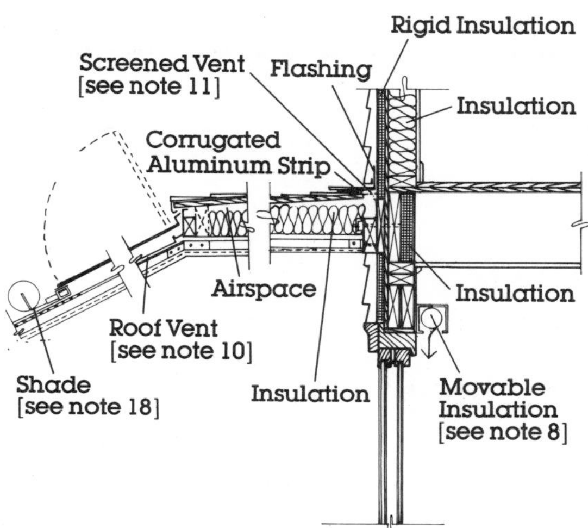

B. Patio Door at Frame Wall

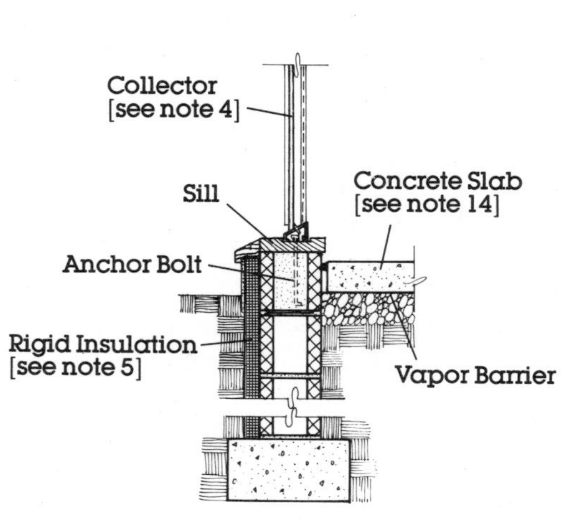

A. Sill at Grade

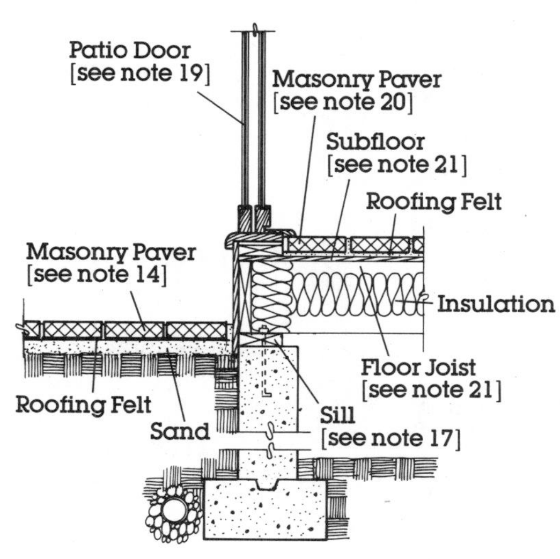

C. Patio Door at Frame Wall

Air Exchange Subsystem

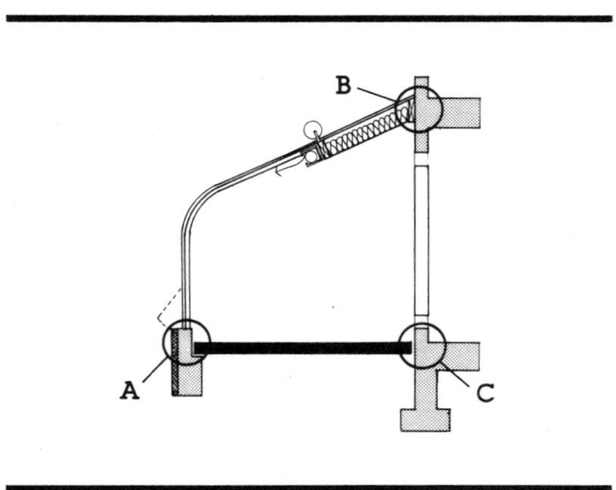

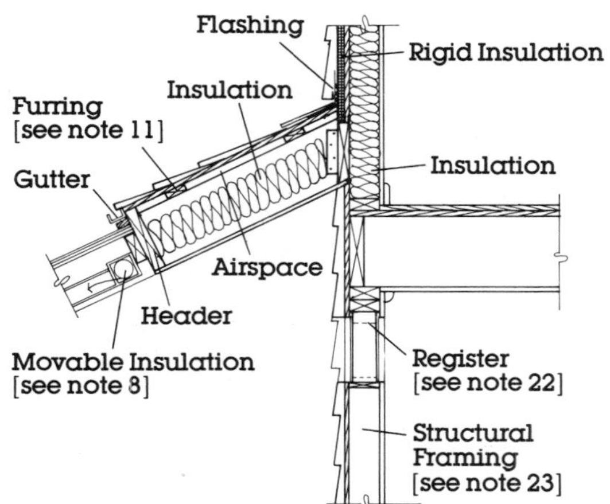

B. Vent at Frame Wall

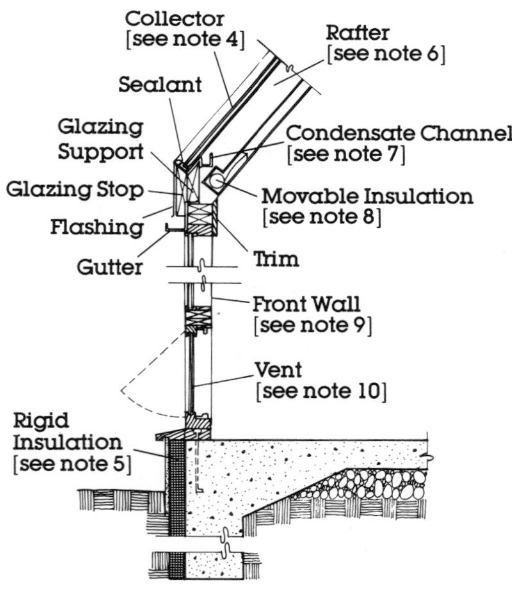

A. Front Wall

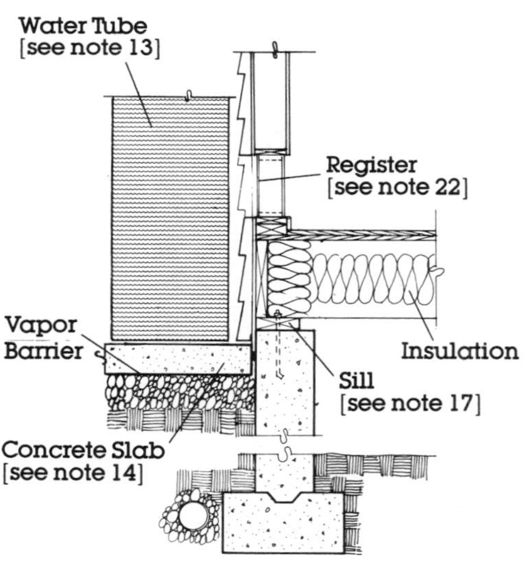

C. Vent at Frame Wall

Thermal Storage Wall Subsystem

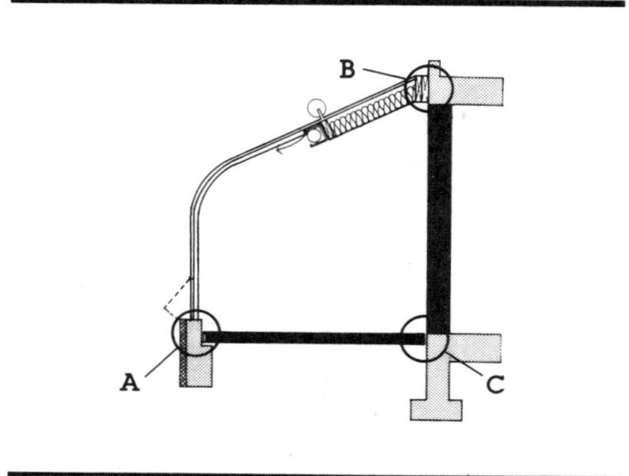

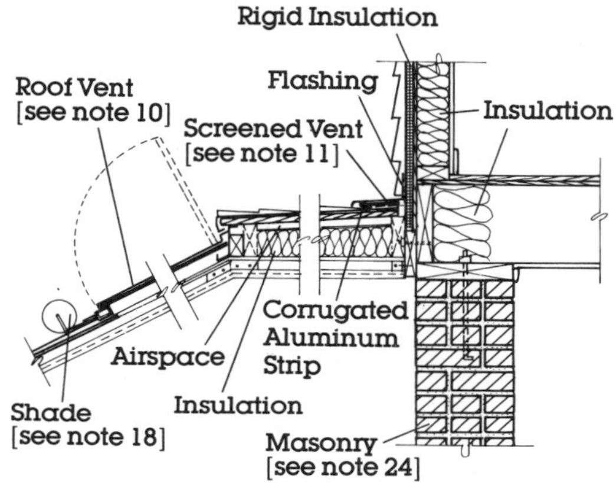

B. Masonry Storage Wall

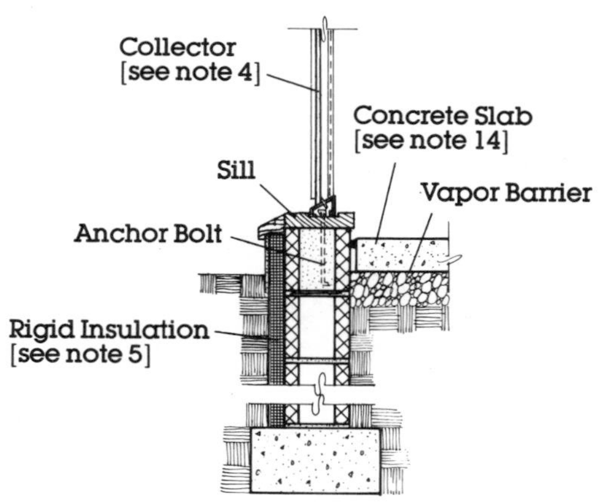

A. Sill at Grade

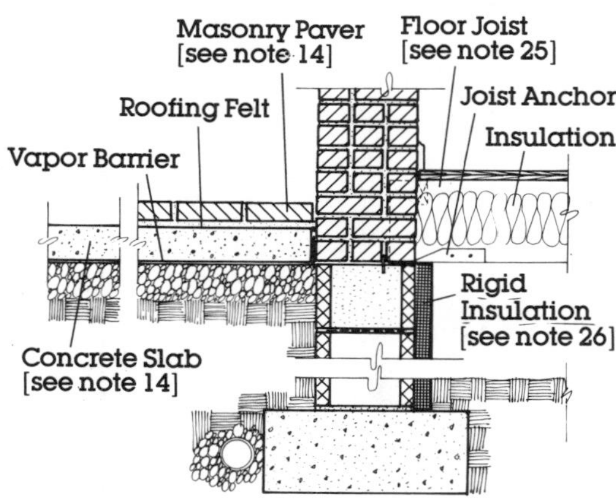

C. Masonry Storage Wall

Remote Storage Subsystem

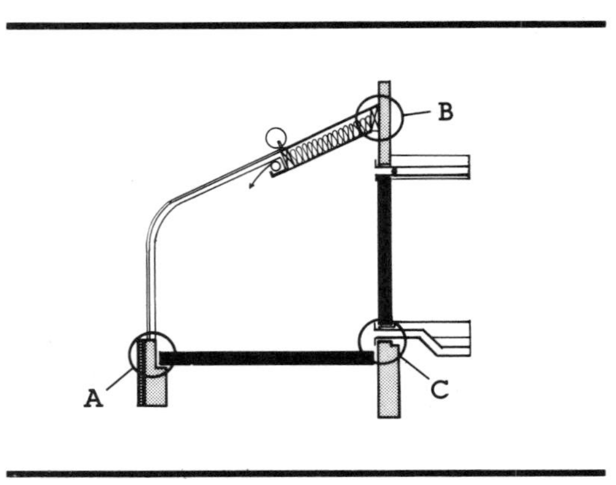

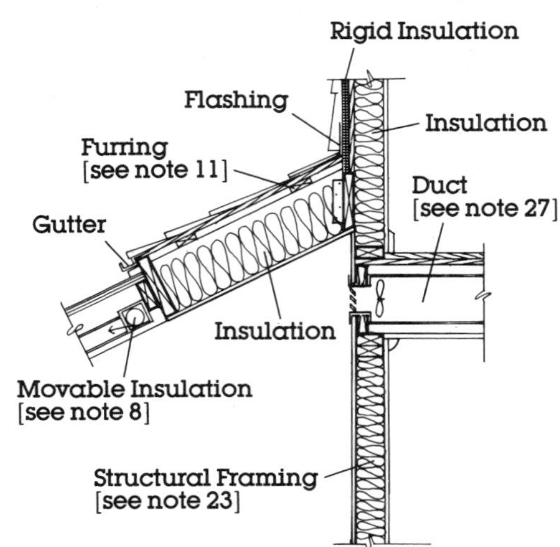

B. Supply Duct at Frame Wall

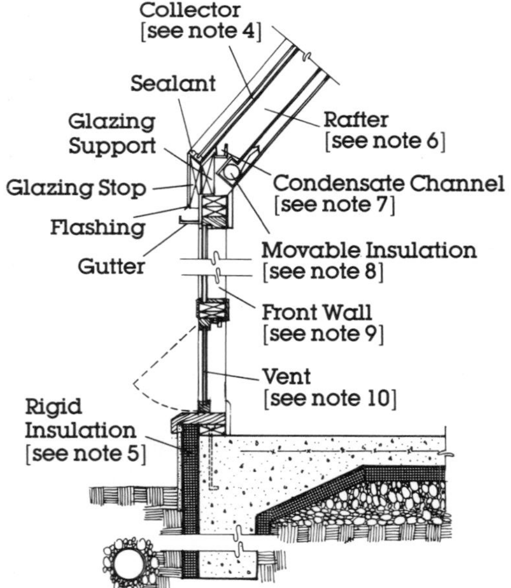

A. Front Wall

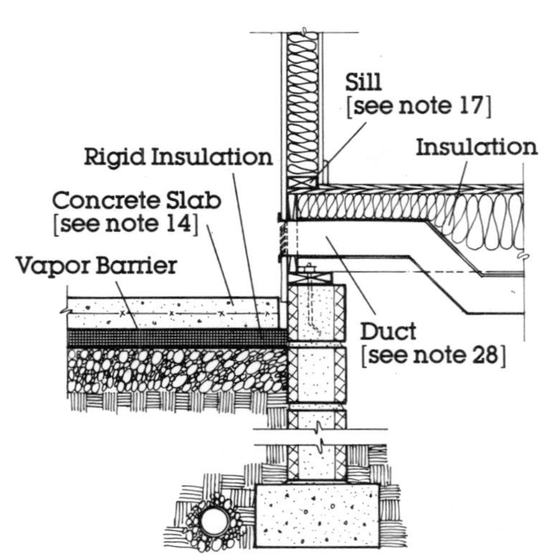

C. Return Duct at Frame Wall

Notes

GENERAL NOTES:

1. All footings must bear on undisturbed soil. Adjust footing size, reinforcing, and depth below grade as required by site conditions and/or local code.

2. Provide adequate drainage in Attached Sunspace floor, particularly where extensive planting is planned.

3. Insulation levels at foundation/floor, opaque walls, and opaque roof section should meet or exceed requirements of local building/energy codes for roof components.

DETAIL NOTES:

4. The glazing material can be glass or plastic. However, double glazing is generally recommended for the collector to ensure optimum thermal performance (see Collector Components in Chapter 7). A ½'' airspace between glazing layers is recommended where feasible. Where corrugated plastic glazing is specified, provide continuous blocking cut to conform to the corrugations. The collector frame can be site-built or prefabricated. Site-built systems are generally framed in wood, which should be treated to resist the effects of moisture. Special attention should be paid in these systems to proper sealing and caulking of the glazing material. Prefabricated systems (e.g., standard greenhouses) are generally framed in metal. The frame specified in any such system should be equipped with thermal breaks, and sill connections should be per manufacturers' specifications and local building codes and/or accepted practice.

5. Foundation and/or front walls may be insulated but must be protected above grade. Cement plaster (stucco) finish applied on wire lath may be used, or other finishes may be specified at builder's option. Consult insulation manufacturer for recommended installation procedures. Insulation should have an R-value that meets or exceeds the requirements of local building/energy codes for foundations.

6. Sizing of framing members is to be determined by weight of glazing material, loading requirements, and desired clear spans. All structural elements must be designed to meet applicable local building codes and/or accepted practice. High humidity levels in Attached Sunspaces used as working greenhouses may necessitate treating wood framing members with preservative. Avoid preservatives harmful to plants.

7. Condensate channel is formed by turning up metal flashing.

8. Movable insulation should be provided to reduce nighttime heat loss during the heating season and daytime heat gain during the cooling season. Because of the limited availability of movable insulation specifically designed for Attached Sunspaces, it may be necessary to site-design and site-fabricate an insulation system for a particular application—for example, an insulating curtain suspended from cables. Where

off-the-shelf movable insulation is specified, prior to detailing the collector component assembly, the insulation manufacturer's specifications should be consulted to ensure that adequate space has been provided for proper installation and operation (see Control Components in Chapter 7).

9. The structure and height of the Attached Sunspace front wall may vary depending upon design and use. The wall may be glazed, insulated opaque frame masonry. Reinforcing, thickness, provisions for weep holes, longitudinal drains, and waterproofing to be determined by site conditions, local building codes, and/or accepted practice.

10. Operable glazing units should be provided to vent the Sunspace to the outside during the cooling season. Intake vents should be as low in the Sunspace wall as possible, while exhaust vents should be high, either near the ridge or high in the gable ends (see Seasonal Operation—Cooling in this chapter). Some greenhouse manufacturers provide roof vents. Side vents are also available and suitable.

11. Opaque roof section must be vented to ensure proper performance of insulation. Vents may be located at the roof/wall connection or at sides of roof construction. Furring is provided as required to allow air to circulate.

12. Sunspace and adjacent living areas open directly to each other for natural air exchange between the two spaces. Headers over opening to be designed for loading and clear span per local building codes and/or accepted practice. For further information, see Open Wall Subsystem in this chapter.

13. Water, stored in appropriate containers, is an effective material for thermal storage. Suitable containers are fiberglass tubes, polyethylene and/or steel drums, and metal culverts. Prefabricated tank systems designed to be set within or on structural wall framing are also available. Care should be taken to ensure that steel and metal containers are lined with corrosion-resistant materials. In all cases, water should be treated with algae-retardant chemicals. For further information on the use of water as thermal storage, including descriptions of appropriate containers, see Storage Components in Chapter 7.

14. Concrete slab-on-grade and any of the wide variety of standard and interlocking concrete masonry unit pavers or standard brick pavers on slab-on-grade are suitable for thermal storage. For further information, including finishes and construction techniques, see Interior Floor Types in Chapter 3.

15. In the cooler climates, it is generally recommended that rigid insulation be placed under masonry floors that are acting as storage components in Attached Sunspaces, particularly when the Sunspace is open to the living spaces. The use of rigid insulation is recommended in all cases to reduce perimeter heat loss.

16. Low wall will not restrict free circulation of air between Sunspace and adjacent living areas. Consult local codes

for structural requirements. For further information, see Open Wall Subsystem in this chapter.

17. Continuous sill sealer is recommended to provide protection against infiltration.

18. Exterior movable shades can be provided for protection against solar heat gain during the cooling season. Such shades are typically aluminum, wood, or fabric such as canvas (see Control Components in Chapter 7).

19. Patio door at adjacent living space may be single or double glazed. Installation is per manufacturers' specifications. For further information, see Direct Gain Subsystem in this chapter.

20. For information on masonry pavers used for Direct Gain storage, including recommended installation procedures, see Interior Floor Types in Chapter 3.

21. Consideration must be given to the additional weight of pavers. Consult structural design loading tables and local codes to determine suitable grade and spacing of joists, and grade and thickness of subfloor. Expansion joints should be provided, as required, at the connection between masonry floors, and stud or masonry walls to prevent cracking.

22. Vents from top of Attached Sunspace distribute heat into adjacent living area, and vents at bottom return cooler air from living space to Sunspace.

23. Insulation is not required in the structural frame wall separating the Attached Sunspace and living spaces. For further information, see Air Exchange Subsystem in this chapter.

24. Alternate storage wall materials can be specified at the builder's option. For further information on Thermal Storage Wall materials and configurations, see Wall Types in Chapter 4.

25. Floor joists must bear on solid masonry not less than 4'' thick. Fire-cut joists as required.

26. It is recommended that wall at crawl space/unheated basement foundation be insulated to prevent heat losses through wall. Insulation levels should meet or exceed building/energy-code requirements for foundation insulation.

27. Air-handling ducts can be connected to vents at the top of the Attached Sunspace to move heated air to the rear of the house, to remote storage, or to a radiant floor system. A small, low-velocity fan can be incorporated into the duct system to ensure proper airflow. For further information on fans, see Distribution Components in Chapter 7.

28. To facilitate proper airflow to remote storage, a cold-air return must be provided. A return-air duct can be connected to a vent at the bottom of the Attached Sunspace, with the cold air supplied from a floor register at the rear of the house, from the cold-air plenum of a rock bed, from the exiting air of a radiant floor system, or from the basement, crawl space, or other unheated underfloor plenum.

TVA Solar House #5
Attached Sunspace

An Attached Sunspace spans the entire south side of this 1,870-sq-ft two-story house designed for a flat or slightly sloping site. The Sunspace combines thermal storage wall and direct gain subsystems to create a space that is well matched to the major living spaces of the home. Circulation and service spaces, including the kitchen, entry, utility room, and bathroom, buffer the colder north wall.

The wall and roof of the Attached Sunspace are constructed of double-glazed operable window units that can be insulated in the evening by a movable curtain. Solid masonry floor pavers are laid mortarless over a 6″ sand bed to retain heat. Heated air is circulated between Sunspace and living space by opening adjoining windows and doors.

The 12″ heavyweight concrete masonry unit thermal mass wall is filled solid with grout and finished with a dark plaster on the greenhouse side to increase solar absorption. An optional woodburning stove next to an interior masonry wall provides auxiliary heat. An opaque section of the Sunspace roof serves as a sunscreen for the Thermal Storage Wall during the cooling season.

During the summer, heat generated within the living space can be vented to an attic and then exhausted to the exterior through louvered vents. The Greenhouse/Sunspace is vented by a wall-mounted fan and operable glazing panels in the collector assembly.

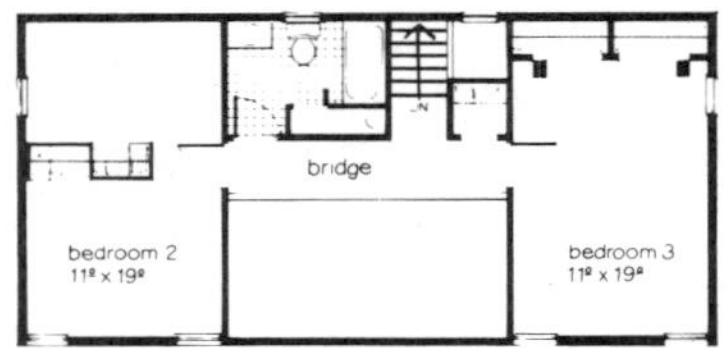

Upper Plan 656 sf

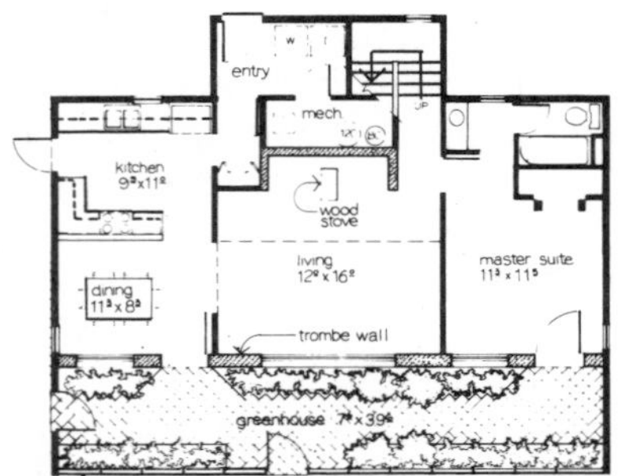

Lower Plan

North

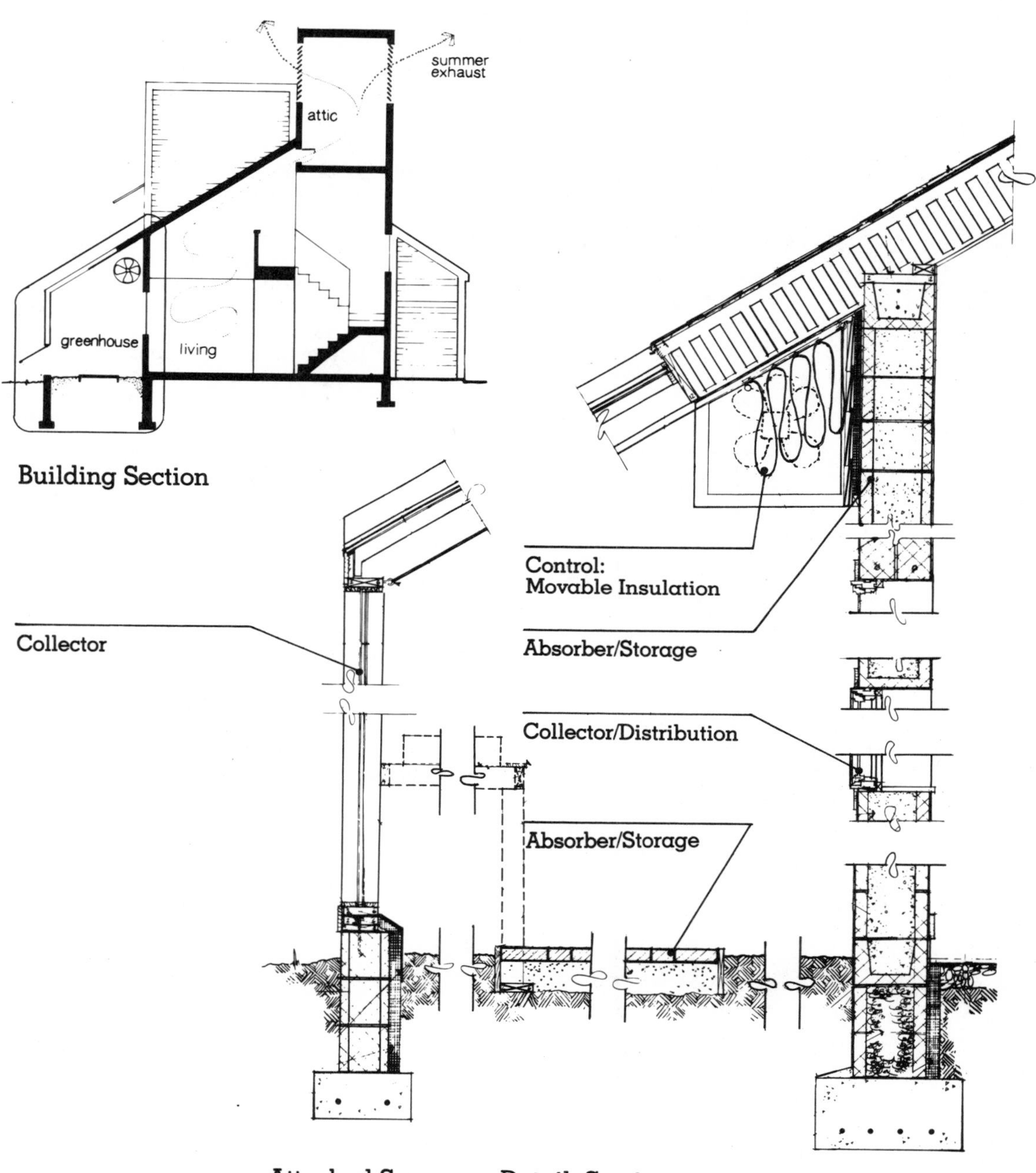

Building Section

Attached Sunspace Detail: Section

PCA Two-Story
Attached Sunspace

Attached Sunspace, patio, and covered porch areas enhance the openness of this 1,800-sq-ft two-story passive house. The garage, air lock, and entry vestibule below, and closets, bath, and stair above created an effective north-wall thermal buffer zone.

Adjacent to the kitchen, the south-glazed Sunspace easily lends itself for use as a breakfast room or working greenhouse. Its concrete slab-on-grade floor provides solar storage. The sliding glass doors are used to control heat flow between the Sunspace and the major living spaces. The great room receives Direct Gain heat, which is stored in the quarry tile floor. The tile rests on 1½″ of concrete to provide the necessary depth of masonry.

During the summer, wing walls and overhangs to the south protect interior spaces by controlling glare and unwanted heat gain.

A double-height space in the great room facilitates heated-air circulation to the upper floors. Supplementary Direct Gain heating is provided by a row of clerestory windows, which also enhance the open feeling of this design. The R-13 exterior wall and R-38 external ceiling and floor insulation help control heat flow through the building envelope throughout the year.

Upper Floor Plan
882 sq.ft.

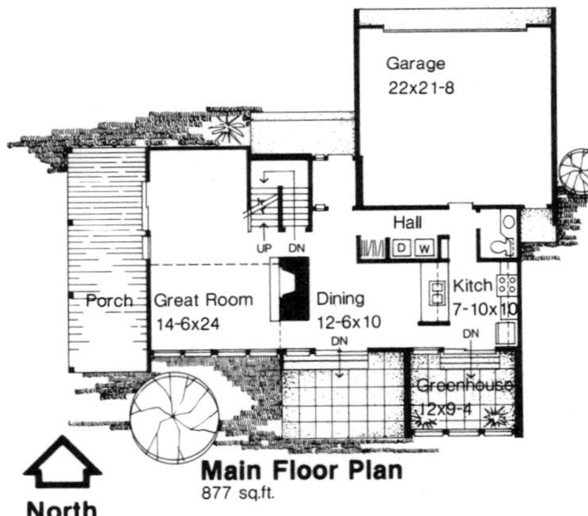

Main Floor Plan
877 sq.ft.

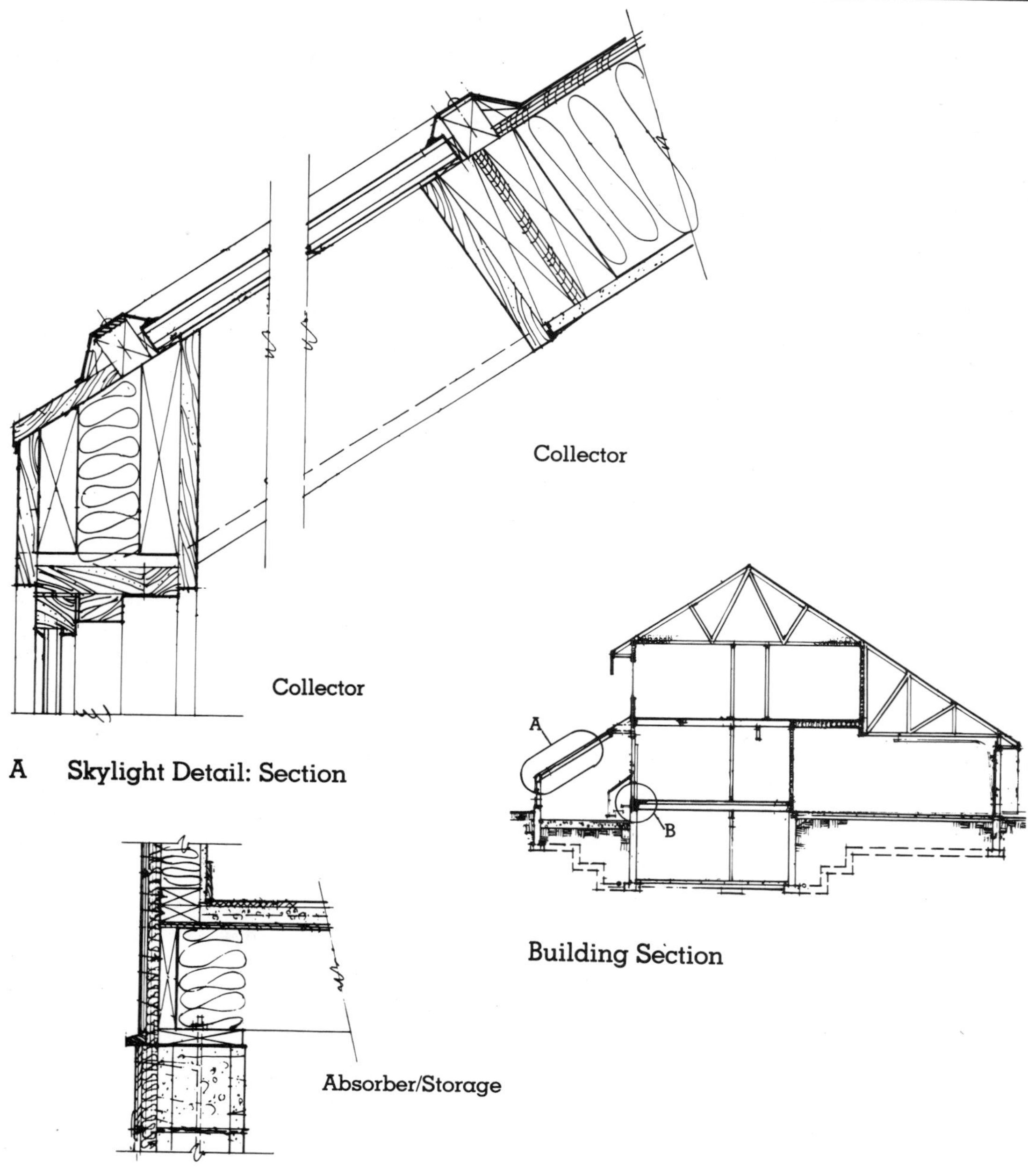

A Skylight Detail: Section

Building Section

B Floor Detail: Section

La Vereda House #4

Attached Sunspace

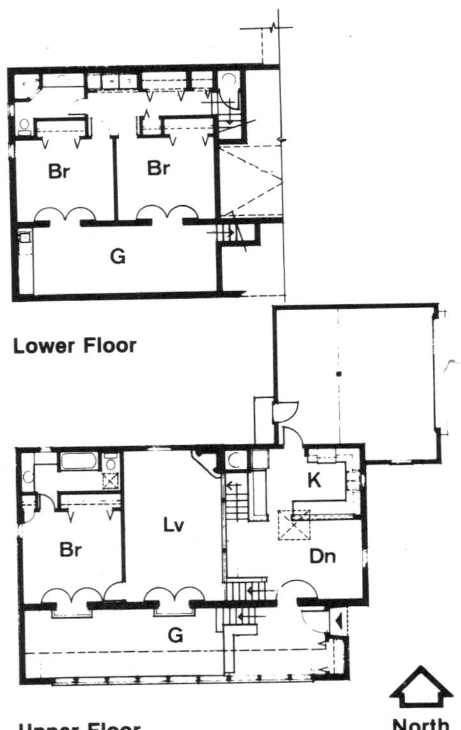

A double-height expanse of greenhouse stretches along the south wall of this split-level home in Santa Fe. Designed for a slightly sloping site, this wood-framed home is partially bermed on the east, west, and north, with trees to the north offering winter wind protection. Heat, which collects in the Sunspace, is stored in an adobe wall and rock bed located underneath the first floor of the home.

A solid roof and projecting overhang shade the adobe wall from the summer sun. The Attached Sunspace is constructed using both tilted and vertical south glazing. In addition to the adobe wall and rock bed, a concrete slab topped with flagstone provides sufficient thermal mass storage.

Doors cut through the adobe wall between the greenhouse and the living spaces can be opened to allow heated air, generated in the Sunspace, to flow throughout the home. During the summer, Sunspace heat is exhausted by wind turbines located in the greenhouse ceiling.

The rock bed contains 810 cu ft of storage beneath the kitchen and dining room. On top of the 1½″- to 2″-diameter rocks, a layer of polyethylene film and a 1″ setting bed of sand provide a base for the 4″ concrete slab, which is finished with tile. A thermostatically controlled fan at the top of the Sunspace draws heated air through ducts to a plenum along the north side of the rock bed. Heat is transferred to the rocks and radiates through the slab. Return air is ducted through a plenum back to the greenhouse, completing the loop. The rock bed is insulated, improving its thermal performance.

The 8″ concrete masonry foundation wall is sheathed in 2″ of rigid R-9 polystyrene with a plaster coating. Ceiling insulation varies between R-30 and, where a combination of rigid polystyrene and fiberglass batt insulation is applied, R-52. Walls are insulated to a level of R-33.

Project: La Vereda Model 4

Builder/Designer
Communico, Inc.
Susan Nichols
Santa Fe, New Mexico

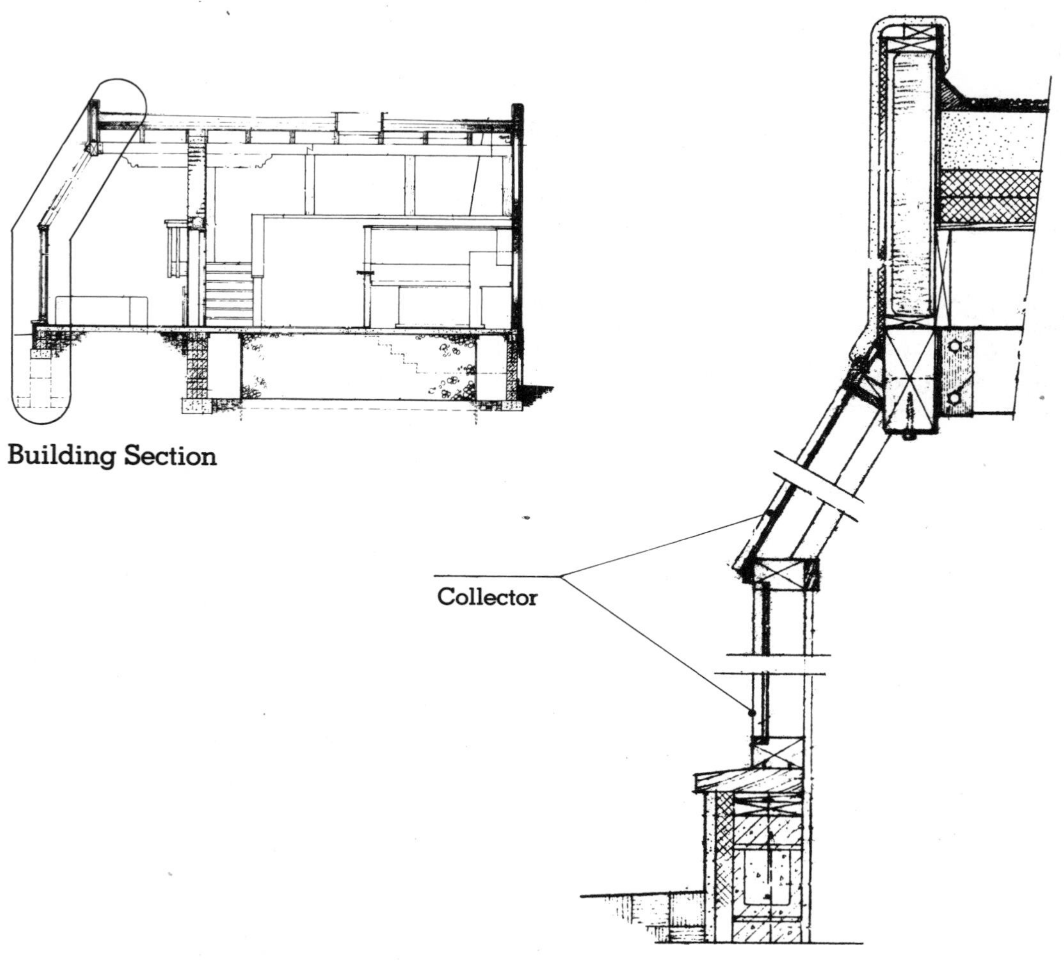

Building Section

A Attached Sunspace Detail: Section

SSEC Atlanta House
Attached Sunspace

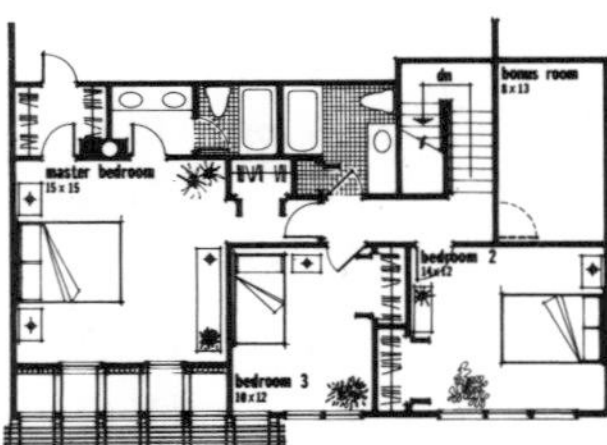

2nd Floor

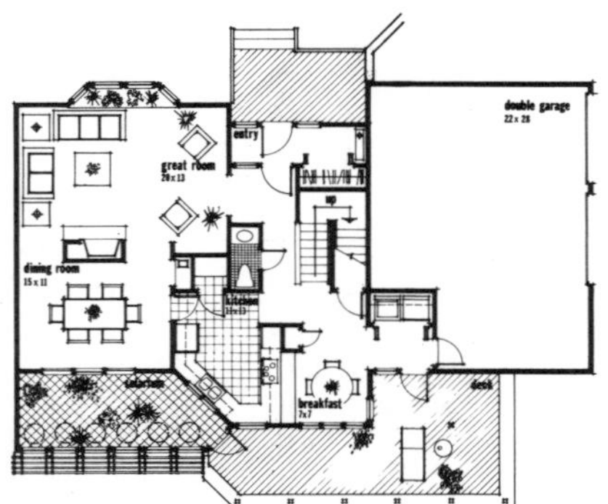

North

1st Floor

A two-story Attached Sunspace incorporates a water wall and phase-change materials for thermal storage in this 2,300-sq-ft home in Atlanta. Living spaces on the upper floor are buffered from heat losses to the north by service and circulation spaces and, on the ground floor, by an air-lock entry vestibule and garage.

The foundation wall is constructed of 8" concrete blocks with exterior rigid insulation, protected above grade with an applied plaster coating. R-38 roof and R-20 wall insulation aid in reducing heat losses.

Sliding glass doors provide both direct access and air distribution from the Sunspace to the dining room. Above the dining room, the main bedroom overlooks the Sunspace through windows inserted between three panels containing additional thermal mass storage in the form of thermal storage rods filled with phase-change materials.

Two thermostatically controlled fans at ceiling level are mounted at either end of the solar porch. One of the fans is opened to the exterior to cool the space in the summer. The other fan operates in the winter, distributing heated air through ducts in the ceiling into the great room.

The triple-height expanse of southern glazing is capped by wooden trelliswork at the first two floors, and with a roof overhang above. Both devices serve as sunscreens to prevent direct solar gain in the cooling season.

The Attached Sunspace contains a water wall of nine 55-gal black drums stacked three by three, providing a significant amount of thermal mass. The Sunspace heat is supplemented by heat generated in a partial basement below. Air heated in both spaces is allowed to rise naturally into the living spaces above.

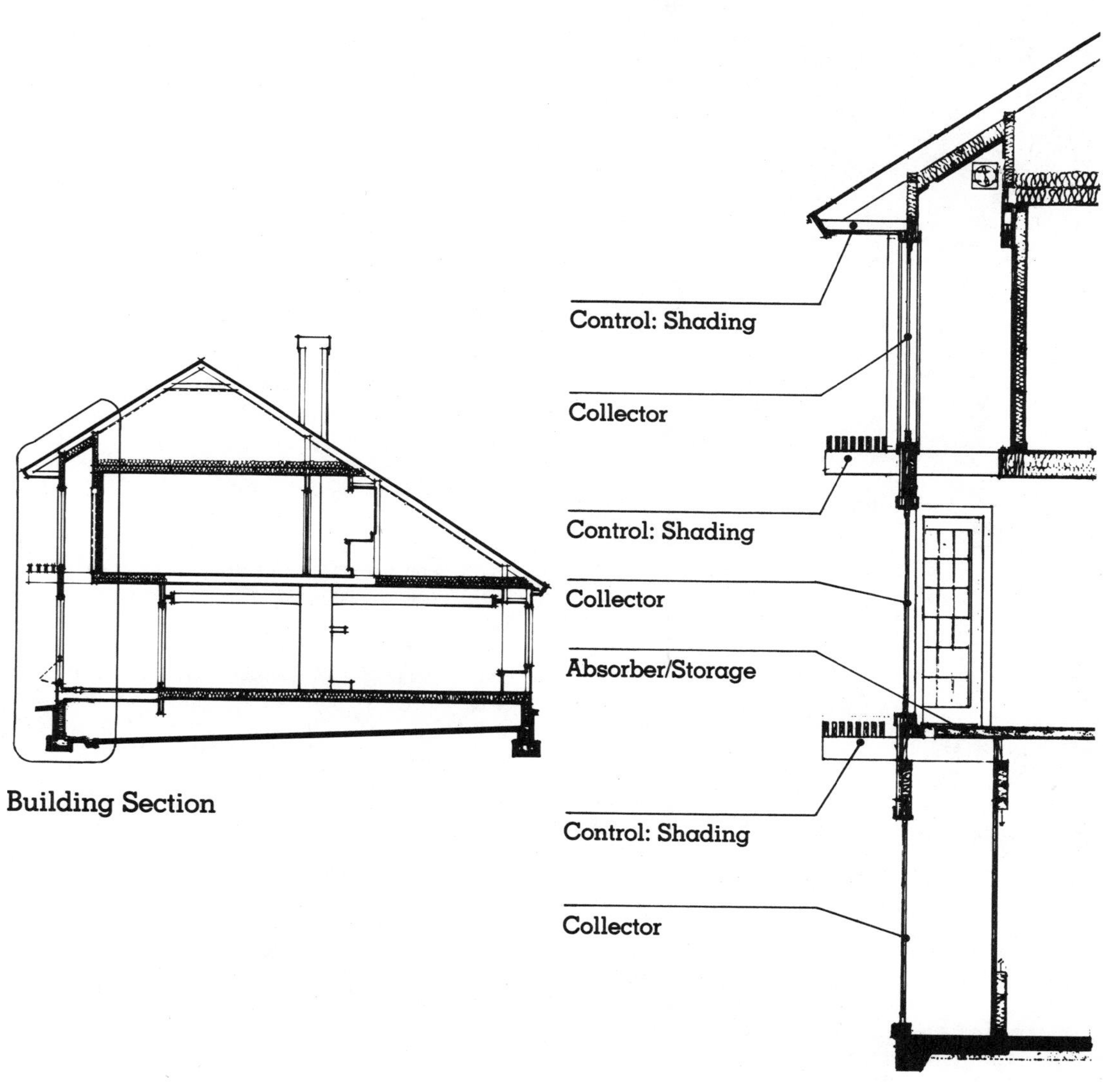

Building Section

Wall Detail: Section

PEG House #1
Attached Sunspace

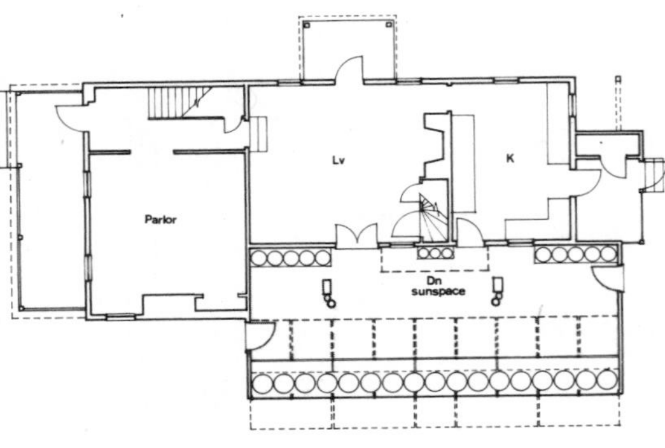

Floor Plan

The 500-sq-ft lean-to greenhouse/dining room addition to the 1,900-sq-ft Hamill house of Princeton, New Jersey, serves the dual function of increasing living space and reducing the home's heating requirements. Thermal storage is afforded by water-filled cylinders, masonry wall construction, and the slate floor, all contained within the Sunspace.

First- and second-floor doors and windows in the masonry wall at the back of the Attached Sunspace can be opened to allow heat flow into the adjacent living spaces. The full-time use of the Sunspace as an added living space requires that the space be well insulated with movable insulation at the glazing.

One hundred square feet of double-glazed vertical collector and 500 sq ft of tilted collector glass over Plexiglas comprise the south wall of the Sunspace. The insulating curtain with an interior reflective coating operates on a track and cable system to reduce nighttime heat loss. Two woodburning stoves provide auxiliary heat.

The solid roof and end walls of the addition offer protection against overheating. Upper glazing panels can be opened with a rod-and-arm mechanism to ventilate the Sunspace, forming a 36′ continuous vent. A shading screen is hung in front of the glazing to reduce the seasonal cooling requirements.

Seventeen water-filled 55-gal storage drums are placed adjacent to the vertical glazing. On the exterior, hinged shutters reflect more light onto the drums when opened during the day. When closed at night, they insulate the lower glazing against heat losses. Additional water storage is provided by ten 18″-diameter and three 12″-diameter fiberglass cylinders, which line the north wall of the Sunspace.

Project: Sunspace Addition

Builder
H. A. Snedeker's Sons, Inc.
Kenneth Snedeker
Kingston, New Jersey

Designer
Harrison Fraker Architects
Harrison Fraker, Peter Brook
Princeton, New Jersey

Sponsor
Private Individual
Princeton, New Jersey

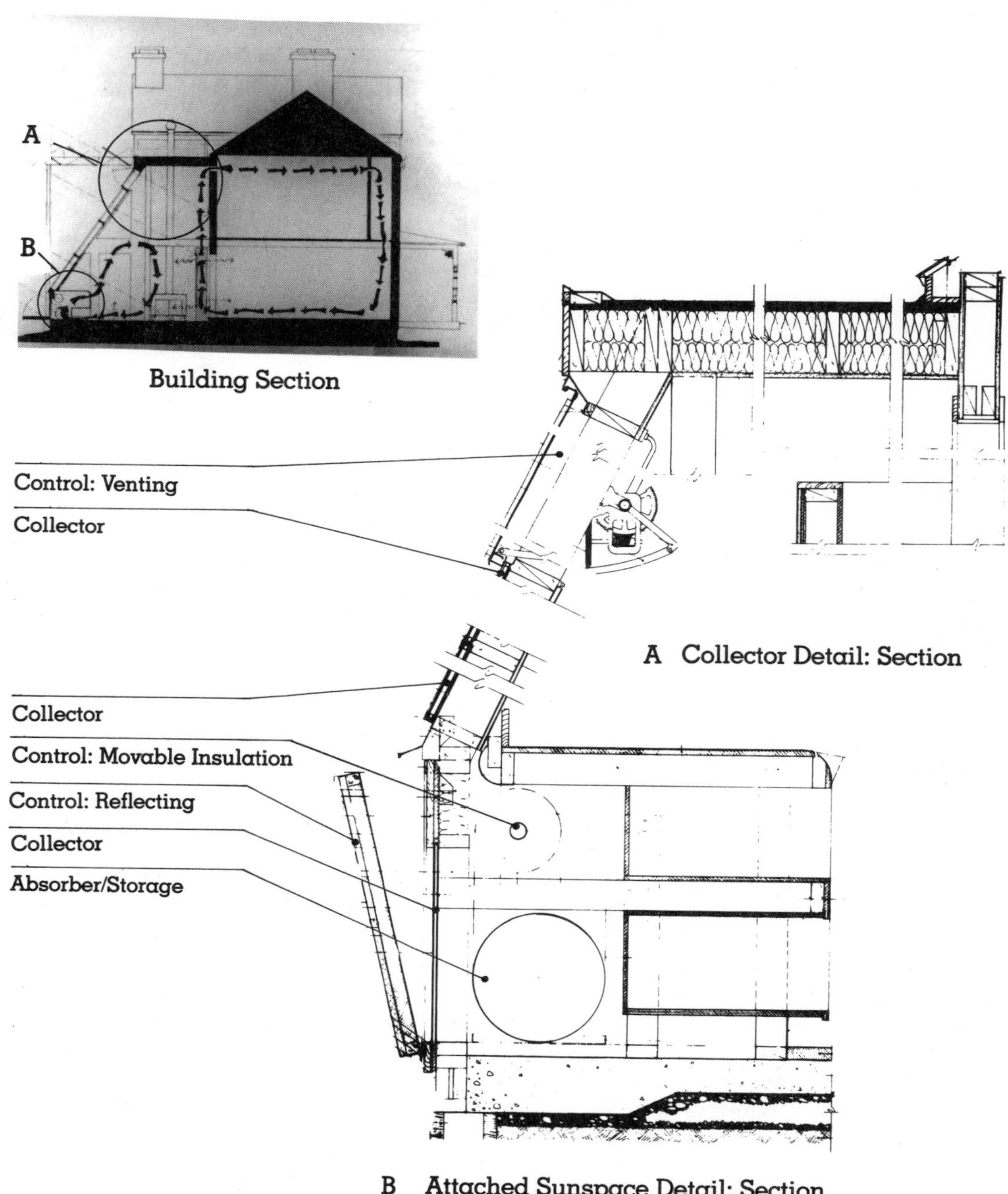

Building Section

A Collector Detail: Section

B Attached Sunspace Detail: Section

Chapter 6
CONVECTIVE LOOP

Design Overview

Convective Loop systems are designed to be used exclusively for heating. They are effective in any location with substantial heating requirements, although they are most appropriate in moderate and severe climates. One major advantage of a Convective Loop system is that it is thermally isolated, and often physically separated, from the living spaces of the home.

As discussed in the previous chapters, Direct Gain, Thermal Storage Wall, and Attached Sunspace systems are potential sources of heat loss during overcast periods and during the night. Control components are typically required in such systems to reduce these losses. Convective Loop systems, however, because they are thermally isolated from living spaces, are not subject to such heat losses, reducing and, in some cases, eliminating, the need for control components.

Isolating solar gains, however, requires that the distribution component and storage component (not required in some Convective Loop configurations) be carefully designed, detailed, and constructed.

The major element in a Convective Loop design is the solar collection panel. These panels, frequently referred to as thermosiphon air panels (TAPs), are similar in appearance and operation to active solar panel collectors. However, unlike active solar panels, which are generally roof mounted, the TAP is typically mounted on the exterior wall in a vertical position (see Figure 6.1) or below the floor line at a steep angle (see Figure 6.2).

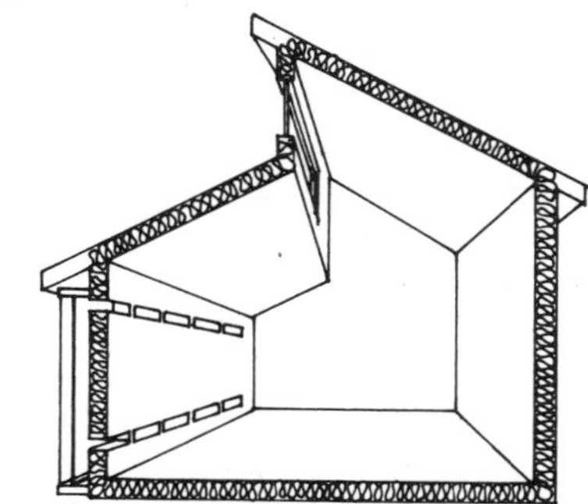
Figure 6.1

Convective Loop systems operate on the basic principle that hot air rises. As in all passive systems, the collector in a Convective Loop admits sunlight, which strikes an absorber surface and is converted to heat. However, in the Convective Loop, this absorbed heat is transferred to the air surrounding the absorber instead of being directly stored by thermal mass. The heated air rises inside the TAP until it passes through a vent located at the top and is distributed either directly into the home or into some form of remote storage medium such as a rock bed. As this air passes out through the upper vent, it is replaced by cooler air entering through a vent located near the bottom of the panel (see Figure 6.3).

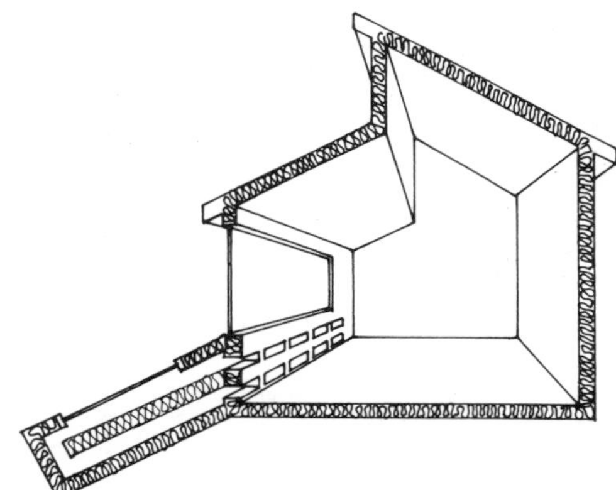
Figure 6.2

All the air movement in such a system is, in principle, by natural convection, although there may be certain designs in which a small fan is required.

Storage is recommended when the amount of heat delivered by the panel represents a substantial part of the total heating requirements of the home. In the more severe climates, this is rarely the case, and only large collector areas will justify the cost of most storage designs.

Seasonal Operation—Heating

During the heating season, sunlight is transmitted through the collector (usually glass or plastic) of the Convective Loop TAP and strikes the absorber surface. The absorber, which is typically a black-colored metallic surface, converts the sunlight to heat. Since metal or other

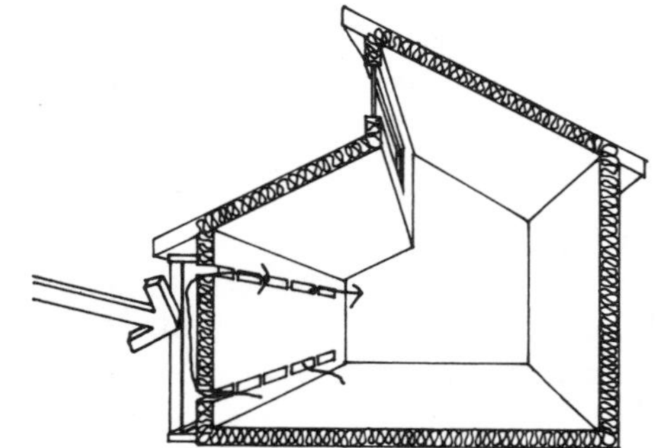
Figure 6.3

materials used as absorbers are good conductors of heat, the air surrounding the absorber will quickly be heated and rise in the panel and through the outlet vent. Cooler makeup air will be drawn from the room into the panel inlet, starting a flow of heat that continues throughout the daylight hours. The heat generated in the Convective Loop panel is distributed either directly into the living spaces or, if required, into remote storage (see Figure 6.4).

This basic process is the same for all Convective Loop system configurations. However, it is in the design, operation, and construction of the components that two basic subsystems can be distinguished. In the discussions that follow, these subsystem types, referred to as vertical and U-tube panel subsystems, are treated separately.

Vertical Panel Subsystem

In this subsystem, the TAP is mounted directly on the south-facing wall, outside the insulation, thermally isolating the panel from the living spaces. Vertical panel systems are usually constructed without storage, delivering heat directly to the living spaces through vents at the top of the panel. Cooler makeup air is drawn into the panel through vents at the bottom of the panel (see Figure 6.3).

Due to the high cost and complexity of construction, remote storage should be considered for the vertical panel configuration only when the system is designed to provide a substantial percentage of the home's total heating requirements.

To prevent warm room air from being drawn back into the panel during cloudy periods or at night, simple backdraft dampers are typically provided at the vents. This system configuration, without storage, is inexpensive to construct, requires little maintenance, and is one of the most effective passive solar designs.

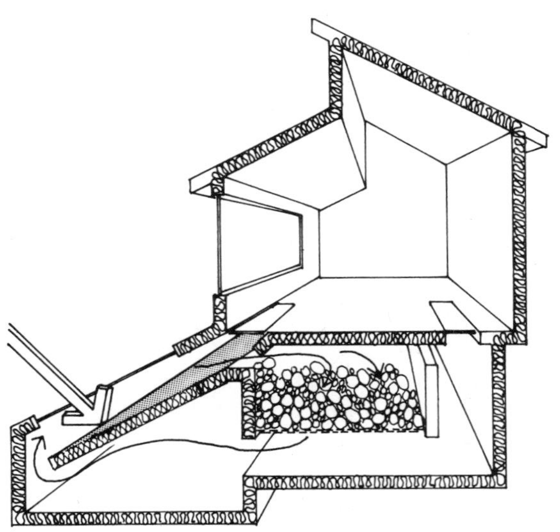

Figure 6.4

U-Tube Panel Subsystem

As its name implies, the TAP in the U-tube subsystem is constructed in the form of a "U" with the cool inlet air drawn down one side of the panel and warm outlet air rising up the opposite side (see Figure 6.5).

This panel configuration has some inherent advantages. If carefully designed and constructed, the panel is self-damping, eliminating the need for control features such as dampers. In the evening or during cloudy periods, the air in the panel cools. This cool air is trapped in the panel and cannot rise up into the living space. Therefore, unlike the TAP design in the vertical panel subsystem, a damper is not required, further simplifying this passive system.

The close proximity of the inlet and outlet vents provides a second advantage for the U-tube. If the vents are located near the floor, distribution to subfloor storage is direct and, consequently, less expensive and simpler to construct than distribution in a vertical panel system. In addition, in the below-floor-level location, less of the south side of the home is covered by the TAP.

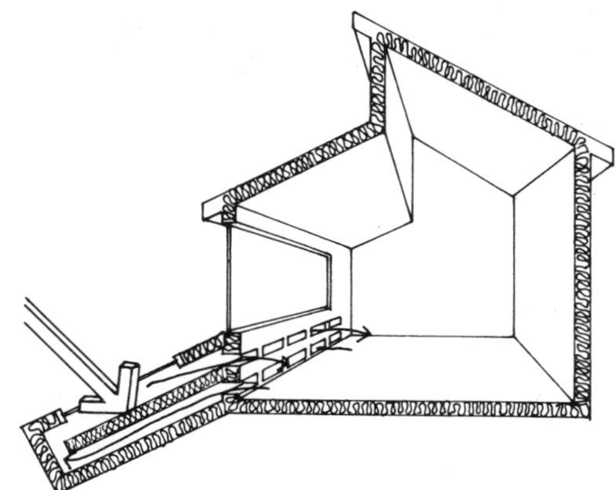
Figure 6.5

The U-tube subsystem can be used either with or without storage. As in the vertical panel subsystem, storage is required only when the panel is to deliver sufficient heat to satisfy a substantial portion of the total home heating needs. The system is "activated" when sun shining on the panel heats the absorber, in turn heating the surrounding air and causing it to rise into the living spaces. Cooler room air is drawn into the inlet vent, completing the loop.

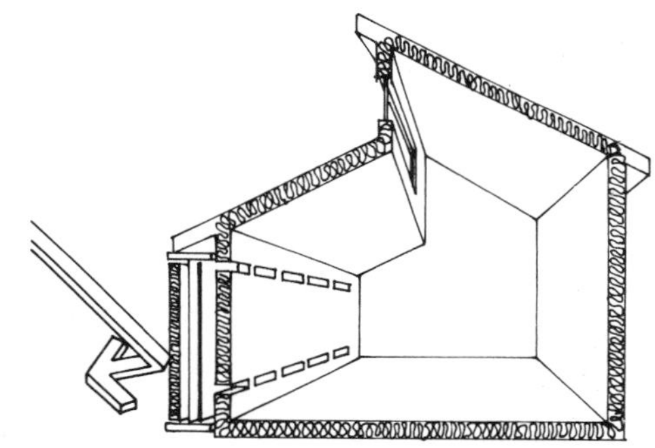
Figure 6.6

The U-tube subsystem without storage is inexpensive to construct, although generally more expensive than the vertical panel subsystem, requires little maintenance, and has a high efficiency relative to the other passive system types.

Seasonal Operation—Cooling

Since Convective Loop systems are designed for heating only, they are inactive during the cooling season. In the vertical panel subsystem, the TAP should be completely isolated from the living space and the storage component by closing the dampers so that any residual heat can be prevented from reaching the home. In addition, it is recommended that the TAP be well shaded or simply covered with an opaque material to prevent unwanted heat buildup in the panel (see Figures 6.6 and 6.7).

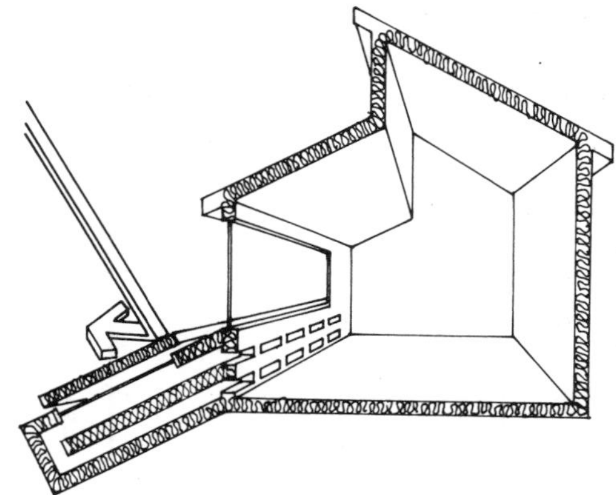
Figure 6.7

An unshaded TAP that is not vented, sometimes referred to as "stagnating," can reach fairly high temperatues. At such temperatures, panel seals are stressed, materials often warp, and the absorber surface can be damaged.

Advantages

- The Convective Loop is the most thermally efficient of all the passive system types.
- Thermal storage is not required if the Convective Loop system is designed to provide less than 30% of the home's heating requirements.
- If designed without storage, the Convective Loop is relatively inexpensive and simple to construct.
- The Convective Loop system is isolated from the living space, allowing the homeowner to completely control heat gains and losses to and from the system.
- The TAP can be physically separated from the home, allowing greater freedom in the design of the home itself.

Disadvantages

- Unless careful attention is paid to its design, the TAP can be an aesthetic liability.
- Unlike the Direct Gain and Attached Sunspace, which can provide other amenities such as light and additional living space, the Convective Loop has only one function—providing heat.
- If mounted on the south-facing wall, the vertical panel subsystem can obstruct light and view. Conversely, the U-tube subsystem, if mounted below the floor line, may impose specific constraints on site grading, landscaping, and use of outdoor areas.

Collector

The collector component of the Convective Loop system, as in the other passive system types, admits solar radiation onto the absorber and traps the heat generated. The collector in a Convective Loop is usually part of a thermosiphon air panel (TAP) whose operation is discussed earlier in this chapter. The TAP is simply an insulated box, glazed on one side, with an absorber plate inside. Air inlets and outlets are provided to remove the heat generated at the absorber. Most TAPs are site-fabricated, although some prefabricated units are available off-the-shelf.

The collector glazing material can be glass or plastic. However, since the temperatures within the TAP can be expected to reach 300°F, materials that will deteriorate at high temperatures should be avoided.

The glazing materials specified may be patio-door replacement glass or plastics available in large sheets, so as to minimize the mullion area in the frame. Single glazing is recommended for all except severe climates, where tempered double glazing should be specified.

Each glazing layer should be set into the frame with sufficient tolerance to allow for 3/16" expansion in both directions, again, due to the high operating temperatures. A generous amount of silicone caulking is advisable to accommodate this expansion while keeping the unit airtight and weathertight. The seals should be periodically checked, since some separation may occur in the first few months of operation.

The frame can be constructed of wood or metal, although for site-built TAPs wood is recommended. In the U-tube configuration, the TAP is often completely exposed, and care should be taken to provide insulation on all sides. In the vertical subsystem TAP, the frame wall of the home forms the backing for the unit, and standard batt insulation thickness is commonly specified. In both cases, exposed wood should be treated to prevent rot. If the frame is metal, it should be separated from the absorber to prevent conductive heat losses to the outside. If the frame is wood, rough-sawn and green wood should be avoided in favor of kiln-dried lumber. Any paint used on the frame should be resistant to high temperatures.

Collector Rules of Thumb

- **Orientation:**
  Orient the Convective Loop collector panel to face south. Orientations up to 20° east or west of true south will not significantly affect performance.

- **Design:**
  The TAP system can be wall mounted, as in the vertical subsystem configuration, or below the floor line for U-tube construction. Floor-line systems should be tilted to an angle of not less than 45°. The optimum tilt is an angle equal to the location latitude plus 10° (see Figure 6.8).

  For example, a tilted panel in Madison, Wisconsin, at 43°NL should be constructed and mounted at a 53° angle (43° + 10°) to the ground.

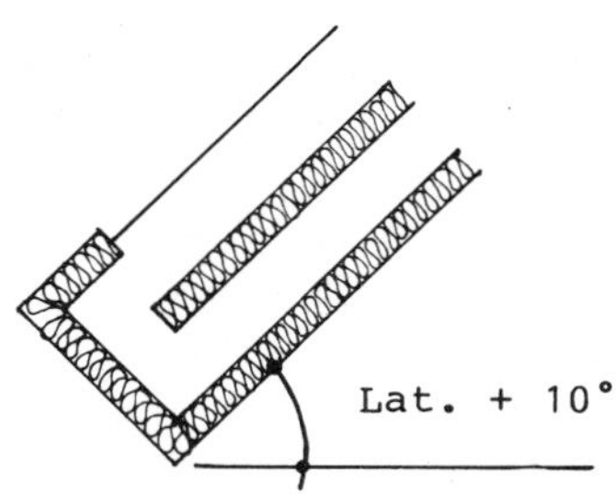

Figure 6.8

- **Area:**
  The collector area should equal from 20-40% of the area to be heated. The height of the collector is typically from 6-18′ but should not be less than 4′.

  For example, if a 300-sq-ft living room is to be heated, the collector area should be from 60-120 sq ft. If 8′ high, the unit would be from 8-16′ wide.

- **Number of Glazings:**
  Single glazing is generally recommended except in severe climates, where double glazing should be specified. Where double glazing is specified, a ½″ airspace between glazing layers is recommended.

Absorber

Unlike the other passive systems, the absorber in a Convective Loop system is a distinct element whose *only* function is to absorb solar radiation and convert it into heat. This heat is in turn transferred to the airstream, which passes either over or through the absorber material. In effect, the absorber works very much like the heating element in a standard convection heater.

The absorber material is typically metal, painted black to effectively absorb solar radiation. The metal surface heats up quickly when exposed to the sun and rapidly transfers this heat to the surrounding air. Paint applied to the metal absorber plate must be able to withstand operating temperatures above 300°F without deterioration. Black enamel paint over a suitable primer or an appropriate selective surface material can be specified (see Absorber Components in Chapter 7).

The absorber plate is typically corrugated metal decking or expanded metal lath, the latter being used where the airstream flows through, rather than over, the absorber material. The actual design, configuration, and material selection of the absorber component will differ for the U-tube and vertical panel subsystem types and, therefore, each is discussed separately below.

Vertical Panel Subsystem

There are three configurations for the absorber plate in the vertical panel subsystem. In the first, described as a back pass absorber, air flows between the absorber plate and the inside wall (see Figures 6.9).

The major advantage of the back pass design is that the airflow is separated from the cold collector component by a dead airspace. Since an airspace is a good insulator, heat losses between the moving airstream and the outdoors will be minimized. In addition, dust and dirt that enters the channel from the home cannot settle on the glazing, which would otherwise eventually reduce the amount of transmitted solar radiation. For these reasons, this configuration is favored by most commercial TAP manufacturers.

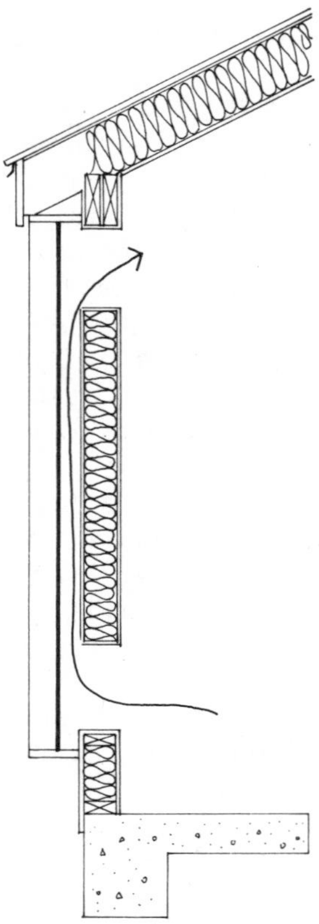

Figure 6.9

In the front pass design, the airstream passes in front of the absorber plate (see Figure 6.10).

The absorber can be fastened directly to the stud wall, simplifying construction and reducing costs. Of the three configurations, however, the front pass is the least efficient, since the moving airstream, which is heated by the absorber, is at the same time losing some of its heat to the cold glazing.

The third and most efficient absorber design is the dual pass. As its name implies,

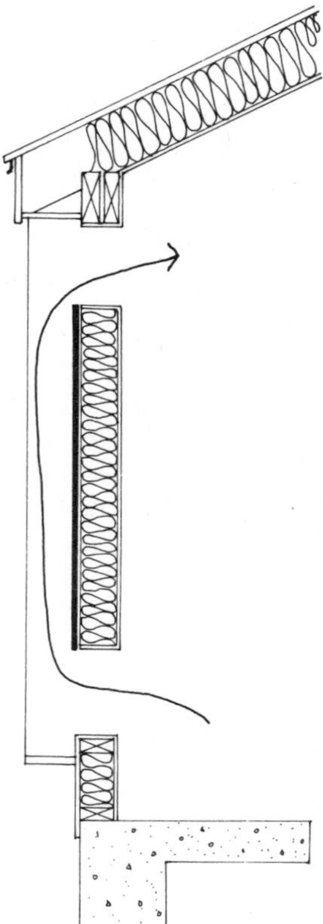

Figure 6.10

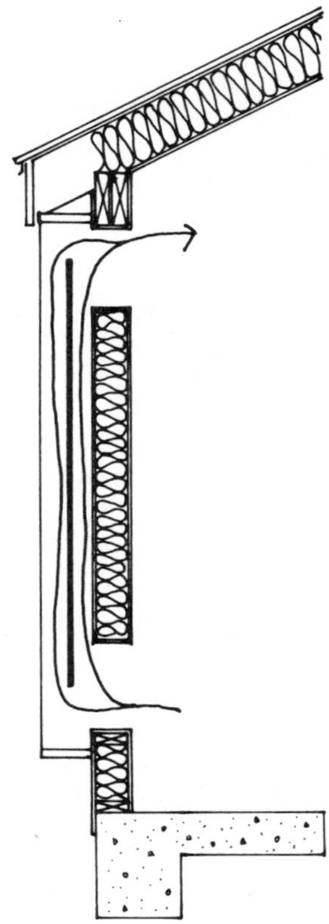

Figure 6.11

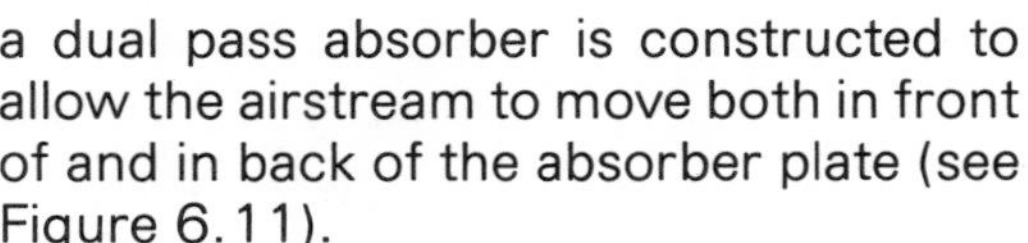

a dual pass absorber is constructed to allow the airstream to move both in front of and in back of the absorber plate (see Figure 6.11).

The effectiveness of this configuration lies in the fact that air is a poor conductor of heat. Therefore, it is advantageous to expose as much of the moving airstream to the absorber plate surface as possible. A dual pass, by allowing air to pass on both sides of the absorber, exposes twice as much surface area to passing air as does either the back or front pass. Of the three configurations, however, the dual pass is the most difficult and expensive to construct.

One practical and effective option for the front and dual pass absorber configurations is to use five or six layers of slit-and-expanded metal lath (typically used for plaster work) as the absorber (see Figure 6.12).

The lath is placed on the diagonal, increasing the heat-transfer surface between absorber and air. The surface behind the absorber is covered with a light-colored or polished surface to reflect any sunlight

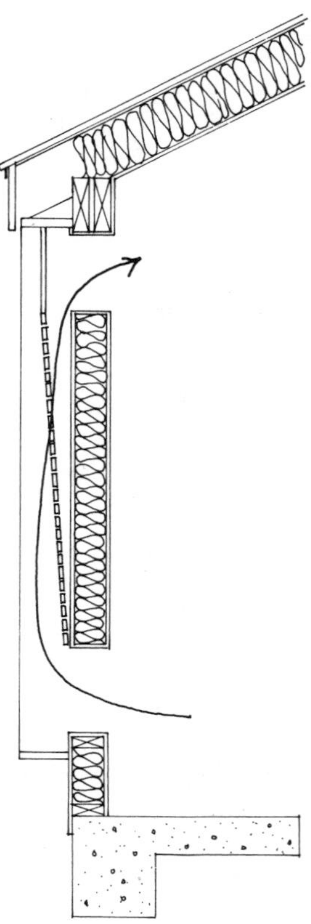

Figure 6.12

that penetrates the lath back onto the absorber surface.

At night in the heating season, the absorber and adjacent airstream will cool below the temperature of the living space. If unimpeded, this cool air will drop, entering the living space through the lower vent. This flow of cool air, called reverse thermocirculation, will rob heat from the living space unless dampers are provided. Dampers can be automatically or manually operated and should be placed on the upper and lower vents. A simple and inexpensive damper can be constructed of thin plastic sheeting and attached along the upper edge of the vent. It is designed to open in the direction of warm-air flow and close at other times, preventing reverse thermocirculation.

U-Tube Subsystem

The TAP in the U-tube subsystem is constructed with a central partition around which the airstream flows, collecting heat from the absorber plate. The absorber is usually found in one of two locations, depending upon the direction of airflow. In the most common configuration, the air flows from behind to in front of the central divider and either directly into the living space or into storage. The absorber is placed on the central divider, much like the front pass configuration in a vertical TAP (see Figures 6.10 and 6.13).

This design can be modified to the diagonal pass-through design discussed earlier

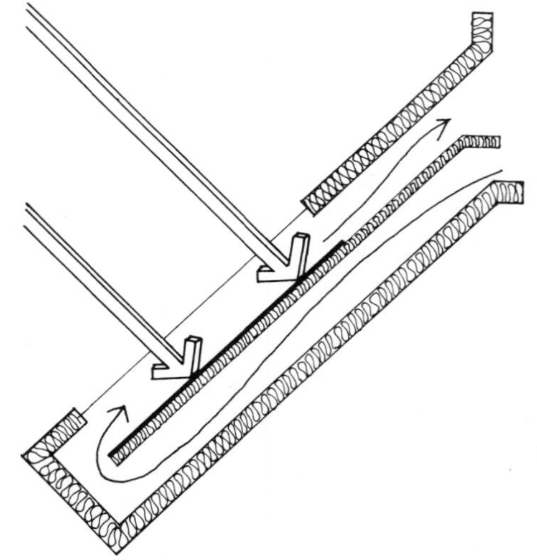

Figure 6.13

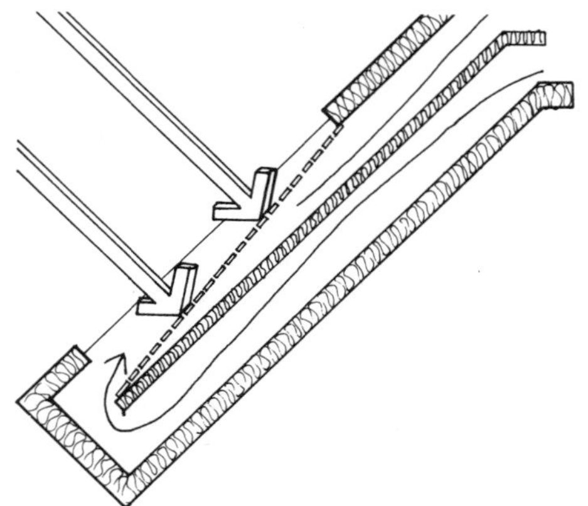

Figure 6.14

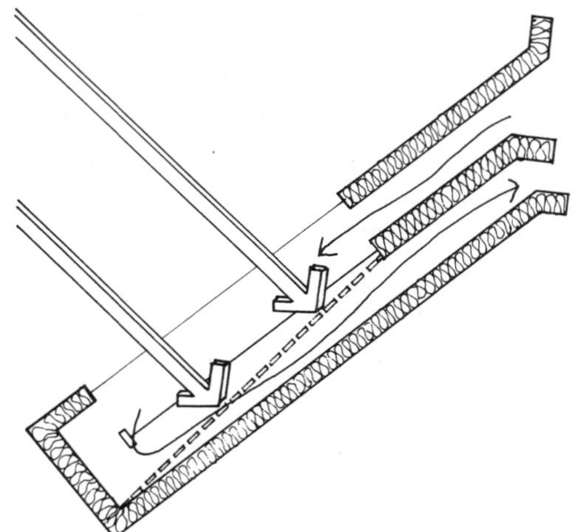

Figure 6.15

in this chapter, in which case the absorber is constructed of slit-and-expanded metal lath (see Figure 6.14).

In the second configuration, known as the reverse flow U-tube, air enters through the top vent of the TAP, flows around the center divider, and out the lower vent. The center divider is glazed in this design, with the absorber placed behind (see Figure 6.15).

This design has lower heat losses to the outside than the front pass, since the air is heated by the absorber *after* it passes the cold glazing. This reverse flow configuration is primarily used in cases where it is advantageous for the lower vent to be the air outlet (see the discussion of slab floor storage in Storage, later in this chapter).

The U-tube TAP, if carefully designed, can be self-damping. When detailing the TAP, the top of the exterior glazing should be below the lowest point of the outlet vent

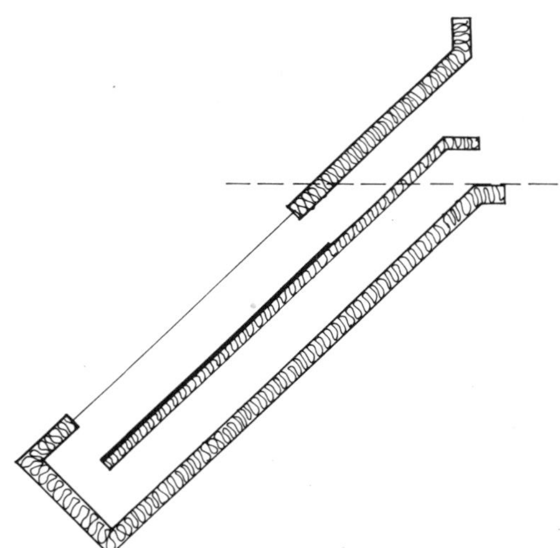

Figure 6.16

(see Figure 6.16). If the glazing is above this level, dampers similar to those described earlier in this chapter for the vertical panel subsystem will need to be specified.

Absorber Rules of Thumb

- **Area:**
  The area of the absorber should be slightly larger than, and directly behind, the collector glazing.

- **Flow Channel:**
  The depth of the flow channel (D) should be $1/20$ the length (L) of the glazing for diagonal absorbers and $1/15$ (L) for flat plate absorbers (see Figure 6-17).

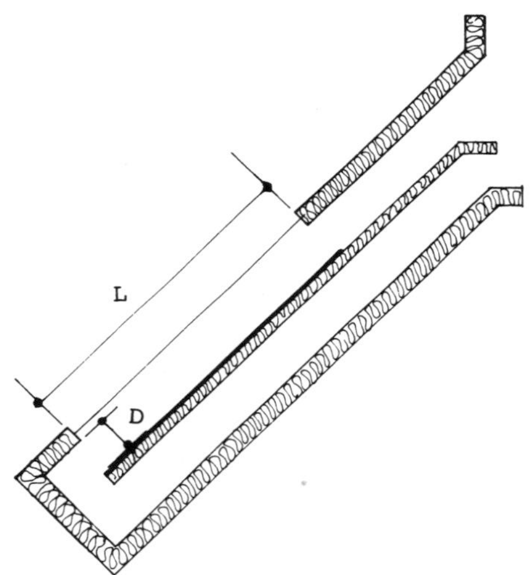

Figure 6.17

For example, if the glazing is 8′ long, the flow channel should be 5″. The channel in a U-tube TAP should be the same depth on both sides of the central divider.

- **Airspace (Back Pass Vertical Panel):**
  The airspace between the absorber and glazing in a back pass vertical TAP should be from ⅝-1¾″.

Storage

Convective Loop systems are generally designed without a storage component, providing heat only during the daytime when the sun is shining. However, if large collector areas are specified, the amount of heat generated in the TAP can, at times, exceed the heating requirements of the home. If storage is provided, this excess heat produced by the TAP can be stored for use at night and during overcast periods.

Since the TAP produces hot air that is distributed to storage, the storage component can be located virtually anywhere in the home. However, two design guidelines should be considered when selecting the storage location: (1) the storage material in most cases will be heavy and will impose structural considerations if placed above the foundation; and (2) the system will operate more effectively if the storage is placed *above* the TAP.

These design guidelines have led to the popularity of under-floor storage with heat provided by a U-tube TAP that delivers warmed air at, or below, floor level.

The two most common storage media are slab-on-grade (radiant floor) storage and rock-bed storage. In slab-on-grade storage, hot air from the TAP is circulated under the floor slab of the home and out through vents at the back of the slab (see Figure 6.18).

As the warm air is passed under the slab, heat is absorbed into the masonry. As in a Direct Gain floor, this heat is slowly released throughout the day and night. The floor should not, therefore, be covered with a

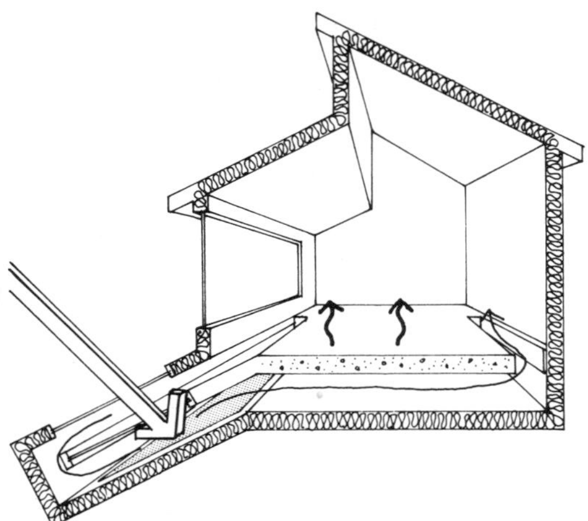
Figure 6.18

carpet or other material with insulating properties. Insulation should, conversely, be placed under the air channel to reduce heat losses to the ground.

The TAP outlet for under-slab distribution is generally the lower vent, and, therefore, a reverse flow panel design should be specified (see Absorber earlier in this chapter). The slab-on-grade storage system is easy to construct and operate and requires little maintenance.

The most common storage component for a Convective Loop system is a rock bed. A rock bed is literally a container filled with rocks, located either under the floor of the major living spaces or in an interior wall.

Rock beds have three modes of operation. In the first mode, heat is generated by the TAP at a time when heating is required in the home. As a consequence, all of the heat from the TAP is delivered directly to the living spaces and the rock bed is bypassed (see Figure 6.19).

In the second operating mode, heat generated in the TAP is not required for immediate use in the living spaces. In this case, the heat is distributed directly into storage for later use. The living spaces are bypassed at these times (see Figure 6.20).

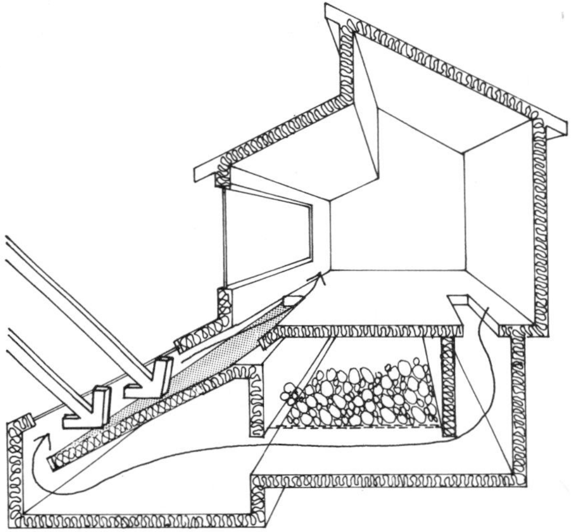
Figure 6.19

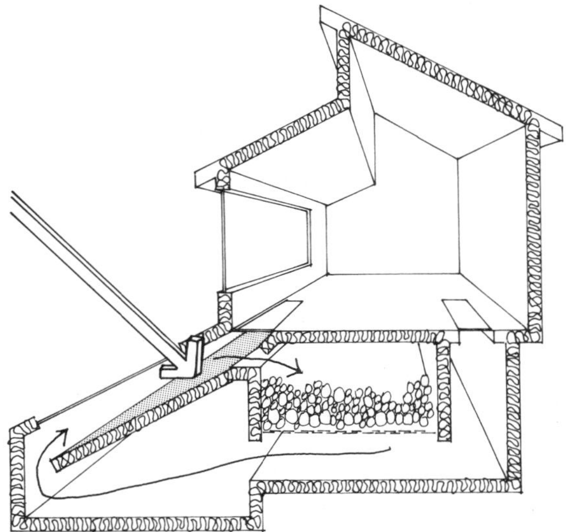
Figure 6.20

In the third mode of operation, heat is required by the living spaces but is only available from storage. This situation is common during the heating season in the evening after a sunny day. Heat from the storage is distributed into the living spaces and the TAP is bypassed (see Figure 6.21).

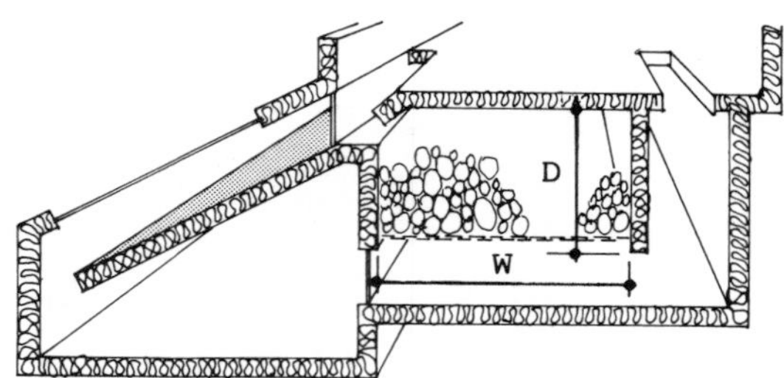

Figure 6.22

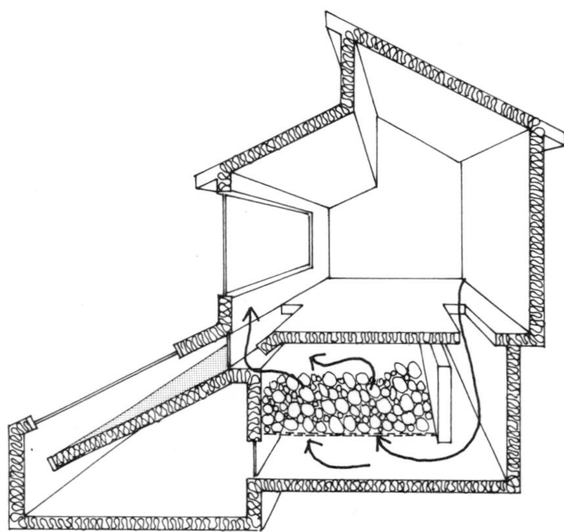

Figure 6.21

Storage Rules of Thumb

- **Area (Rock Bed):**
  The depth (D) should equal about 4′ for rocks 4-6″ in diameter and 2′ for rocks 1-4″ in diameter. The cross-sectional area (D × W) should equal about one-third of the collector area (see Figure 6.22).

- **Volume:**
  The rock storage bed should contain 2-3.5 cu ft of storage area per square foot of collector glazing area. For example, 150 sq ft of collector would require 300-525 cu ft of rock storage. If the cross-sectional area of the bed is 50 sq ft (1/3 × 150), and the depth, using 4″ rocks, is 4′, the overall dimension of the rock bed might be 4′ × 12.5′ × 10′.

Distribution

In Convective Loop systems, the distribution component moves air between TAP, living space, and, if required, storage. For systems designed without storage, distribution is required only when the TAP is not located adjacent to the living spaces. This is rarely the case, and most Convective Loop systems without storage are designed to allow the TAP to deliver heated air directly into the living space through vents.

If remote storage is provided, hot air coming out of the TAP should be delivered with as little loss of heat through the distribution system as possible.

Generally, the distribution system consists of insulated ducts connecting TAP, storage, and living space. Air movement is by natural convection, although in some instances the use of a small fan, placed in the opening of the duct, is advisable. A fan is particularly recommended in designs where the TAP outlet vent is located above the storage component.

In a popular vertical panel subsystem design with storage, heated air is taken off the top of the TAP and distributed to storage through insulated ducts that run between the floor joists or roof trusses. If storage is provided under the foundation, the vertical runs are enclosed in an interior partition wall. The ducts feed horizontal insulated plenums, which distribute the heated air evenly throughout the rock bed. Heat flow into the living space can be by radiation if the rock bed is adjacent to a slab floor above, or by convection through floor registers for rock storage under a frame floor. After passing through the rocks, the air is collected in a return plenum and distributed back to the TAP through return ducts.

In the U-tube subsystem, as noted above, the TAP is frequently adjacent to storage. Distribution between the TAP and storage is through a continuous vent at the top and a shared plenum at the bottom. Air movement between storage and living space is through floor vents. At times when heat is provided directly to the living spaces from the TAP, the heated air is distributed through floor registers adjacent to the TAP. Return air, from the north side of the heated living space, is distributed back to the TAP through ducts that circulate the air into a continuous plenum under the rock storage feeding the TAP inlet vents.

Distribution Rules of Thumb

- **Design**:
  In all cases, the distribution system should be designed to be free of any obstructions that would impede airflow. Duct runs should be designed to be direct and to contain as few turns as possible.

- **Vent Openings**:
  Vent openings between the TAP and living space should be the same width as the flow channels. The use of continuous linear diffusers (available from most mechanical equipment distributors) is recommended for vents used for direct distribution.

Control

Shading

To avoid excessive heat buildup in the TAP during the cooling season, it is recommended that a covering be placed over the collector glazing. A sheet of opaque material, such as plywood or rigid insulation, which is roughly the size of the TAP, secured at the edges, will prevent solar heat from reaching the absorber.

If a vertical TAP is used, the facade of the home should be designed to allow the collector to be completely covered without obstructing the view out of the windows. Simple overhangs and other conventional shading devices are *not* recommended, since they will not sufficiently reduce the potential heat buildup to prevent deterioration of the TAP materials.

Insulating — Fixed Insulation

Fixed insulation is generally specified for three parts of a Convective Loop system: the thermosiphon air panel (TAP), the storage components and the distribution component. In the TAP construction of the vertical panel subsystem, insulation is placed between the studs on the inside of the TAP to the same levels as those specified for the rest of the home. The insulation and studs should be covered with a smooth material, such as gypsum board, which allows the air to flow unimpeded through the TAP.

In a U-tube subsystem, the TAP is frequently exposed on all sides and will require a complete insulation "wrap" to reduce heat losses from the sides and back. In addition, the central divider should be constructed of an insulating material to reduce heat flow between the air channels. In both cases, a rigid insulation, such as urethane, which can withstand high operating temperatures, is recommended. Polystyrene is emphatically *not* recommended in TAP construction.

The storage component should be insulated to reduce heat losses to the ground during the heating season. The type, location, and selection of insulating materials will vary, depending upon the design and operation of the storage component (see Convective Loop details that follow).

If ducts are used to distribute warm air from the TAP, they should be insulated to reduce heat losses. Insulation is particularly important if the ducts run in an unheated space, such as a basement or crawl space.

Control Rules of Thumb

Insulating

- **Fixed Insulation:**
  R-8 insulation is recommended in the construction of the TAP in a U-tube subsystem design. The central divider should be insulated with R-5 to R-8 insulation. Ducts used for warm-air distribution should be wrapped with R-4 insulation.

Convective Loop Details

The drawings on the following pages illustrate common construction detailing for Convective Loop systems. The details are divided into four detail pages and are intended to simply suggest a few options among the many Convective Loop system configurations possible.

The first two details illustrate typical construction features for the vertical panel and U-tube panel subsystems, respectively. The vertical panel subsystem is shown in both back and dual pass configurations. Panel constructions can be used both with and without thermal storage. The next two details illustrate the primary storage options: rock storage and radiant floor. Both storage designs can be adapted for use with either panel option. Note that the storage details can also be used with Attached Sunspace.

The details are followed by a set of construction notes, referenced from the details. The notes contain guidelines, troubleshooting tips, and other information useful in constructing a Convective Loop system.

CONVECTIVE LOOP
Wall Collector

BACK PASS

DUAL PASS

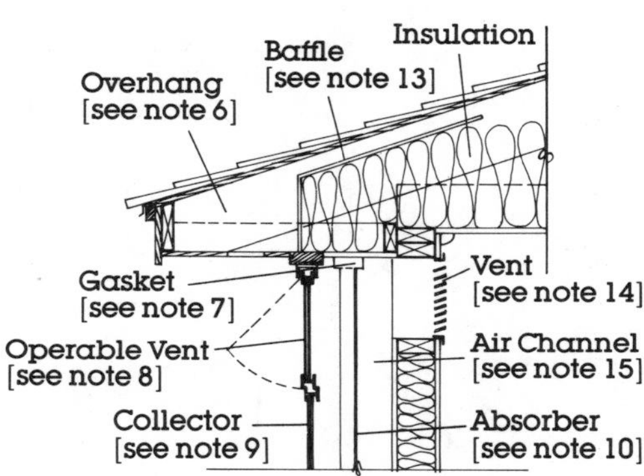

Back Pass Collector at Roof

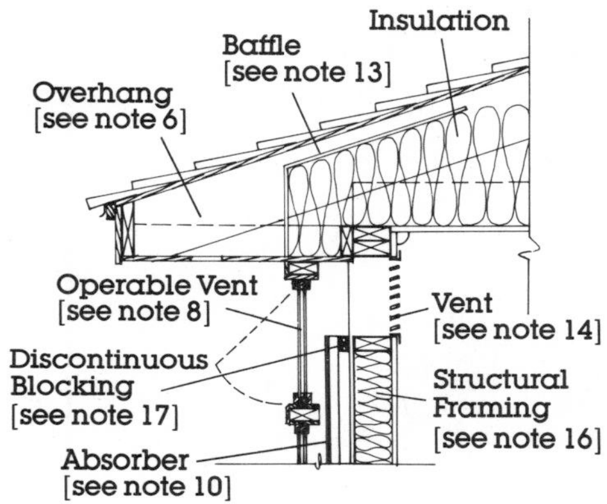

Dual Pass Collector at Roof

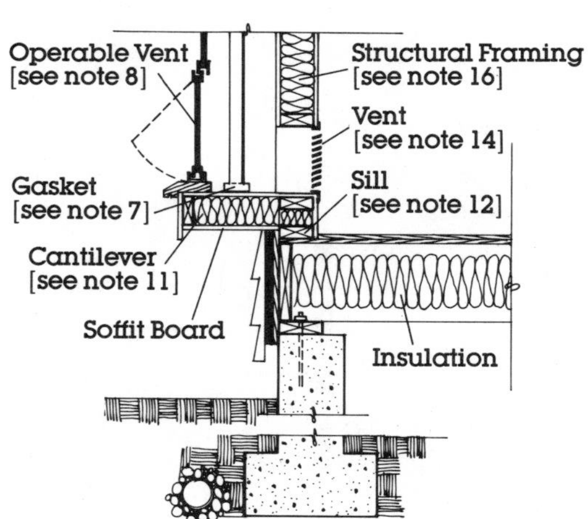

Back Pass Collector at First Floor

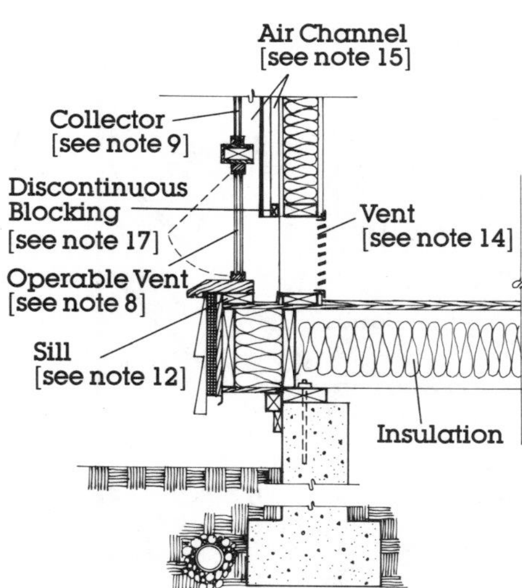

Dual Pass Collector at First Floor

Below-the-Floor Collector

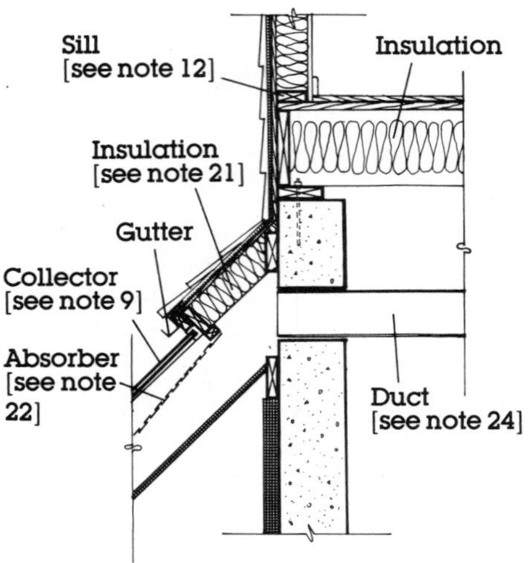

U-Tube Collector at Supply Duct

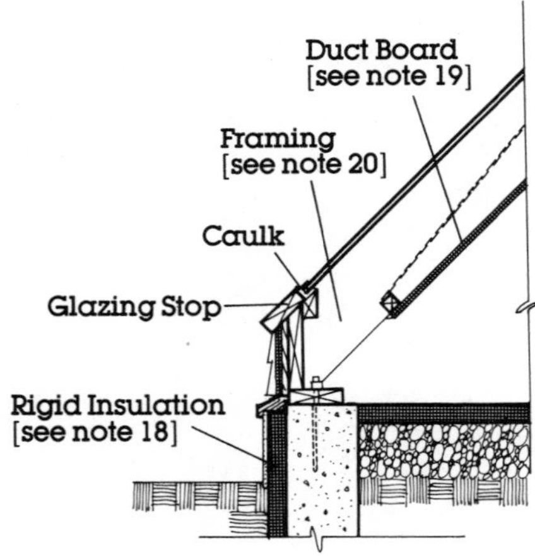

U-Tube Collector at Grade

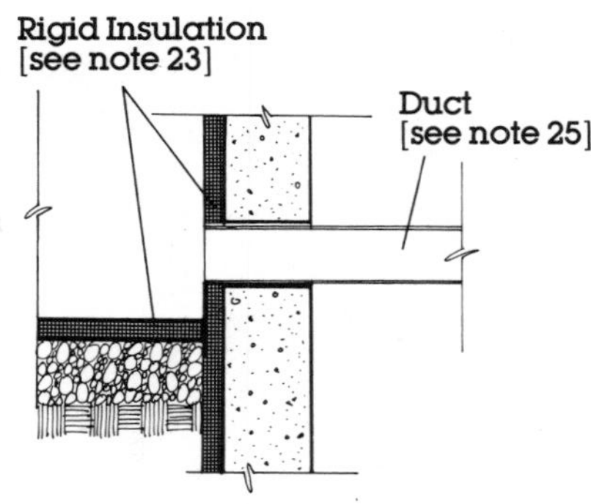

Return Duct at Foundation

Collector Notes

GENERAL NOTES:

1. All footings must bear on undisturbed soil. Adjust footing size, reinforcing, and depth below grade as required by site conditions and/or local code. Where groundwater problems exist, provide sufficient granular fill to prevent water penetration at slab-on-grade. Reinforcing in concrete slab-on-grade construction is not recommended except where required by local conditions and/or code. Expansion joints should be provided, as required, at the connection between concrete or masonry floors, and stud or masonry walls to prevent cracking.

2. Sizing of headers is to be determined by local building codes and/or accepted practice.

3. Insulation levels at foundation, floors, walls, and ceiling must meet or exceed the requirements of the local building/energy codes.

4. Other roof configurations may be used at builder's option. Consult local building codes for restrictions and special requirements concerning roof/wall connections, especially in seismic and/or high-wind areas. The opaque roof must be vented to ensure proper performance of the insulation. Vents may be located at the sides of the roof or at the ridge, depending upon the roof construction. An airspace is provided to allow for circulation.

5. Provide gutters and downspouts as required.

DETAIL NOTES:

6. Overhang at collector should be sized to optimize solar gains during heating season and prevent overheating during cooling season.

7. A strip gasket, such as EPDM, at the top and bottom of the corrugated-metal absorber plate will provide an airtight seal at the edges. Care must be taken to ensure that the gasket material is able to withstand the high temperatures generated in the space.

8. Portions of the collector should be operable to allow venting of the Thermal Storage Wall during periods of excessive heat gain. Awning-type or other operable windows may be specified and should be equipped with demountable insect screens that can be removed during the heating season.

9. Double glazing is generally recommended to ensure optimum thermal performance. A ½'' airspace between glazing layers is recommended where feasible. The glazing material can be glass or plastic (see Collector Components in Chapter 7). The collector unit should be demountable or externally operable to aid in maintenance and cleaning. The collector frame may be detached from the frame wall and designed to support the glazing, glazing frame (if preframed), roof, and other related loads, or it may be attached to and supported by the frame wall. Framing should be kept to a minimum, since it reduces the effective collector area and may cast shadows, both of

which will affect the thermal performance of the wall. All structural framing and connections shall be designed per local building-code restrictions and/or accepted practice. Frames should be continuous at edges.

10. The absorber should be a lightweight material, such as corrugated roof metal, providing maximum heat transfer and maximum surface area. For further information, see Absorber Components in Chapter 7.

11. Collector assembly is supported on a cantilever. Care must be taken to ensure that insulation levels meet or exceed requirements of local building/energy codes. Sizing of members is to be determined by loads and local building codes and/or accepted practice.

12. Continuous sill sealer is recommended to provide protection against infiltration.

13. Provide baffles where necessary to maintain 1'' minimum airspace for venting attic.

14. Operable vents at top of adjacent wall area are provided to allow heated air to enter living spaces, while low vents draw cooler room air into collector assembly, completing the "loop." Heated air may also be ducted from vents at top to remote storage for later use. For further information, see Storage and Distribution, earlier in this chapter.

15. It is generally recommended that the air channel depth should be 1/15 the length of the glazing for flat plate absorbers and 1/20 the length of the glazing for diagonal mesh absorbers. For further information, see Absorber Rules of Thumb in this chapter.

16. Stud framing should be as required to support roof and other loads. Insulation levels should meet or exceed requirements of local building/energy codes. The air channel side of the stud wall should be finished with a smooth-surfaced sheathing to augment airflow.

17. Discontinuous blocking is provided to allow free airflow on both sides of absorber plate.

18. Cement plaster (stucco finish applied on wire lath) over exterior insulation may be used. Other finishes may be specified at builder's option, but rigid insulation must be protected where exposed above grade.

19. The air channel is created by nailing a rigid sheathing material to the underside of the framing members. It is recommended that an insulative material such as duct board be specified.

20. Framing must be sized to support the glazing system.

21. Opaque, insulated framing should be provided to reduce heat loss at the top of the collector assembly. The length of the opaque portion should be sufficient to extend below the bottom of distribution ducts from the collector assembly.

22. Three to five layers of expanded wire mesh are recommended for the absorber. Other materials may be specified at builder's option. For further information, see Absorber Components in Chapter 7.

23. Rigid insulation is recommended to control heat loss and to reduce the introduction of ground vapor and odors into the airflow. In areas with drier soils and low water table, a 6-mil polyethylene sheet may suffice.

24. A blockout in the foundation must be provided to allow for distribution of heated air to living spaces and/or remote storage. Lintels should be designed and reinforced as required by local codes and/or accepted practice. Ductwork should be insulated to prevent heat loss. For further information, see Distribution in this chapter.

25. A blockout in the foundation must be provided for the return air ducts bringing cooler air from living spaces and/or remote storage, completing the "loop." Lintels should be designed and reinforced as required by local codes and/or accepted practice. For further information, see Distribution in this chapter.

CONVECTIVE LOOP

Remote Storage

(Vertical-Feed Rock Bed)

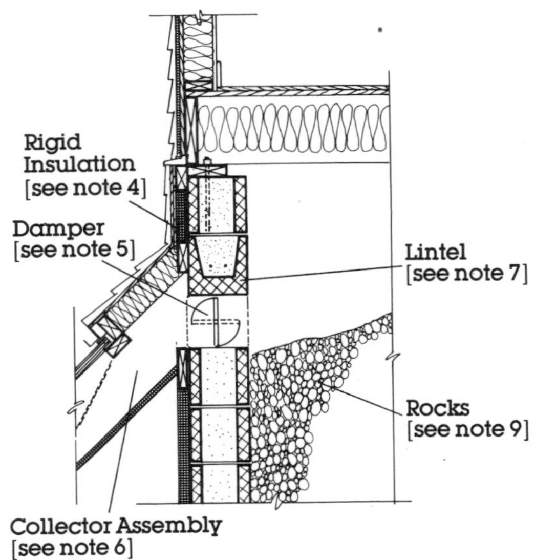

Rock Bed at Panel Supply

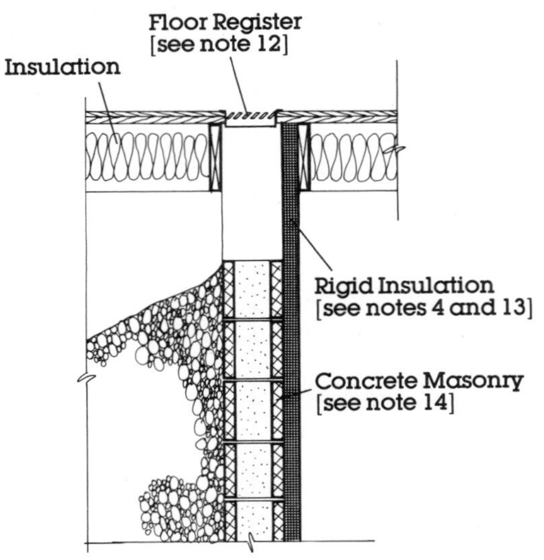

Rock Bed at Floor Register

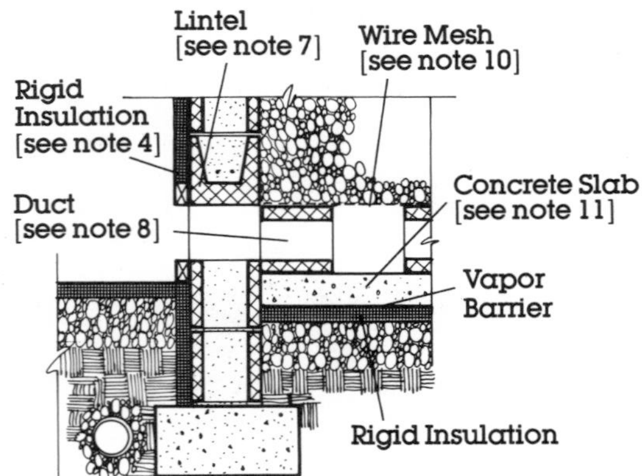

Rock Bed at Panel Return

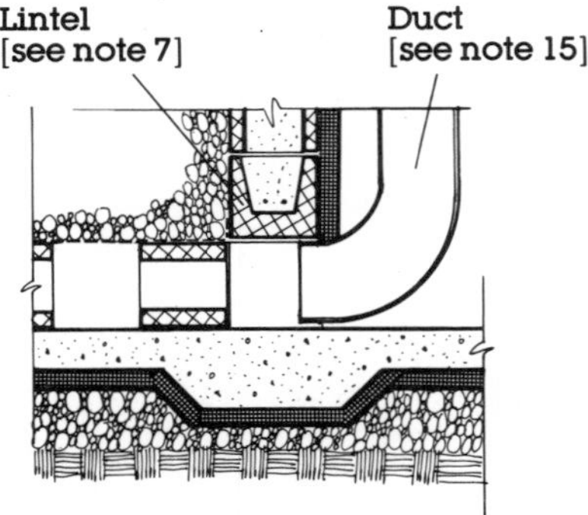

Rock Bed at Return

Remote Storage

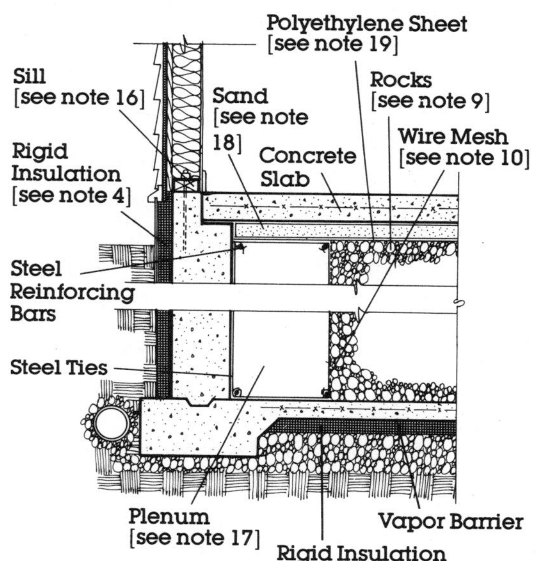

Horizontal-Feed Rock Bed at Return Plenum

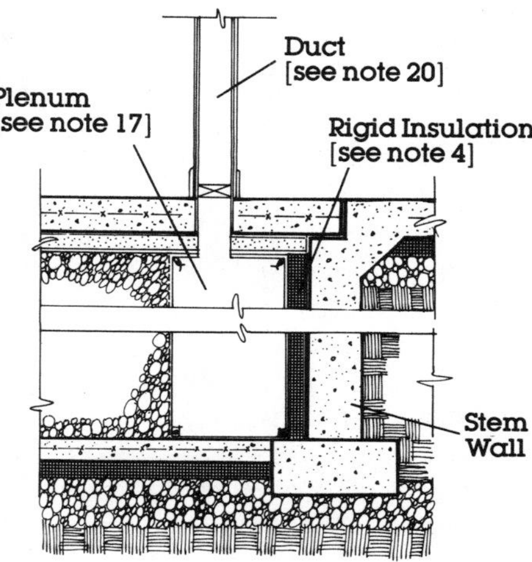

Horizontal-Feed Rock Bed at Supply Plenum

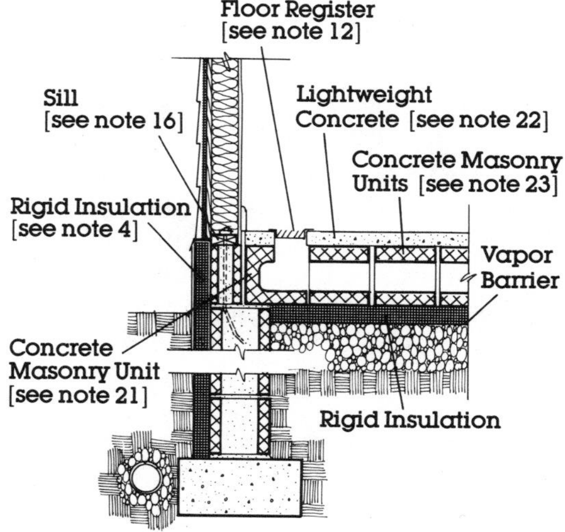

Radiant Floor at Return Plenum

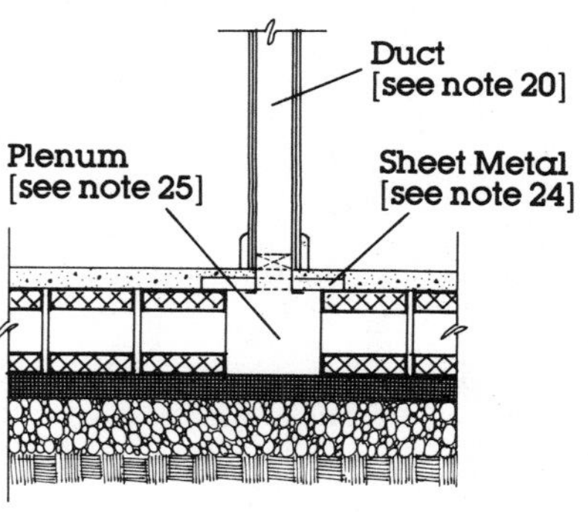

Radiant Floor at Supply Plenum

Remote Storage Notes

GENERAL NOTES:

1. All footings must bear on undisturbed soil. Adjust footing size, reinforcing, and depth below grade as required by site conditions and/or local code. Where groundwater problems exist, provide sufficient granular fill to prevent water penetration at slab-on-grade. Reinforcing in concrete slab-on-grade construction is not recommended except where required by local conditions and/or code. Expansion joints should be provided, as required, at the connection between concrete or masonry floors, and stud or masonry walls to prevent cracking.

2. Sizing of headers is to be determined by local building codes and/or accepted practice.

3. Insulation levels at foundation, floors, walls, and ceiling must meet or exceed the requirements of the local building/energy codes.

DETAIL NOTES:

4. Rock-bed insulation must be well sealed to restrict moisture. Joints should be butted and sealed with duct tape or silicone caulk. Polyethylene sheet (6 mil) may also be specified on the outside face of insulation as an added protection. Insulation in contact with rocks should be protected from damage with a thin sheet of masonite or plywood. All rigid insulation exposed above grade should be protected.

5. Dampers may be plywood, sheet metal, or steel plate, hinged or set in a pivot. The use of a reversible motor and heat sensor can greatly improve effectiveness in controlling the flow of heated air. Care must be taken to insulate and weatherstrip the damper to control nighttime cold-air flow.

6. The collector assembly can be designed for any of a number of systems, including below-the-floor Convective Loop. Remote rock-bed storage can also be designed for use with Attached Sunspace systems. For further information, see Chapter 5: Attached Sunspaces.

7. A blockout must be provided for the movement of air into and out of the rock bed. Lintels should be designed and reinforced as required by local codes and/or accepted practice.

8. Concrete masonry units are used to form the return plenum. Cores must be aligned and spaced 8'' apart to allow for the free movement of air from the rock bed into the plenum.

9. Washed river rock or gravel may be specified for use in a rock bed. Care must be taken to ensure that rock size is consistent and that rocks are well cleaned and dry. For further information on sizing rocks and recommended installation procedures, see Storage Components in Chapter 7.

10. Expanded wire mesh is recommended for use in confining the rocks. The mesh should be continuous or overlapped at least 12'' at joints. It is recommended that a 3'' lip be provided at each edge of the rock bed.

11. A concrete slab is generally recommended in areas with high ground-moisture levels to act as a base for the plenum. In areas with dry soil and stable soil conditions, the use of rigid insulation and vapor barrier will suffice.

12. The floor register should be operable so that it can be opened to allow for heat to be delivered to the living space when available or completely closed when heat is not required.

13. Rigid insulation is used to create ductwork to floor registers.

14. The enclosure walls of the rock bed must be designed and reinforced to withstand the lateral loads imposed by rocks.

15. A duct is provided to carry cool air from the living spaces and rock bed back to the collector assembly.

16. Continuous sill sealer is recommended to provide protection against infiltration.

17. Provide supply and return plenums in the form of a box frame fabricated out of steel reinforcing bars covered with expanded wire mesh. For further information, see Storage Components in Chapter 7.

18. A layer of sand is provided to protect against the formation of an airspace above rocks due to settlement.

19. A polyethylene sheet is laid over the rocks to prevent sand from filtering into the rock bed.

20. The duct is bringing heated air from the collector assembly, Attached Sunspace, or other collector. It is recommended that the ductwork be insulated and well sealed. Ducts can run within stud partitions or other furred-out spaces.

21. A concrete masonry unit U-block set on its side can be effectively used as a plenum. The leg of the block should be removed in conjunction with slab blockouts to provide exit for airflow.

22. A lightweight concrete topping finish is applied over concrete masonry units. Other finishes may be used at builder's option. However, care must be taken not to use materials that will impede the heat flow of the radiant floor.

23. The radiant floor storage system employs concrete masonry units laid with cores aligned. Heated air is delivered by duct and blown through the cores, heating the blocks. Heat stored is then released into the adjacent living space through radiation.

24. Corrugated sheet metal is laid over the natural plenum, formed by removing a row of concrete masonry units, as a base for the lightweight concrete.

25. A row of concrete masonry units has been removed to provide a plenum for distribution of heated air throughout the radiant floor storage system.

SERI #1—Lobato House

Convective Loop

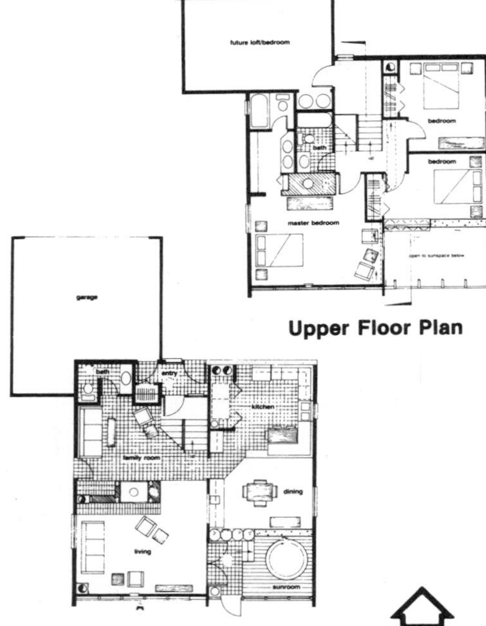

Upper Floor Plan

Main Floor Plan

North

This 2,300-sq-ft home in Boulder, Colorado, was designed under a program sponsored by the Solar Energy Research Institute (SERI) and employs a Convective Loop system for passive solar heating. Heated air from a two-story thermosiphon air panel and an Attached Sunspace is stored in a rock bed and the thermal mass within the living space. The distribution of heated air within the system is augmented by small fans.

Partial earth berming, air-lock entry vestibules, and the garage act as thermal buffers on the minimally glazed north side of the home. To the south, a double-height Attached Sunspace is well integrated into the plan and serves as a focal point for recreational activities.

The 216 sq ft of south-facing thermosiphon air panels in the living and bedroom areas are glazed with patio-door replacement glass. Sunlight transmitted through these collectors is absorbed by a black aluminum screen in the airspace between the glazing and the stud wall. This absorbed heat is transferred to the air passing up through the airspace. Induction fans circulate the heated air through insulated ducts into a rock storage bed, which stores the heat for eventual radiant release through the floor slab.

Located at the southeast corner of the house, the double-height Sunspace provides additional passive heating. Vertical and sloped south-wall glazing allows sunlight to strike water storage containers that lie along the north wall. The water storage is comprised of four water-filled fiberglass tubes, each holding 450 gal of water. Additional thermal mass is afforded by 12″ of exposed aggregate concrete wall. Beneath the tubes, brick pavers on a concrete slab further supplement the storage mass. Heated air from the Sunspace is transferred to adjacent spaces through operable doors and windows.

Project: Heritage I (SERI #1)

Builder
Heritage Construction
Tom Fowler
Boulder, Colorado

Designer
Rudolph Lobato & Assoc.
Rudy Lobato
Longmont, Colorado

Sponsor
Solar Energy Research Institute,
Denver Metro Home Builders Assoc.
Denver, Colorado

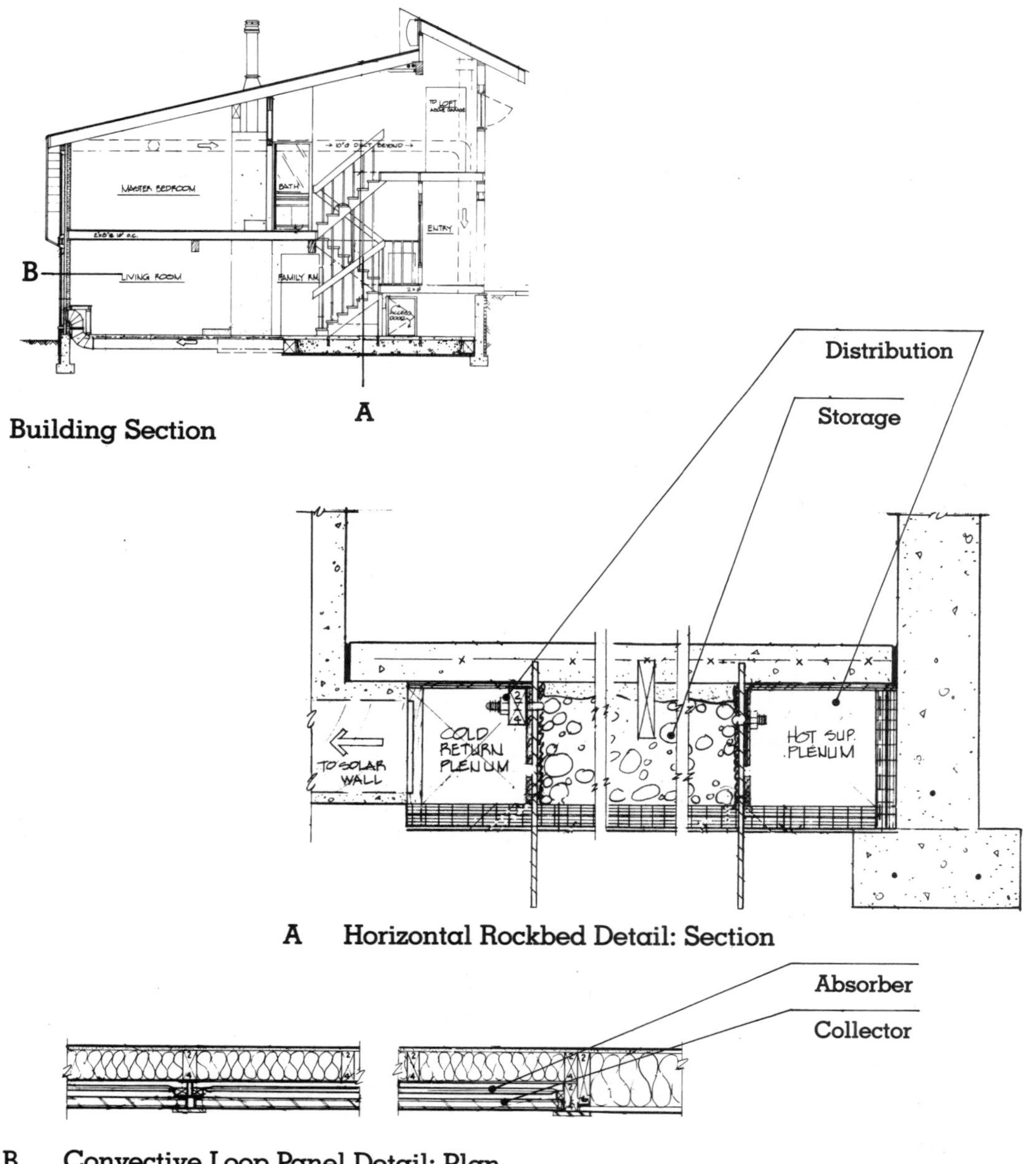

Building Section

A Horizontal Rockbed Detail: Section

B Convective Loop Panel Detail: Plan

Ashelman House
Convective Loop

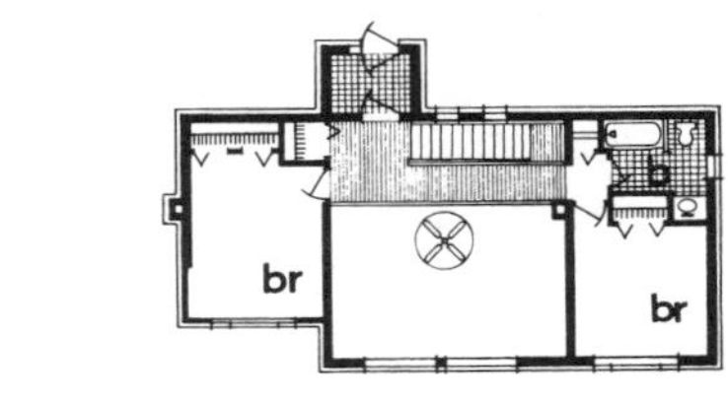

Upper Floor

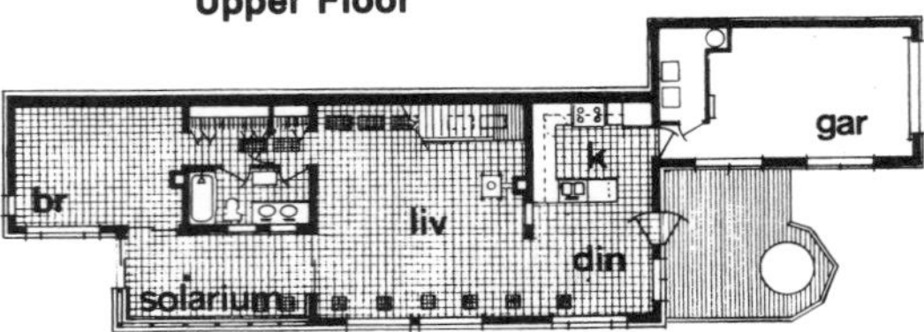

Lower Floor

North

A Convective Loop system provides passive solar heating to the 2,155-sq-ft Ashelman house in Martinsburg, West Virginia. Built into a south-facing slope, the home is buffered on the north by a service area and an air-lock entry. The house is fully exposed on the south, where thermosiphon air panels below the first floor collect solar heat.

An Attached Sunspace (solarium) and an open-air sun deck flank the south-facing living spaces on the lower level. Above, bedrooms are connected by a bridge from the main entry overlooking the double-height Direct Gain living room. Masonry construction throughout, including precast 4′ concrete panels supporting the sod roof, tile over the slab floors, and five 55-gal water-filled storage drums placed in the Sunspace, contribute to the thermal mass of the home.

The 12′ × 32′ tilted array of air panels is constructed using one layer of tempered glass over five layers of expanded metal lath painted flat black to increase solar absorption. The assembly lies within an 8″ airspace and is backed by 2″ of urethane insulation. Four inches of rigid Styrofoam insulate the cold-air return channel behind the panels.

A rock bed containing 4″- to 8″-diameter rocks provides backup storage for the solar heat collected in the panels. Panel-heated air rises into the rock-bed storage area beneath the home. Manually controlled vents in the living-space floor are opened to allow warm air to rise directly into the living quarters if required. Cooler return air is ducted into the rock-bed storage through floor vents on the north side of the home, completing the "loop."

Roof overhangs and intermediate trellises shade the south wall during the cooling season.

Removable insulation over the panels has been added since construction to significantly reduce temperatures in the panels during the summer.

Project: Sunrise

Builder/Designer
Randall Ashelman-Natural Sun Homes
Randall Ashelman
Berkeley Springs, West Virginia

Sponsor
U.S. Department of Housing and Urban Development
U.S. Department of Energy
Washington, D.C.

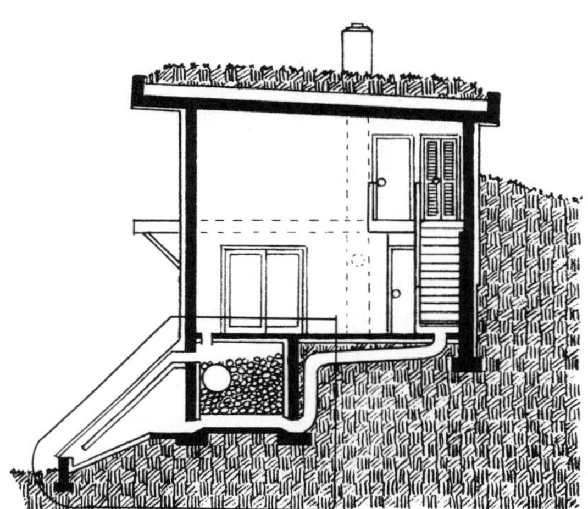

Building Section

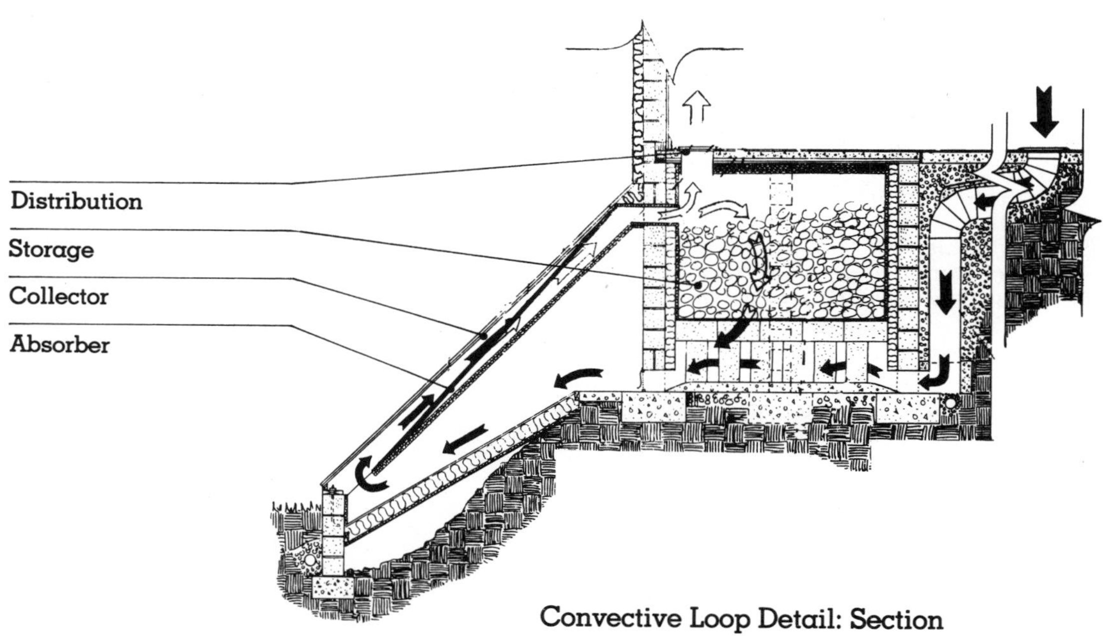

Convective Loop Detail: Section

TEA House
Convective Loop

The TEA House located in Massachusetts utilizes thermosiphoning air panels along its south wall, creating a Convective Loop system for passive solar heating. This design illustrates the simplicity with which a standard design can be modified to supplement heating requirements with a passive solar system. It is particularly appropriate for application to existing structures because the air panels can be installed on almost any south wall, regardless of the configuration of the floor plan.

The panels consist of clear glass covering corrugated aluminum absorber plates, which are painted black. In the back pass design, as it is called, top and bottom vents in the insulated wall allow air to move between absorber plates and the gypsum-board finish. Sunlight passing through the collector glazing strikes the aluminum plates, where it is absorbed as heat. The heat is transferred through the plate to warm the air in the airspace. This heated air rises through the channel and passes into the house through the upper vent. At the same time, denser, cool air from the home enters the channel through the lower vents to replace the displaced heated air.

This thermosiphoning effect naturally circulates heat through the house in a Convective Loop during the day. A 4-mil polyethylene backdraft damper over the bottom grill restricts nighttime heat loss from reverse thermosiphoning. Additionally, insulated vent covers on hinges can be latched closed at night to further reduce heat loss, as well as during the cooling season to prevent heat gain. This system design does not require storage and, as a result, provides heat only during the daytime. The panels can be covered with an opaque panel during the summer to reduce the possibility of its rapid deterioration.

Project: T.E.A.

Designer
T.E.A. Foundation, Inc.
1 West St.
Keene, NH 03431

From the Easy to Build Solar series of construction manuals by T.E.A. Other manuals include: *State of the Art, Sunspace, Retrotrombe,* and *Batch Heater.* Each manual is available for $9.95 (post. incl.) from T.E.A. and includes blueprinted plans as well as information.

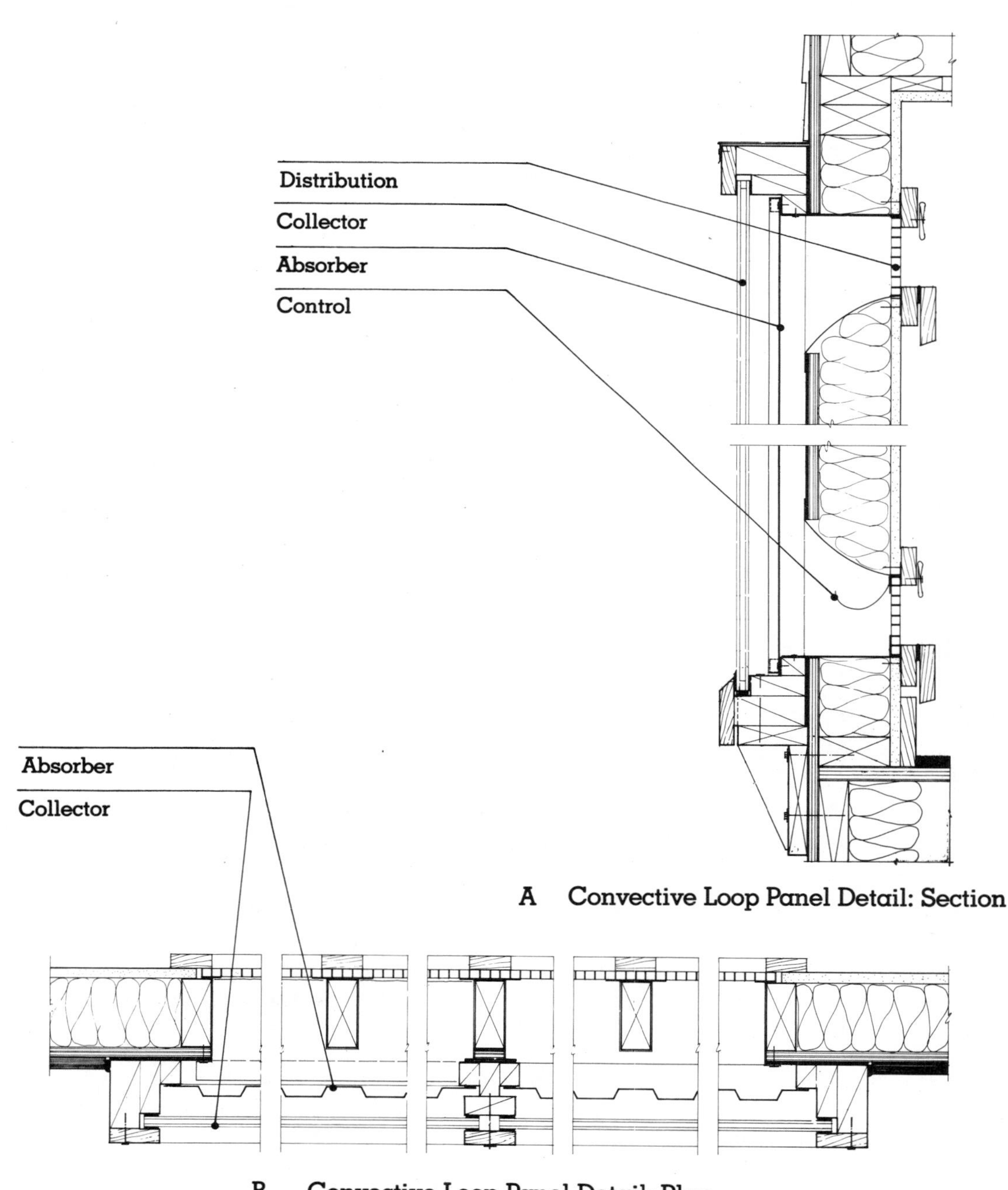

A Convective Loop Panel Detail: Section

B Convective Loop Panel Detail: Plan

Chapter 7
MATERIALS

Introduction

This chapter presents construction and thermal performance information that will aid in specifying the materials illustrated in the construction details in Chapters 3 through 6. For the most part, the materials designated in these details are intended to represent typical, rather than specific, products.

Once a particular detail has been selected and material dimensions determined using the rules of thumb, specific materials or products can be specified using the information in this chapter. In most cases, the materials illustrated are available from a number of manufacturers. In instances where unique or specific products are discussed, individual manufacturers are identified and their products listed.

To simplify this material selection and specification process, the chapter is divided into subsections by component type, including: (1) collector components, (2) absorber components, (3) storage components, (4) control components, and (5) distribution components. Each component section is further subdivided with discussions focusing on issues relevant to that subject area.

The collector components section includes: (1) a review of the basic characteristics of passive solar collectors; (2) a discussion of the types of glazing materials available for use in passive installations, including a series of tables that offer general performance comparisons between a number of specific glazing products;

(3) a description of the major framing systems used with collectors; and (4) a brief discussion of caulking and sealants appropriate for passive solar applications.

The section on absorbers covers a general discussion of the function of the absorber, the types of products available to use as absorbers, and their performance characteristics. Two tables accompany the discussions. The first lists specific manufacturers of selective-surface products and general absorber materials. The second provides absorptivities of various common building materials.

The storage components section is divided into six subsections covering the various storage materials presented in the construction details (Chapters 3 through 6): (1) concrete masonry, (2) concrete, (3) brick, (4) water, (5) phase-change materials, and (6) rock. Each of these subsections includes, where relevant, a general description of the basic composition and typical dimensions of the materials; a discussion of relevant construction issues, including finishes (exterior and interior), color, and texture; and a definition and discussion of the specific physical properties that affect the thermal performance of each material. Tables are presented that provide data, of a general nature, with regard to that material type and, where appropriate, specific products.

The discussion of the distribution components provides a brief review of the applicability of distribution to the basic system types and general guidelines useful in component design.

The control components section is subdivided into three sections covering shading, reflecting, and insulating components. Each section outlines the function of these components and the techniques for their proper operation in the basic passive systems.

Control components tend to vary in effectiveness and installation procedures by manufacturer. Tables are, therefore, provided that list general control products and their comparative performance characteristics.

Collector Components

Introduction

The collector is that portion of a passive solar system that admits solar energy and traps the heat it generates. This heat is absorbed by other elements in the system and is then stored for later use or distributed directly to the living spaces. The collector thus forms an integral part of every passive heating system.

The collector can be as simple as a standard window unit or as elaborate as a large, site-built Sunspace with both vertical and tilted glazing. In all cases, the collector is composed of three basic elements:

- Glazing—Composed of some form of glass or plastic that allows radiant energy to strike the absorber or storage component while restricting the outflow of the heat energy.

- Collector Frame—That supports the glazing material.

- Caulking and Sealants—That ensure that the entire installation is airtight and weathertight and able to operate at peak efficiency.

These elements can be combined in a large variety of configurations whose performance in passive solar installations is dependent upon: the orientation and tilt of the collector; the thermal properties of the specific glazing material employed (see tables that follow); the *net* area of the material being used for collection (exclusive of frame); the size and type of frame being used; the effectiveness of the caulking and/or sealants being used; and the overall configuration of the collector (glazing material and frame).

Types of Collectors

There are an almost limitless number of possible designs for passive solar collectors, from simple, off-the-shelf windows to solid walls of glass or plastic that can cover the entire south side of a home. Some of the more common configurations and their possible applications are reviewed in Figure 7.1.

It must be remembered that the glazing material in these configurations does not need to be glass. Any one of a variety of materials (some of which are identified below) can be used if the basic system operating conditions, outlined in the prior chapters, are kept in mind.

As a general rule in any passive application, using large individual units• to cover a given collector area (e.g., one large picture window as opposed to several smaller units) will decrease the overall amount of framing and will thus increase the net collector area exposed to the sun. From the standpoint of energy conservation, this arrangement will also reduce the amount of perimeter framing exposed to the environment and consequently will reduce the amount of heat loss due to infiltration through the framing. Similarly, using a unit with less crack length, as in a single-hung versus a double-hung window, will also reduce infiltration. Considerations such as these must be weighed against other criteria such as cost, ease of maintenance, aesthetic preference, and availability to arrive at the best collector design for a given application.

Glazing Material

At present, glass is the most commonly used glazing material. Other materials, such as acrylics, polycarbonates, fiberglass reinforced polyester, and polyethylene, can substitute directly for glass if they are in panel or sheet form, or they can be combined with glass or another more rigid material if they are thin films.

Information regarding the thermal properties and physical characteristics of the commonly used passive solar glazing materials is presented in Tables 7.1 through 7.5.

The tables are not intended to be comprehensive, and certain specific products have been left out because of a lack of sufficient data. Rather, the information presented is intended to serve as an indicator of general product performance. More detailed information, including specific product sizes and availability, should be obtained directly from the manufacturer before any product is actually specified.

The tables are relatively self-explanatory. They are organized by glazing material classification, with some commonly used specific products listed under each as examples.

Tables 7.3 and 7.4 present a variety of relevant physical characteristics for each product. The technical data presented in these two tables are provided primarily

Collector Types **Figure 7.1**

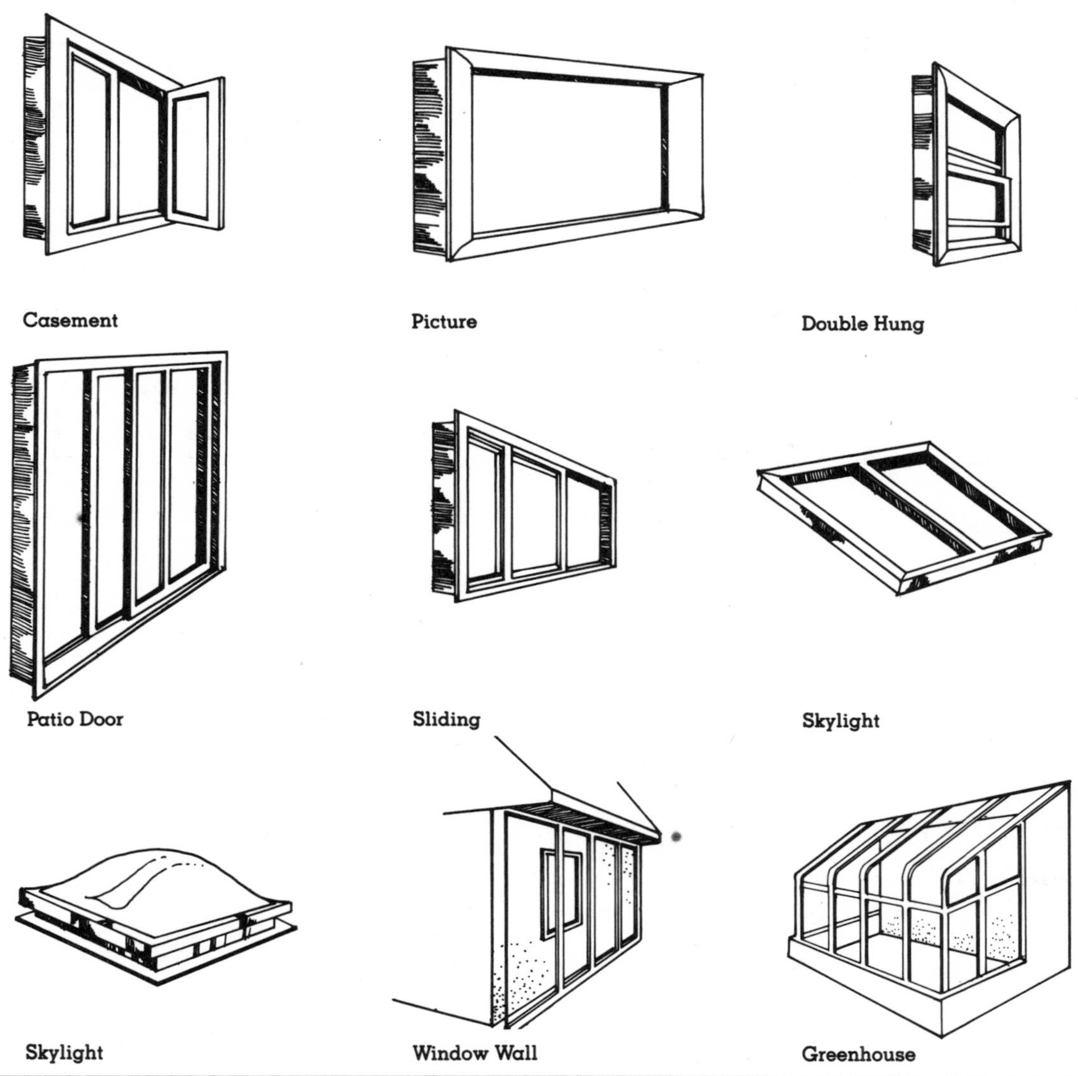

Casement
Picture
Double Hung
Patio Door
Sliding
Skylight
Skylight
Window Wall
Greenhouse

for use with the various analytical tools described in Appendix C.

When assessing a material's suitability for a specific situation, special attention should be paid to a product's "weatherability" and its maximum operating temperature. Thermal Storage Walls, and Convective Loop panels in particular, can generate extremely high working temper-

(continued on page 197)

Glazing: Applications **Table 7.1**

	Direct Gain			Thermal Storage Wall	Attached Sunspace
	Windows	Clerestories	Skylights		
GLASS					
Single Glazed Float Glass	*	*		*	*
Double Glazed Float Glass	*	*		*	*
ACRYLIC					
Plexiglas (Rohm & Haas)	*	*	*		*
Lucite (Du Pont)	*	*	*		*
Exolite (CY/RO)	*	*	*		*
POLYCARBONATE					
Lexan (G.E.)	*	*	*	*	*
Tuffack Twinwall (Rohm & Haas)	*	*	*	*	*
Cyrolon SDP (CY/RO)	*	*	*	*	*
FIBERGLASS REINFORCED POLYESTER					
Lascolite (Lasco Industries)	*	*	*	*	*
Filon w/Tedlar (Vistron Corp.)	*	*	*	*	*
Sunlite Premium II (Kalwall)	*	*	*	*	*
POLYESTER FILM					
Mylar Type W (Du Pont)	*	*			*
Flexiguard(3M)	*	*			*
POLYETHYLENE FILM					
Monsanto 602 (Monsanto)	*	*			*
Teflon (Du Pont)	*	*			*
POLYVINYL FLUORIDE FILM					
Tedlar (Du Pont)	*	*			*

Glazing: Characteristics Table 7.2

	Fixed	Operable	Transparent	Translucent
GLASS				
Single Glazed Float Glass	*	*	*	
Double Glazed Float Glass	*	*	*	
ACRYLIC				
Plexiglas (Rohm & Haas)	*	*	*	
Lucite (Du Pont)	*	*	*	
Exolite (CY/RO)	*	*		*
POLYCARBONATE				
Lexan (G.E.)	*	*	*	
Tuffack Twinwall (Rohm & Haas)	*	*		*
Cyrolon SDP (CY/RO)	*	*		*
FIBERGLASS REINFORCED POLYESTER				
Lascolite (Lasco Industries)	*			*
Filon w/Tedlar (Vistron Corp.)	*			*
Sunlite Premium II (Kalwall)	*			*
POLYESTER FILM				
Mylar Type W (Du Pont)	*			*
Flexiguard(3M)	*			*
POLYETHYLENE FILM				
Monsanto 602 (Monsanto)	*		*	
Teflon (Du Pont)	*		*	
POLYVINYL FLUORIDE FILM				
Tedlar (Du Pont)	*		*	

Glazing: Physical Properties — Table 7.3

	Thickness	U-Value	Solar Trans.	Infrared Trans.	Maximum Operating Temperature
GLASS					
Single Glazed Float Glass	.090 -.125	1.13	.91	.01	400
Double Glazed Float Glass	.625	.52 -.65	.77 -.85		
ACRYLIC					
Plexiglas (Rohm & Haas)	.125	1.06	.90	.02	180-200
Lucite (Du Pont)	.125	1.06	.92	.02	180-200
Exolite (CY/RO)	.630	.58	.83	.02	160
POLYCARBONATE					
Lexan (G.E.)	.125	1.05	.81 -.89	.02	250-270
Tuffack Twinwall (Rohm & Haas)	.220	.62	.77		195
Cyrolon SDP (CY/RO)	.630	.58	.74		240
FIBERGLASS REINFORCED POLYESTER					
Lascolite (Lasco Industries)	varies	.73	.86		200-300
Filon w/Tedlar (Vistron Corp.)	varies	.95	.86		200-300
Sunlite Premium II (Kalwall)	.040	.55	.87	.02	200-300
POLYESTER FILM					
Mylar Type W (Du Pont)	.003 -.014	1.05	.85	.16 -.32	350
Flexiguard(3M)	.007		.89	.095	275
POLYETHYLENE FILM					
Monsanto 602 (Monsanto)	.004		.85	.70	140
Teflon (Du Pont)	.001		.96	.57	300
POLYVINYL FLUORIDE FILM					
Tedlar (Du Pont)	.004		.95	.30	225

Glazing: Physical Properties Table 7.4

	Thermal Expansion	Ease of Handling	Weather-ability	Estimated Lifetime	Shading Coef.
GLASS					
Single Glazed Float Glass	.47	P	E	25+	1.0
Double Glazed Float Glass	.47	P	E	25+	.79 -.90
ACRYLIC					
Plexiglas (Rohm & Haas)	4.10	E	E	10-20	
Lucite (Du Pont)	3.90	E	E	10-20	
Exolite (CY/RO)	4.00	E	E	10-20	
POLYCARBONATE					
Lexan (G.E.)	3.75	E	F	15-17	
Tuffack Twinwall (Rohm & Haas)	3.30	E	F	5-7	
Cyrolon SDP (CY/RO)	4.00	E	F		.83
FIBERGLASS REINFORCED POLYESTER					
Lascolite (Lasco Industries)	1.60	VG	G	10-20	
Filon w/Tedlar (Vistron Corp.)	2.30	VG	G	10-20	
Sunlite Premium II (Kalwall)	1.36	E	G	20	
POLYESTER FILM					
Mylar Type W (Du Pont)	1.50	F	F	4	
Flexiguard(3M)		F	G	10	
POLYETHYLENE FILM					
Monsanto 602 (Monsanto)		F	P	1	
Teflon (Du Pont)	30.00	P	E	25	
POLYVINYL FLUORIDE FILM					
Tedlar (Du Pont)	2.80	F	E	10-20	

Glazing: Advantages and Disadvantages Table 7.5

	ADVANTAGES	DISADVANTAGES
GLASS	o excellent selective transmission o transparent o good weatherability o heat, air-pollution and u-v resistant o low thermal expansion o easily obtained	o expensive o breaks easily o heavy o often difficult to install in large sheets
ACRYLIC	o high optical clarity, strength and weatherability o lightweight and easy to handle o impact resistant o insulation and transmission similar to glass	o expensive o prone to surface abrasion o buckles and cracks if not properly installed o significant expansion and contraction characteristics o will sag at high temperatures
POLYCARBONATE	o very high impact strength o similar to acrylics, but less solar transmission	o scratches easily o not rigid o becomes brittle and changes color after prolonged exposure to the sun
FIBERGLASS REINFORCED POLYESTER	o low cost, easy to handle o high strength and durability with coatings such as Tedlar o available flat or with corrugations	o solar transmittance reduced when u-v coatings are added o will yellow and 'blossom' without coatings o medium lifespan
POLYESTER FILM	o low cost o high surface hardness	o medium lifespan o u-v degradation if not coated o relatively high longwave transmittance
POLYETHYLENE FILM	o very inexpensive o light, flexible, easy to install o good inner glazing properties	o short lifespan (less than one year) o poor resistance to long-wave radiation o wind and temperature can cause sag
POLYVINYL FLUORIDE FILM	o excellent weatherability and strength o high solar transmittance o can be bonded to fiberglass as a u-v resistant screen	o expensive o available only in thin films o relatively high longwave transmittance

atures (150-300°F) and can put severe strain on certain collector materials.

Wood frame units are generally more expensive than metal units and, if not sufficiently protected, are more prone to weathering effects such as shrinkage and swelling. They are, however, better insulators than their metal counterparts, which are typically poor insulators unless provided with thermal breaks. Metal units are generally stronger and more weather-resistant.

Commercially available frame and glazing assemblies have the advantage of being readily available off-the-shelf and are typically of a consistent quality. However, in those passive installations that require some degree of nonstandard construction, commercially available units may pose problems that a site-built system designed for a specific installation will be able to avoid.

For example, the collectors on Thermal Storage Walls require exterior operation for periodic cleaning and seasonal ventilation. Few commercially available collectors can be operated in this fashion because their hardware is designed for interior operation.

If such units are not adapted for special use, they will need to be mounted with their interior face (and finish) on the outside, exposed to the weather, in order to be operable from the exterior. Patio doors with key locks can operate successsfully in these conditions, but for most other off-the-shelf collectors it will be necessary to make special arrangements with the manufacturer to install the necessary hardware on the outside of the stock products. In the case of site-built collector assemblies, such maintenance and operability problems should be considered prior to designing and building the framing system.

Frame

The framing system is the second basic component of a passive collector, and like the glazing material, it will affect the overall thermal performance of the installation. There are two broad categories into which framing systems can be divided: those that are commercially available and, to varying degrees, precut and preassembled; and those that are site-built.

Commercially Available

Commercially available frames can be either wood (preassembled, or precut and site-assembled). They are fabricated with their collection material (usually glass) already installed or precut for easy assembly, and they are most commonly used in Direct Gain and some Attached Sunspace and Convective Loop applications. These units are modular and can be used to develop overall collector systems of virtually any size or configuration. Fixed units can be combined with operable ones to allow for ventilation as well as solar heat gain.

Site-Built

An alternative to using commercially available units is to design a system to be constructed on-site, using wood framing specially fabricated for the job along with available glazing materials such as patio-door replacement glass, acrylic panels, or polycarbonate sheets. This type of "one-off" approach is particularly appropriate in applications such as Thermal Storage Walls for which commercially available framing systems are generally not available.

Care must be taken to ensure that proper fasteners are used to attach the glazing to the frame (glazing manufacturers' specifications must be referenced), and it is recommended that the collector material be held in place by a wood stop or metal extrusion screwed to the wood mullion. A generous amount of space (½'' is recommended) should be left between the edge of the sheet of glazing and the framing mullion to allow for the expansion and contraction caused by the high temperatures that are frequently generated. Green wood should not be used in a site-fabricated mullion system, and any wood used should be treated with a wood sealant. If wood is used as a face piece to hold the collector material in place, it should be at least ¾'' thick to avoid curling at the edges.

Commercially Available and Site-Built Composite Systems

In certain applications, it may be advantageous to insert preframed, commercially available glazing units into a specially designed, site-constructed collector frame. For example, preframed patio doors might be installed in a larger Thermal Storage Wall collector frame that has been site-constructed of wood and attached directly to the wall. The choice of an appropriate glazing and frame system combination will depend on local conditions, aesthetic preference, and cost.

Sealants and Caulking

All collector installations need to be carefully caulked and sealed. In the case of nonvented Thermal Storage Walls, where temperatures can be in excess of 150°F, or Convective Loop panels, which experience temperatures up to 300°F, they may need to be recaulked after the first several months of operation to ensure efficient future performance. The seal should be airtight but able to be broken in case the panel must be removed.

The major construction sealants used in collector construction and their performance characteristics are listed in Tables 7.6 and 7.7 to aid in selecting appropriate products. As mentioned earlier in this chapter, these tables are intended to provide guidance and should not supersede experience with particular products or construction performance.

Absorber Components

Introduction

Generally located on the surface of the storage material, the absorber converts solar radiation into heat. When sunlight strikes a material, it is, in part, reflected, effectively reducing the amount of solar radiation that can be absorbed. Consequently, absorber materials are usually dark, minimizing heat losses due to reflection. In the process of converting sunlight to heat, the temperature of the absorber rises. If the absorber forms the surface of the storage component, as in a Thermal Storage Wall, the heat moves into and is stored in the thermal mass. In this case, the surface of the wall itself is considered the absorber.

Typically, the surface is painted a dark color to enhance its ability to absorb solar radiation and convert it to heat. This is the case in Direct Gain, Thermal Storage Wall, and Attached Sunspace systems, where storage and absorber are the same material, and the absorber is simply the south-facing surface of the storage material. In such cases, a coat of dark paint will aid in its absorption func-

Caulking Table 7.6

Name*	Cost	Adhesion	Elongation (%)	Life (yrs)	Cont. Service Temp. (°C)	Resistance to UV
Urethane (1 part)	$2-3	Excellent	300-450	20+	-50 to 135	Good
Silicone	$3-4	Good (excellent w/ primer)	100-200	20+	-60 to 205	Excellent
Acrylic Terpolymers	$3-4	Excellent		20+	85 maximum	Very Good
Hypalon (1 part)	$2	Very Good	15-50	15-20	-20 to 110	Good
Polysulfide (1 part)	$2-4	Excellent	200-350 (2 part)	20+	-50 to 120	Very Good
Butyl Rubber	$1-2	Very Good	75-125	7-10	-40 to 135	Fair
Acrylic Latex	$1-2	Excellent except w/ metal	25-60	2-14	-30 to 100	Fair
Oil	$1	Fair - Good	5	1-7	-25 to 85	Poor

Reprinted from Energy Efficient Building and Rebuilding: The Profit Opportunities (see Bibliography reference No. 22).

* Specific product material properties may vary from those listed in this table. Consult manufacturers' literature for specific information.

tion. Certain storage materials, such as colored concrete, brick, or textured concrete masonry units, are naturally dark and do not require a coat of paint.

Materials used to enhance the absorber performance, such as paint applied to masonry in a Thermal Storage Wall and a Direct Gain application, may be exposed to temperatures of 200°F or higher. Care must be taken to ensure that paints or other material applied to the absorber can withstand these extreme temperatures without degradation. If the storage is remote from the absorber, or if storage is not provided, the absorber transfers its heat to the surrounding air. In this case, the absorber is often a metal plate, a material well suited to quickly transferring heat to a moving airstream. The entire plate is the absorber and provides no other system function. This type of absorber is used in Convective Loop systems. Again, care must be taken to use paints to coat the metal surface that can withstand the high temperatures (up to 300°F in Convective Loop panels) without degrading.

For these applications, corrugated-metal roofing is a suitable absorber material. Depending upon the type of absorber plate required (see Chapter 6: Convec-

Caulking Table 7.7

Name*	Shrinkage Free	Cure Time (days)	Paint	Thinner	Use Primer On
Urethane (1 part)	Excellent	14	If desired	MEK, acetone, lacquer thinner	Not required
Silicone	Excellent	5	Possible Difficulties	Paint thinner Naptha, toluol	Not required
Acrylic Terpolymers					Not required
Hypalon (1 part)	Good	60-180	If desired	Toluene, xylene, xylol	Not required
Polysulfide	Excellent	4-7	If desired	TCE,toluene MEK	Special primer on all but metal
Butyl Rubber	Fair	7	Recommended	Paint thinner Naptha	Not required
Acrylic Latex	Fair	3	Recommended	Water	Pourous surfaces
Oil	Fair	120-360	Necessary	Paint thinner	Pourous surfaces

Reprinted from Energy Efficient Building and Rebuilding: The Profit Opportunities (see Bibliography reference No. 22).

* Specific product material properties may vary from those listed in this table. Consult manufacturers' literature for specific information.

tive Loop), expanded metal lath can also be used. Usually three to five layers of lath are specified to allow the air-stream to circulate *through* the absorber, thereby coming in contact with the largest possible area of absorber and enhancing heat transfer.

Thermal Properties

The performance of the absorber component depends partially on two properties of the material used for the surface finish of the absorber: absorptivity and emissivity. The absorptivity is a measure of the material's ability to absorb solar radiation while minimizing losses due to reflection. As a result, white and light colors are poor absorber-finish colors. Emissivity is a measure of the material's ability to emit the collected solar energy and is associated with a loss of solar heat otherwise usable by the passive system.

Therefore, the surface of the absorber facing the sun should have a high absorptivity and a low emissivity.

Most materials typically specified as finish material have both high absorptivities and high emissivities. However, there are a few products currently available that exhibit high absorptivity and low emissivity characteristics.

These products, referred to as selective-surface materials, are typically available

Characteristics of Absorber Materials — Table 7.8

Material	Product Name and Manufacturer	Absorptivity	Emissivity
1. Exposed and unfinished concrete	———	0.65	0.87
2. Flat black Latex paint (nonselective)	Various	0.96	0.87
3. High-temperature absorber coating (nonselective)	Black Absorber Paint Solar Components, Inc.	0.95	0.95
4. Coated nickel foil, 6"-wide rolls (selective) with pressure-sensitive backing	Maxorb Solarfoil	0.97	0.10
5. Black chrome-coated foil, 24"-wide strips (selective)	Berry Foil Berry Solar Products P.O. Box 327 Edison, NJ 08817 (201) 549-3800	0.95	0.10
6. Copper substrate with coated surface (selective) and with pressure-sensitive backing	Sunsponge	0.95	0.10
7. Aluminum substrate with coated surface (selective) and with pressure-sensitive backing	Sunsponge II	0.91	0.15
8. Aluminum (polished and unfinished)	———	0.12	0.09

in sheets designed to be adhered to the face of the storage component.

Table 7.8 illustrates these properties for a variety of materials. The absorptivity of concrete, if left unfinished, is 0.65, indicating that 65% of the solar radiation striking the surface of, for example, a Thermal Storage Wall is absorbed. Simply painting the concrete with a flat black paint can increase the absorptivity to 0.96. However, the emissivity is also fairly high at 0.87.

Materials 4 through 7 on the table are currently available selective-surface materials. As can be seen from the table, the absorptivity of these materials is approximately the same as the black paint, but the emissivities are much lower, which is an indication of their superior performance as absorber-finish materials.

In general, where standard materials, such as black paint, are used as finish materials, a high absorptivity is usually associated with a high emissivity. Table 7.9 lists absorptivities for standard materials. If selective-surface materials are not specified, this table can be used as a guide to the comparative performances of readily available finishes as absorbers.

Storage Components

Introduction

Once the collectors of the system are properly selected, designed, located, and

Table 7.9

SOLAR ABSORPTANCE OF VARIOUS MATERIALS*

Material	Absorptance
Optical flat black paint	.98
Flat black paint	.95
Black lacquer	.92
Dark gray paint	.91
Black concrete	.91
Dark blue lacquer	.91
Black oil paint	.90
Stafford blue bricks	.89
Dark olive drab paint	.89
Dark brown paint	.88
Dark blue-gray paint	.88
Azure blue or dark green lacquer	.88
Brown concrete	.85
Medium brown paint	.84
Medium light brown paint	.80
Brown or Green lacquer	.79
Medium rust paint	.78
Light gray oil paint	.75
Red oil paint	.74
Red bricks	.70
Uncolored concrete	.65
Moderately light buff bricks	.60
Medium dull green paint	.59
Medium orange paint	.58
Medium yellow paint	.57
Medium blue paint	.51
Medium Kelly green paint	.51
Light green paint	.47
White semi-gloss paint	.30
White gloss paint	.25
Silver paint	.25
White lacquer	.21
Polished aluminum reflector sheet	.12
Aluminized Mylar film	.10
Laboratory vapor deposited coatings	.02

This table is meant to serve as a guide only. Variations in texture, tone, overcoats, pigments, binders, etc. can vary these values.

*Data obtained from: Passive Solar Design Handbook, Volume Two: Passive Solar Design Analysis by J.D.Balcomb et al. (see BIBLIOGRAPHY reference no. 38)

constructed, a great deal of heat in the form of solar energy will enter the passive solar system. The amount of heat that can be utilized and the measure of comfort afforded in the living space is largely a function of the storage characteristics of the system. In the winter, for example, the sun shines for only a few hours during the day, while the majority of home heating needs are at night. A well-designed, well-constructed storage system will make the sun's energy available in the living space, protecting against overheating during the day and providing needed heat in the evening.

The discussions that follow present basic information on materials that can be used as passive solar storage components. A storage component can generally be constructed of any of a variety of materials that exhibit a high capacity to store heat. The most commonly used residential construction materials that fit this criterion are those in the masonry family, water in suitable containers, phase-change materials, and rocks.

There are advantages and disadvantages inherent in each material discussed below. For example, both water and rocks are effective storage materials. Although these materials are inexpensive, the container or tank necessary to effectively utilize them may be difficult to design and/or may be expensive.

The possibility of using masonry materials as structural components of the building provides the potential of combining the cost of structure and storage materials. This has resulted in establishing masonry as the most popular storage material in passive solar homes constructed to date.

Recent product advances have established phase-change materials (PCMs) as a promising storage material. PCMs store heat chemically, rather than depending on thermal mass, as is the case with masonry, rocks, and water. As a result, storage can be relatively lightweight and thinner than other storage materials.

Table 7.10 presents a comparison of the major storage-component materials. It is meant only as a quick reference guide for use in selecting a material in the early design stages. Each of these material types, appropriate for use in a passive design, is discussed below in detail.

In each discussion, the following information is presented: a definition of the material and a discussion of typically available shapes and sizes; a review of construction issues specifically related to the materials used in passive solar applications; and an analysis of the thermal performance characteristics of the material, including charts summarizing specific thermal properties that can be used with the various analysis methods presented in Appendix C.

Concrete Masonry

Concrete masonry units (concrete block, concrete brick, and concrete block paving units), referred to as CMUs, are composed primarily of portland cement, graded aggregates, and water. They are fabricated into specific shapes from machine-molded, very dry, no-slump concrete. The resulting elements are then exposed to an accelerated curing procedure and subsequently dried to reduce the moisture content to specified levels. Concrete masonry has traditionally been used both as a structural material in above- and

Characteristics of Primary Thermal Mass Storage Materials **Table 7.10**

Material	Typical Thickness (in.)	Volume to Store 100 Btu's (ft³)	Weight to Store 100 Btu's (lb)	Comments
Concrete Masonry Unit	12-18	43.3	4,545	Can be structural
Concrete	12-18	33	4,762	Can be structural
Brick	16-18	40.65	5,000	Can be structural
Water	6-12	16 (133 gal)	1,000	Inexpensive, locally available, container required
PCM	2-4	0.16	12.2*	Large potential, may be expensive, container critical to performance
Stone	N/A	72	5,000	Inexpensive, locally available, container required

*Based on PCM used in Texxor Heat Cells

below-grade walls and as an exposed finish material on walls and floors.

Concrete masonry construction is based on a modular dimensional system whose base module has been established as 4″ vertically and 4″ horizontally (nominal measurements). Standard dimensions for CMUs are equal to the nominal size minus the thickness of one ⅜″ mortar joint. Thus, the common 8″-high by 16″-long concrete block is actually 7⅝″ × 15⅝″ (see Figure 7.2).

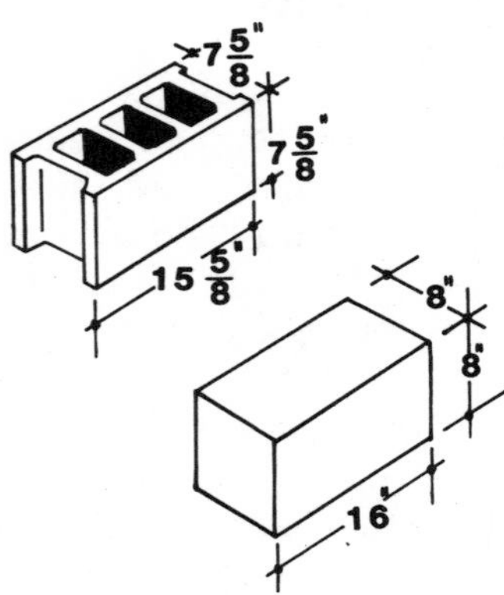

Figure 7.2

Regardless of their dimensions, CMUs are manufactured as either hollow or solid units. Due to their reduced weight and ease of handling, hollow units are used more frequently than solid ones. However, in most passive solar construction, it is recommended that concrete masonry units used for storage be solid, or hollow with grout-filled cores, to increase the overall efficiency of the passive system (a specification for the grout to be used in CMU cores is included later in this chapter).

A sampling of the large variety of sizes and shapes of concrete masonry units available in the market today is presented in Figures 7.3 through 7.12.

Concrete Masonry Units **Figure 7.3**

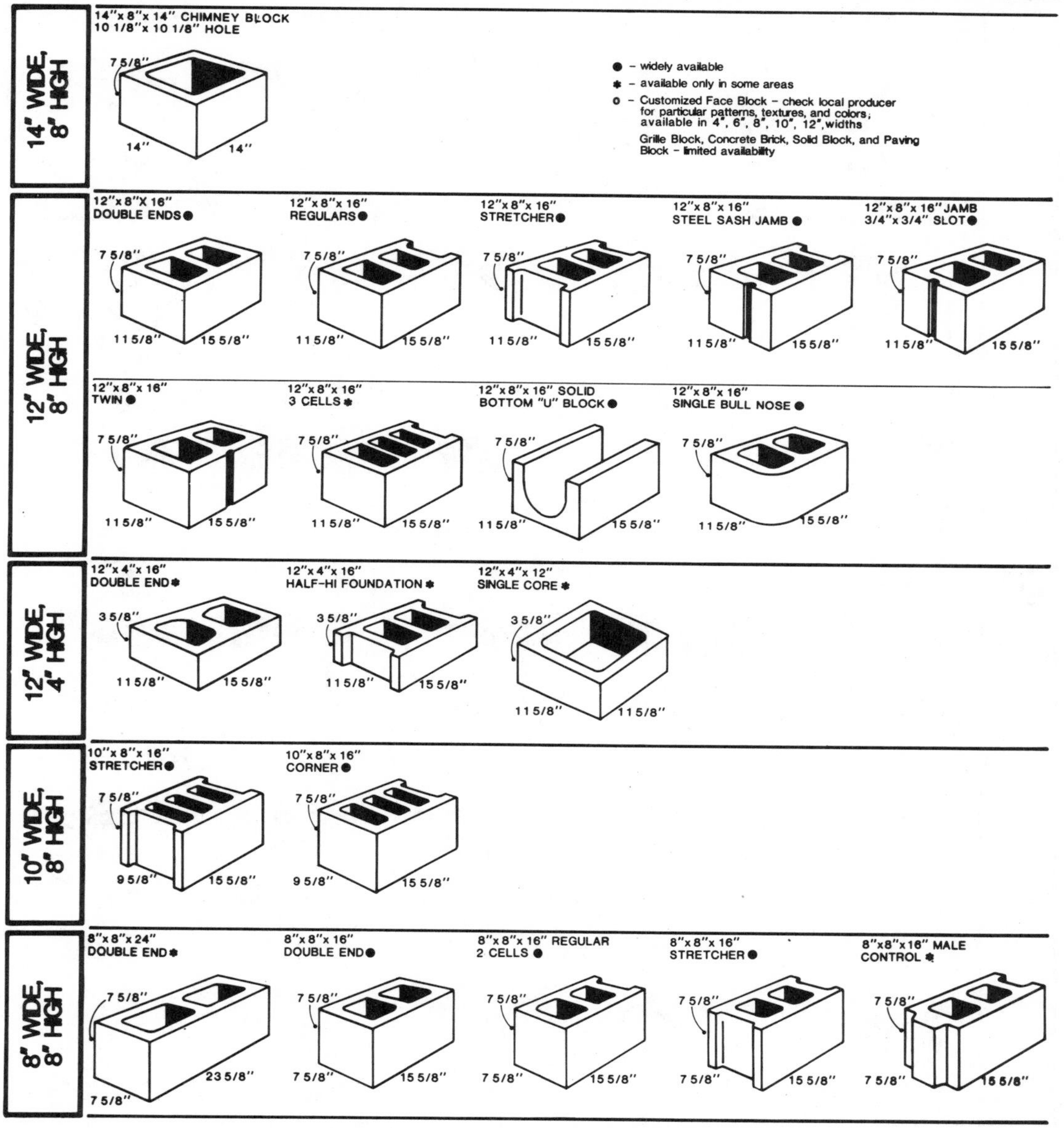

● – widely available
✱ – available only in some areas
o – Customized Face Block – check local producer for particular patterns, textures, and colors; available in 4", 6", 8", 10", 12", widths

Grille Block, Concrete Brick, Solid Block, and Paving Block – limited availability

14" WIDE, 8" HIGH

14"x 8"x 14" CHIMNEY BLOCK 10 1/8"x 10 1/8" HOLE — 7 5/8", 14", 14"

12" WIDE, 8" HIGH

12"x 8"X 16" DOUBLE ENDS● — 7 5/8", 11 5/8", 15 5/8"

12"x 8"x 16" REGULARS● — 7 5/8", 11 5/8", 15 5/8"

12"x 8"x 16" STRETCHER● — 7 5/8", 11 5/8", 15 5/8"

12"x 8"x 16" STEEL SASH JAMB ● — 7 5/8", 11 5/8", 15 5/8"

12"x 8"x 16" JAMB 3/4"x 3/4" SLOT● — 7 5/8", 11 5/8", 15 5/8"

12"x 8"x 16" TWIN ● — 7 5/8", 11 5/8", 15 5/8"

12"x 8"x 16" 3 CELLS ✱ — 7 5/8", 11 5/8", 15 5/8"

12"x 8"x 16" SOLID BOTTOM "U" BLOCK ● — 7 5/8", 11 5/8", 15 5/8"

12"x 8"x 16" SINGLE BULL NOSE ● — 7 5/8", 11 5/8", 15 5/8"

12" WIDE, 4" HIGH

12"x 4"x 16" DOUBLE END ✱ — 3 5/8", 11 5/8", 15 5/8"

12"x 4"x 16" HALF-HI FOUNDATION ✱ — 3 5/8", 11 5/8", 15 5/8"

12"x 4"x 12" SINGLE CORE ✱ — 3 5/8", 11 5/8", 11 5/8"

10" WIDE, 8" HIGH

10"x 8"x 16" STRETCHER● — 7 5/8", 9 5/8", 15 5/8"

10"x 8"x 16" CORNER ● — 7 5/8", 9 5/8", 15 5/8"

8" WIDE, 8" HIGH

8"x 8"x 24" DOUBLE END ✱ — 7 5/8", 23 5/8", 7 5/8"

8"x 8"x 16" DOUBLE END● — 7 5/8", 7 5/8", 15 5/8"

8"x 8"x 16" REGULAR 2 CELLS ● — 7 5/8", 7 5/8", 15 5/8"

8"x 8"x 16" STRETCHER ● — 7 5/8", 7 5/8", 15 5/8"

8"x8"x16" MALE CONTROL ✱ — 7 5/8", 7 5/8", 15 5/8"

Concrete Masonry Units **Figure 7.4**

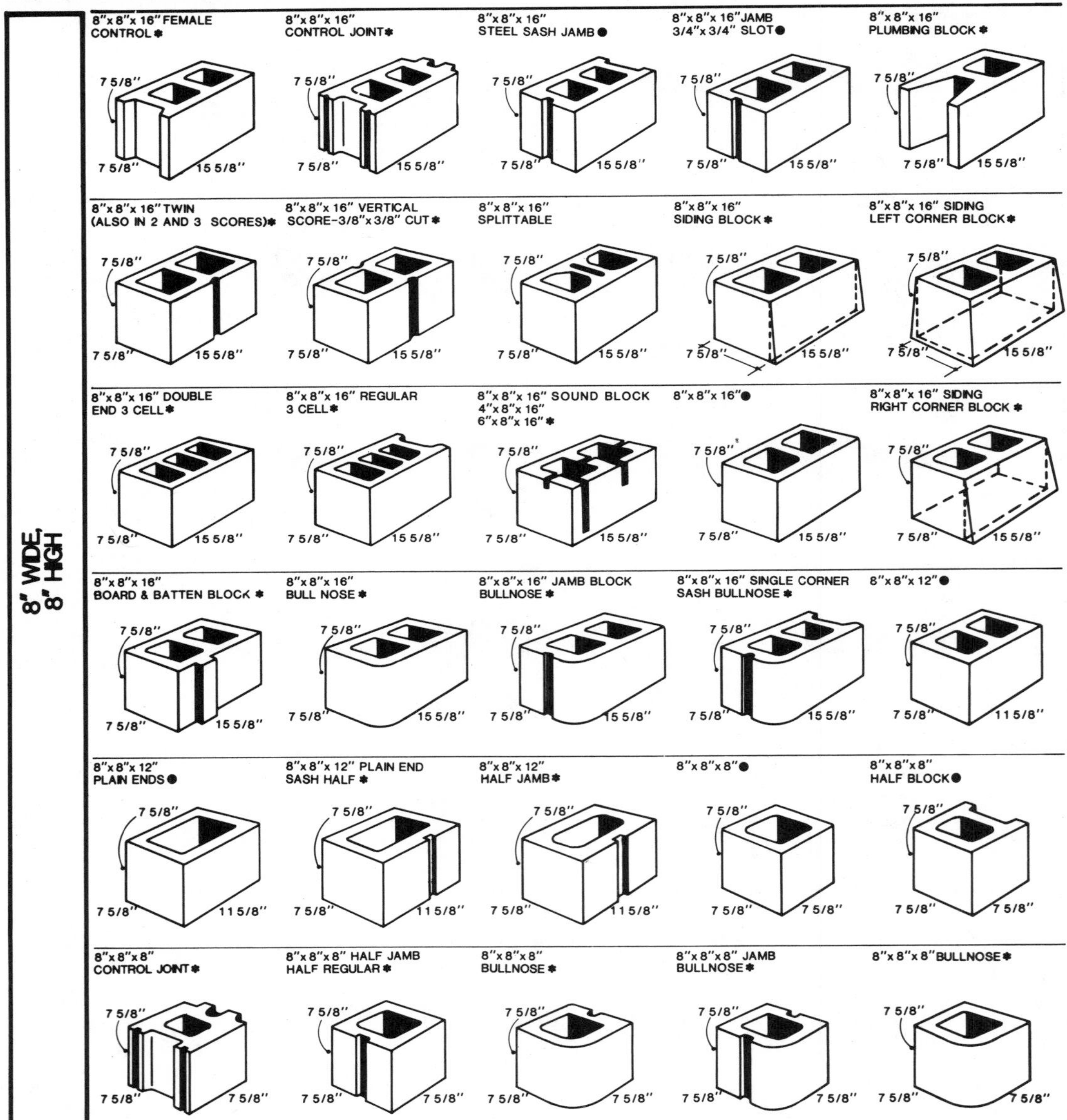

Concrete Masonry Units

Figure 7.5

8" WIDE, 8" HIGH

8"x8"x8" JAMB BULLNOSE ✱

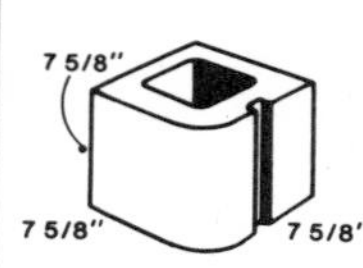

8"x8"x8" SIDING BLOCK ✱

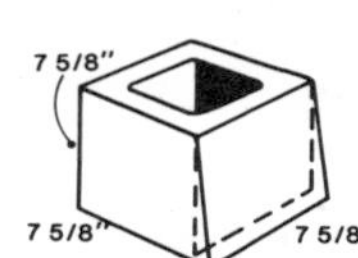

8" WIDE, 4" HIGH

8"x4"x16" HALF HI 3 CELL ●

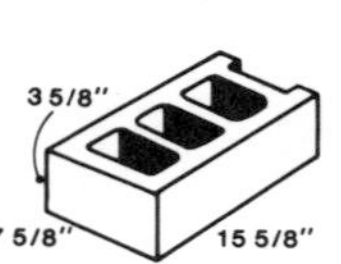

8"x4"x16" HALF HI DOUBLE END ●

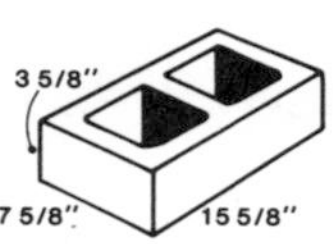

8"x4"x16" HALF HI ●

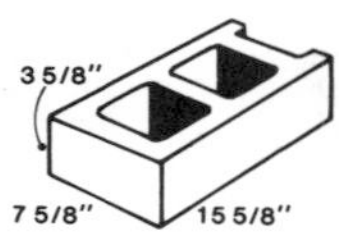

8"x4"x16" HALF HI STRETCHER ●

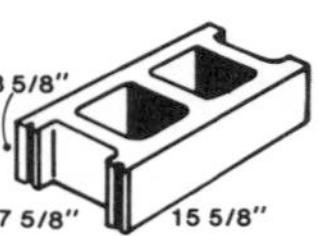

8"x 4"x 16" JAMB HALF HI ✱

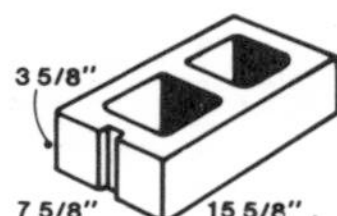

8"x4"x16" HALF HI SASH ✱

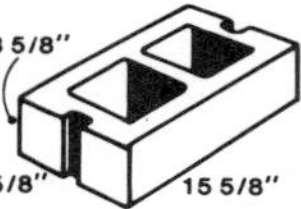

8"x4"x8" HALF HI ●

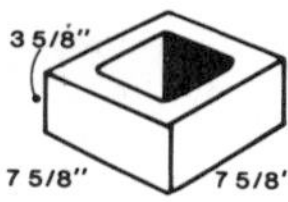

8"x4"x8" HALF HI ✱

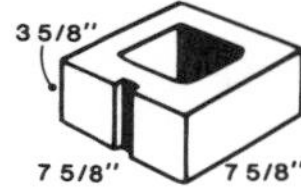

8"x4"x12" HALF HI ✱

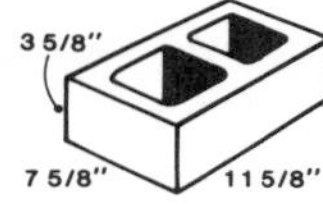

8"x4"x16" HALF HI FURRED JAMB ✱

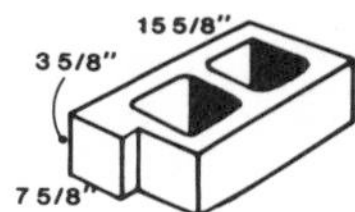

8"x4"x16" ASHLAR HALF HI I-CELL ✱

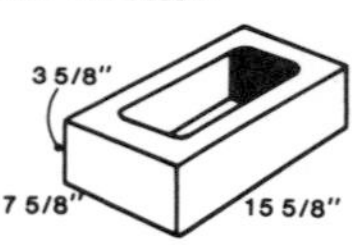

8"x4"x8" HALF HI SPLIT FACE ✱

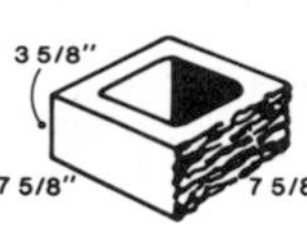

8"x4"x16" HALF HI SEMI-SPLIT FACE ✱

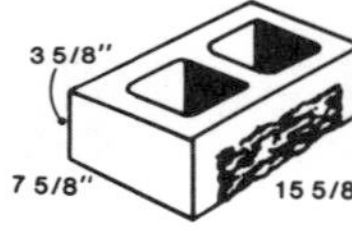

8"x4"x16" (SINGLE) HALF HI, DOUBLE TWIN, 4" ✱

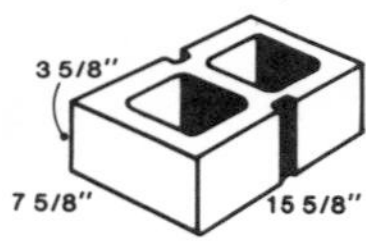

8"x4"x16" HALF HI BULLNOSE ✱

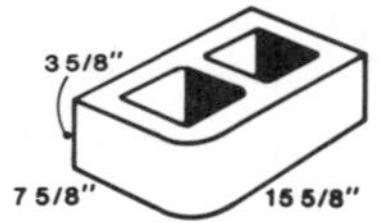

8"x4"x8" JAMB BULLNOSE HALF HI ✱

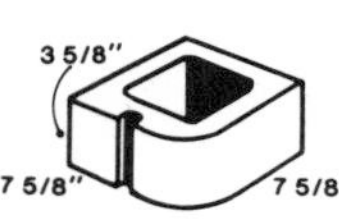

8"x4"x8" HALF HI BULLNOSE ✱

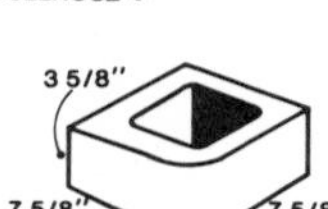

8"x4"x16" CORED SPLIT BLOCK ✱

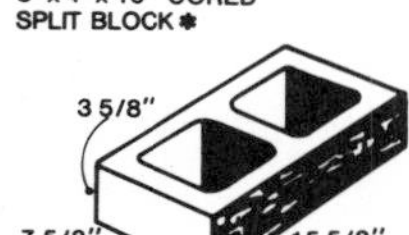

6" WIDE, 8" HIGH

6"x8"x16" DOUBLE ENDS ●

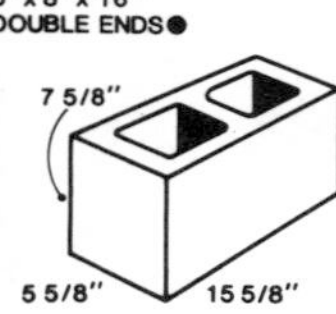

6"x8"x16" REGULARS AND PARTITIONS ●

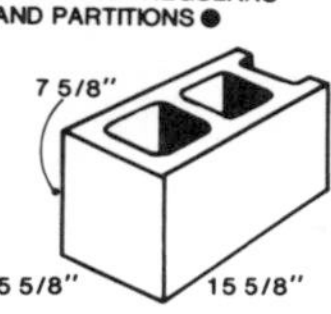

6"x8"x16" STRETCHER ●

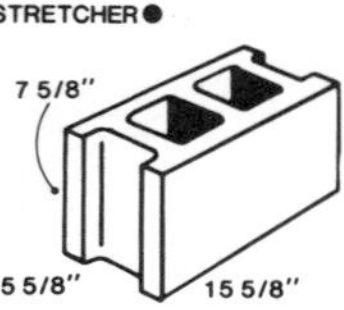

6"x8"x16" JAMB 3/4"x3/4" SLOT ●

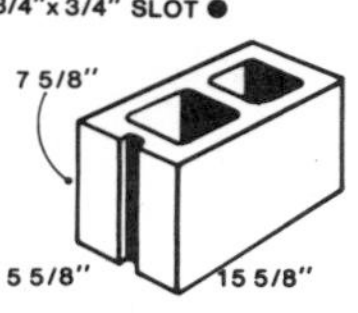

6"x8"x16" TWIN (ALSO WITH 2 AND 3 SCORES) ✱

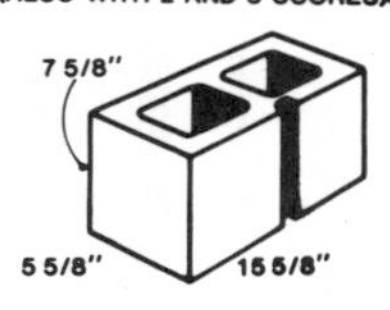

Concrete Masonry Units **Figure 7.6**

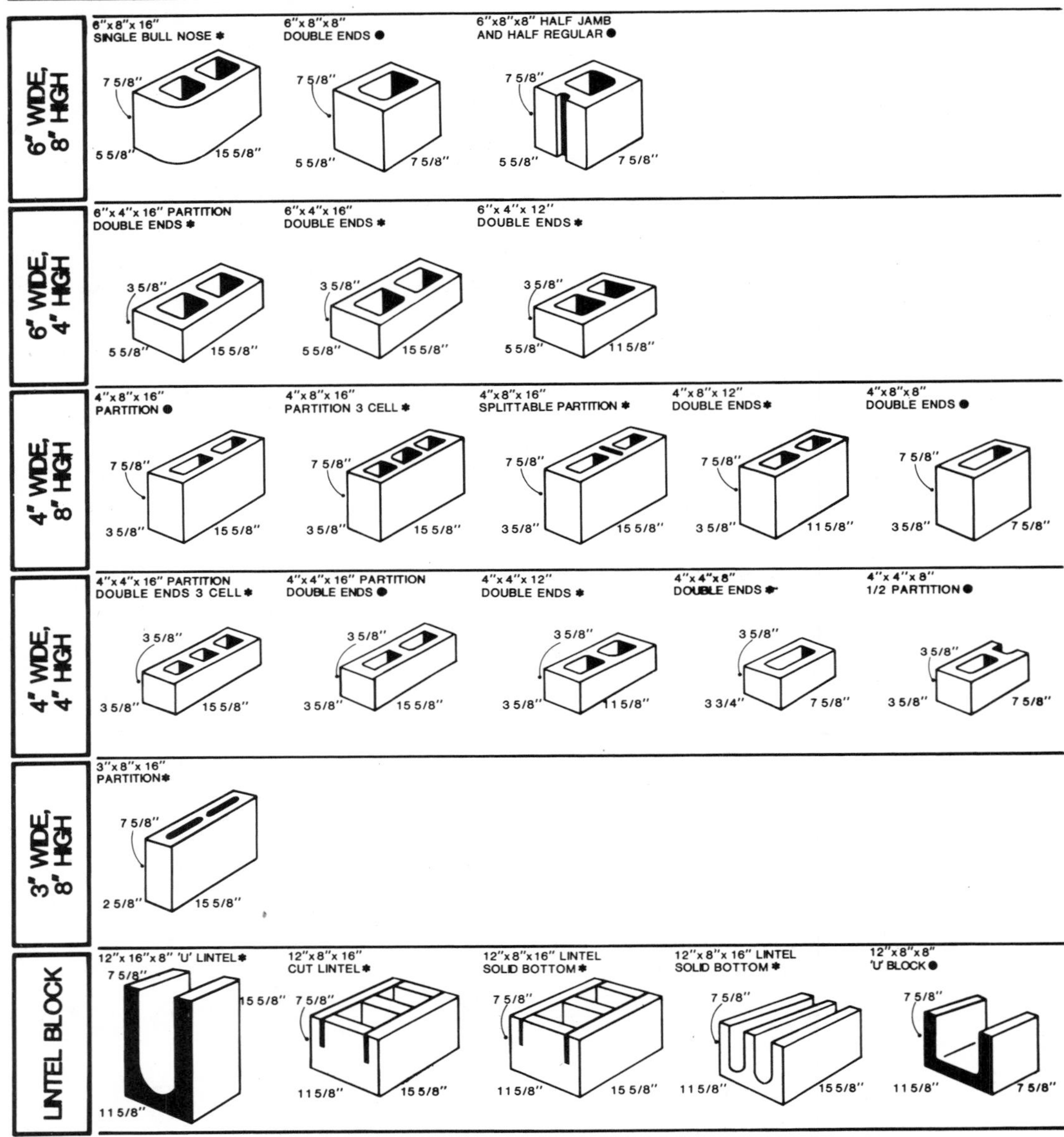

Concrete Masonry Units **Figure 7.7**

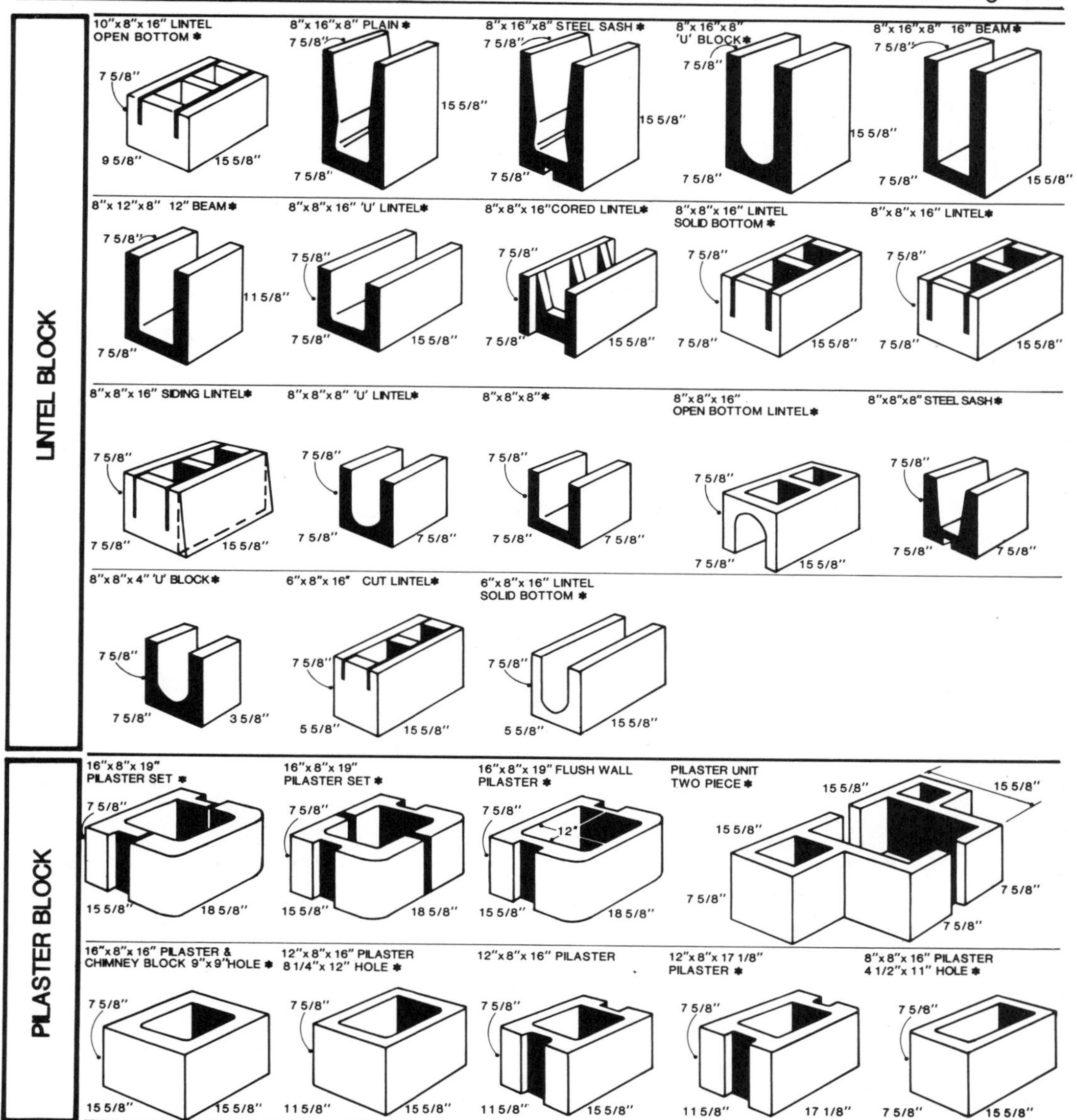

Concrete Masonry Units **Figure 7.8**

PILASTER BLOCK

8"x8"x16" NO.37 PILASTER ✱

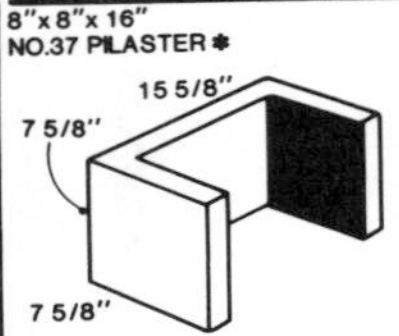

COLUMN AND FLUE BLOCK

16"x8"x16" COLUMN STYLE NO.1 ✱

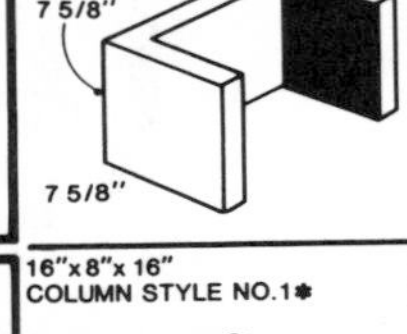

16"x8"x16" FLUE BLOCK ✱

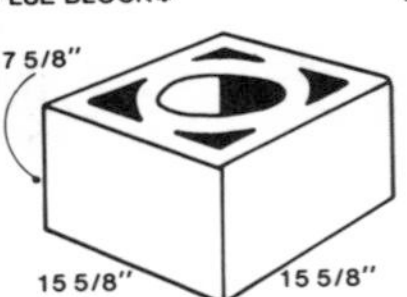

16"x4"x16" FLUE BLOCK ✱

CUSTOMIZED FACE BLOCK

8"x8"x16" VERTICAL PATTERN o

7 5/8" 7 5/8" 15 5/8"

8"x8"x16" VERTICAL PATTERN o

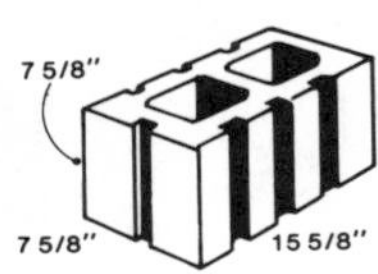

8"x8"x16" VERTICAL PATTERN o

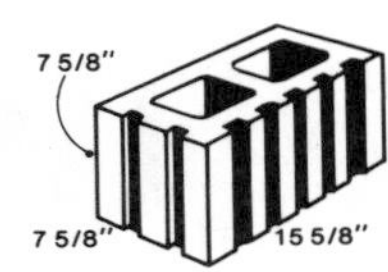

8"x8"x8" VERTICAL PATTERN o

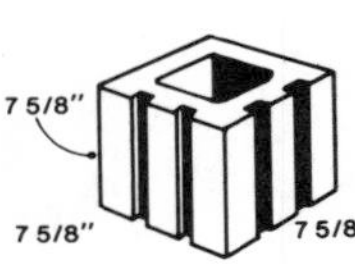

8"x8"x16" FLUTED o

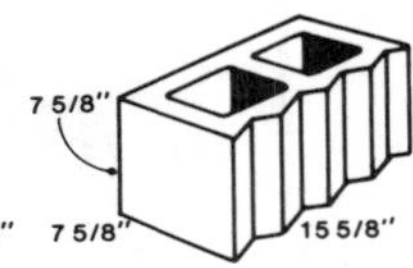

6"x8"x16" ROUND FLUTED o

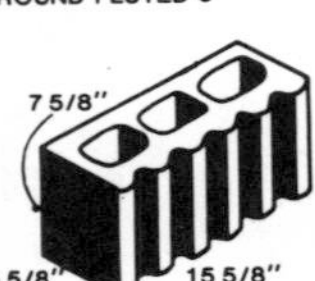

8"x8"x16" FLUTED & SPLIT o

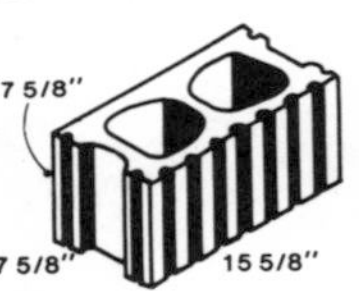

8"x8"x16" FLUTED & SPLIT o

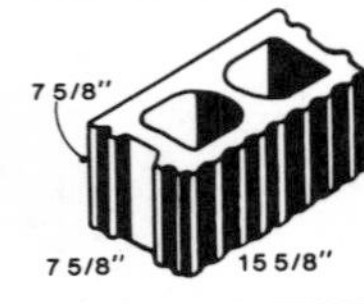

4"x8"x16" FLUTED & SPLIT o

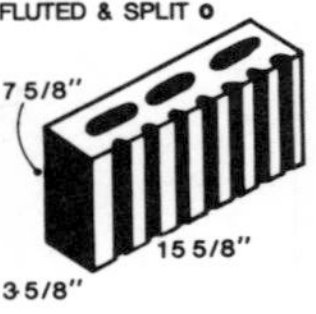

8"x8"x16" 3 SCORE o

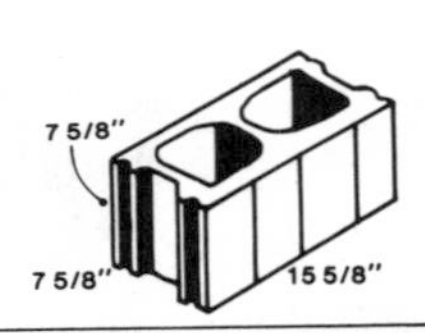

8"x4"x16" SLUMP BLOCK o

3 5/8" 7 5/8" 15 5/8"

4"x4"x16" SLUMP BLOCK o

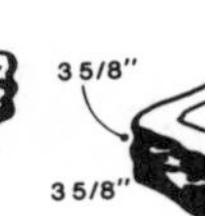

4"x3"x16" SLUMP BLOCK o

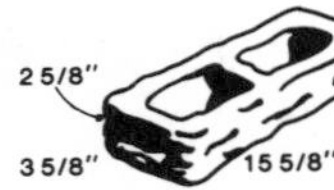

4"x2"x16" SLUMP BLOCK o

4"x2 1/4"x16" SPLIT FACE o

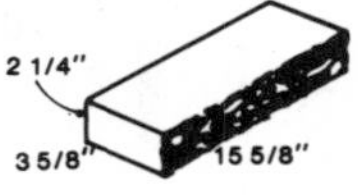

12"x8"x16" SPLIT FLUTED o

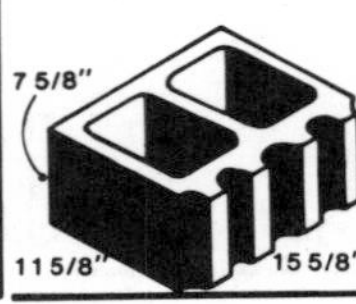

8"x8"x16" SPLIT o

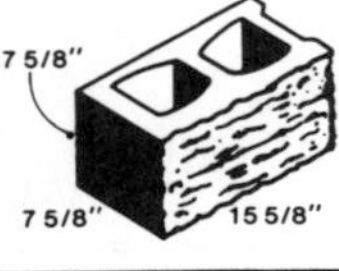

8"x4"x16" SPLIT o

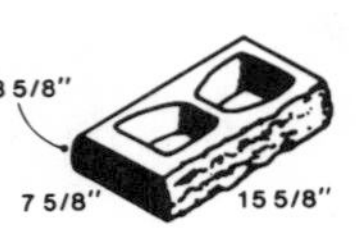

8"x4"x16" SPLIT o

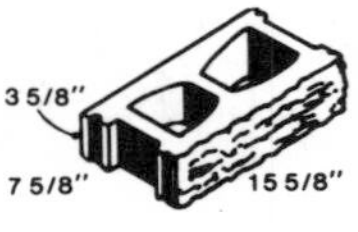

Concrete Masonry Units **Figure 7.9**

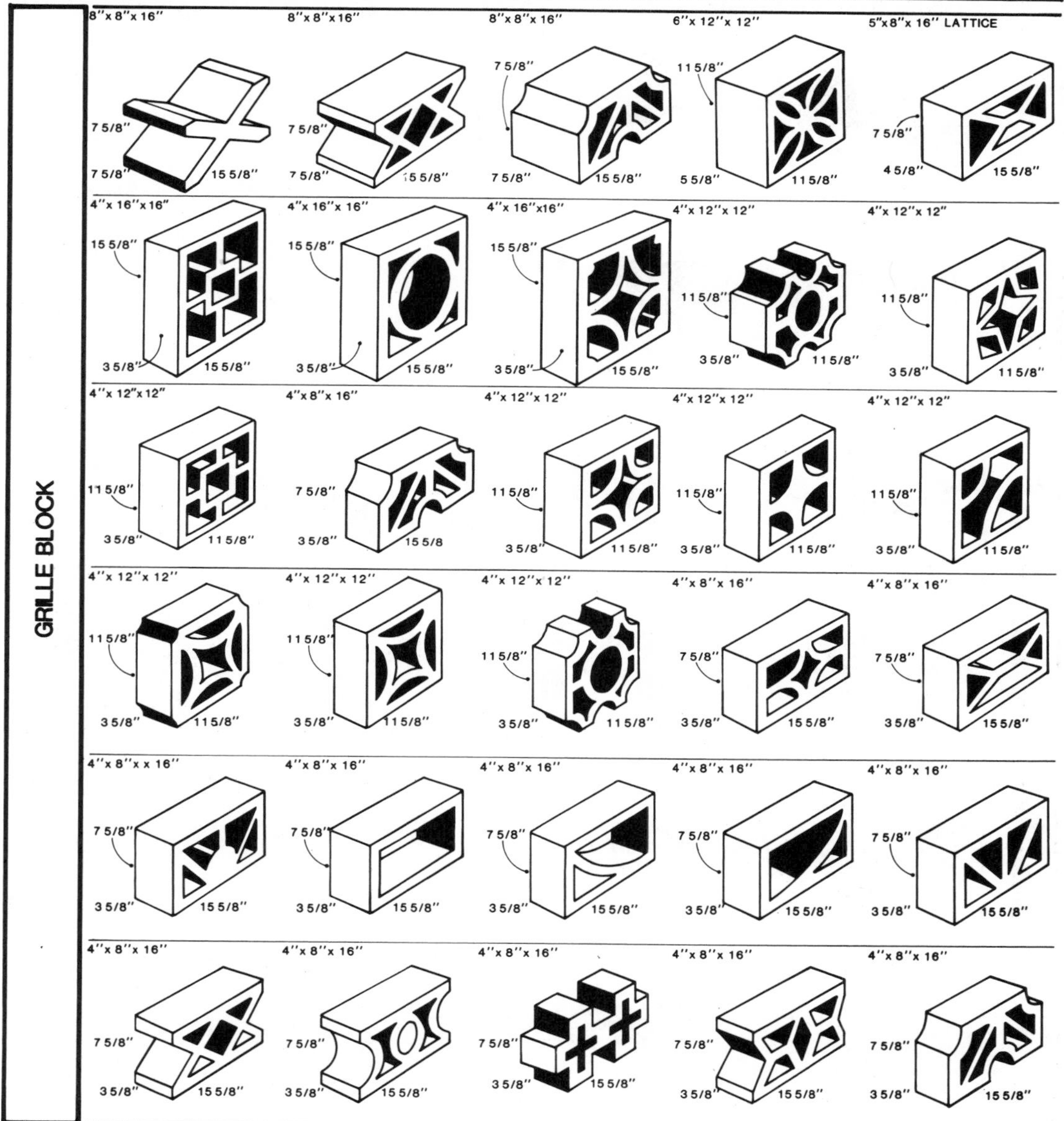

Concrete Masonry Units **Figure 7.10**

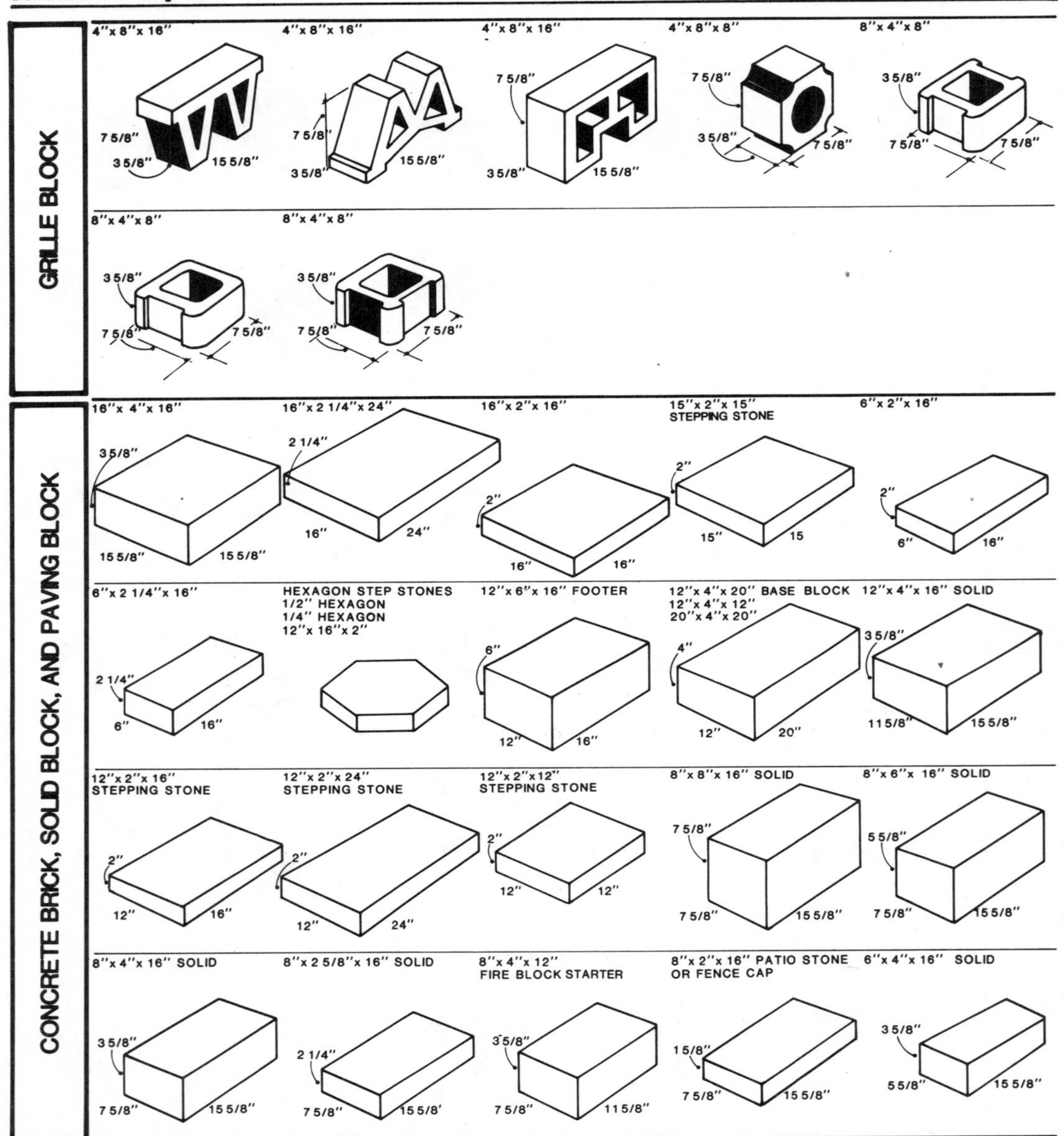

Concrete Masonry Units **Figure 7.11**

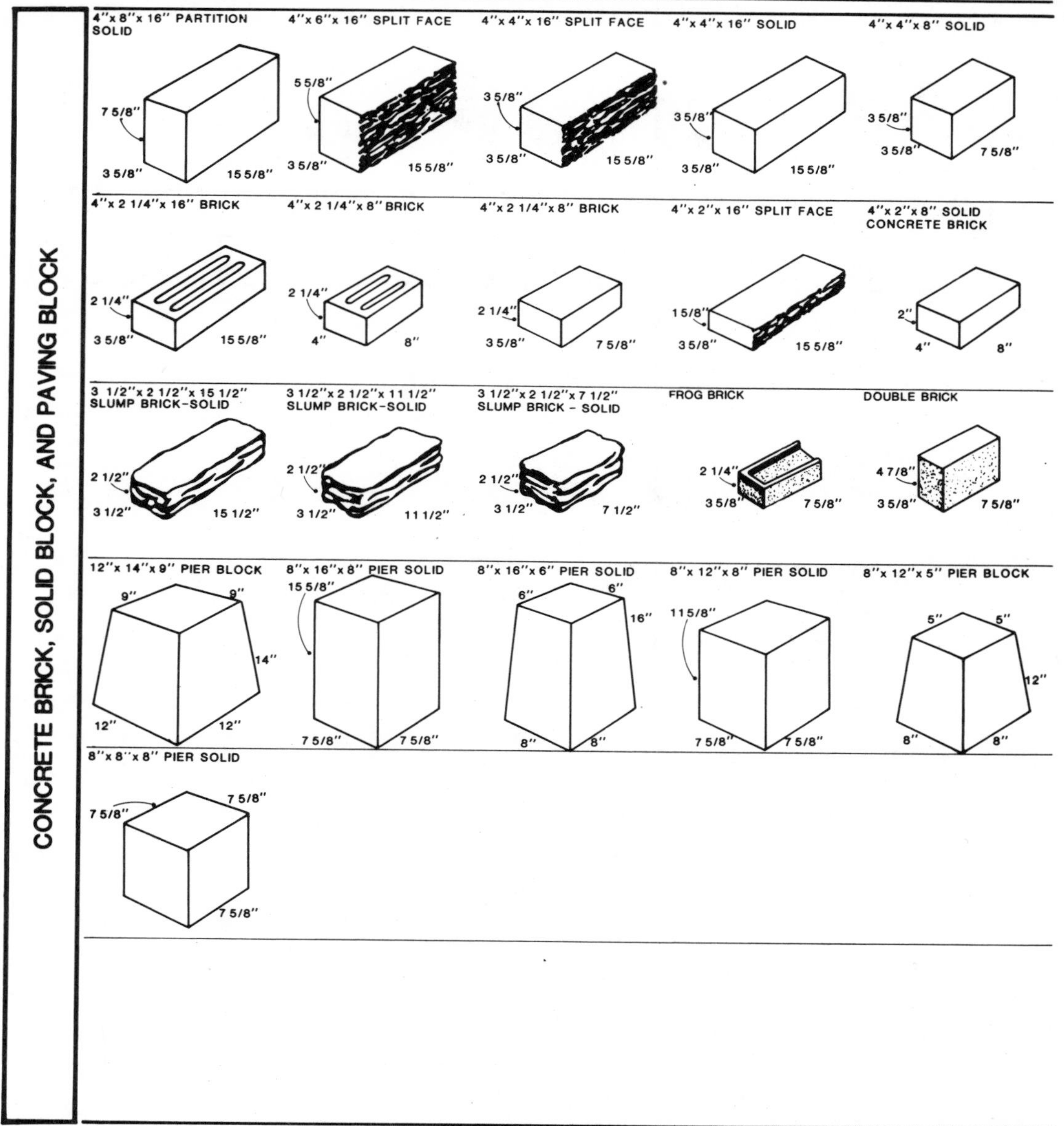

Concrete Masonry Units **Figure 7.12**

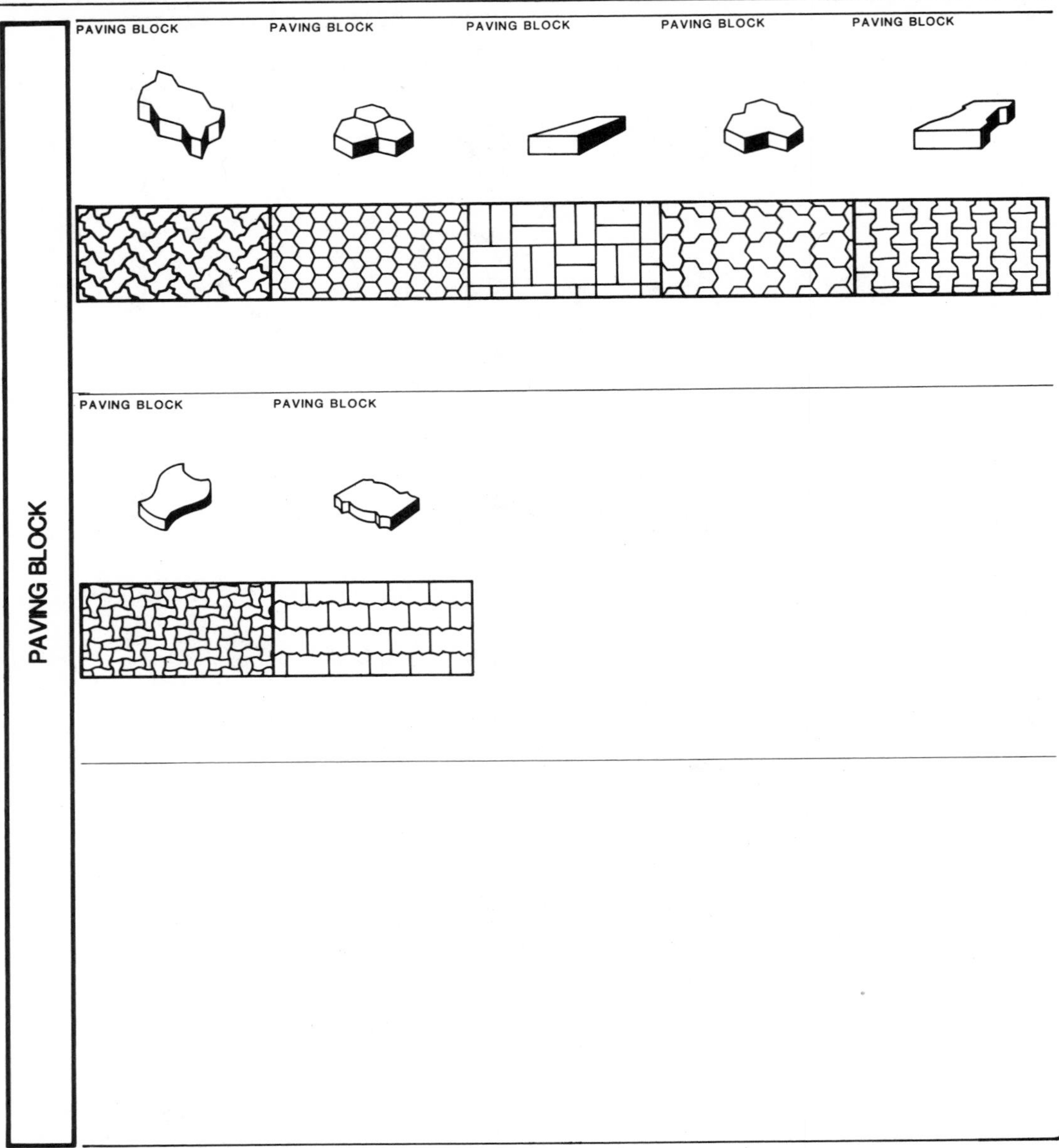

Construction

In most passive solar installations, construction with concrete masonry will follow standard practice. The only major departures from typical construction are: (1) placing insulation on the *outside* of a CMU being used for Direct Gain storage (e.g., an exterior CMU wall); (2) the use of solid or grout-filled, as opposed to hollow, units to store heat; and (3) the use of heavyweight rather than lightweight block. These changes require only minor modifications to the building process, which are described in the details in Chapters 3 through 6.

Finishes

In Thermal Storage Wall and Attached Sunspace construction, the exterior face of a CMU wall is used to absorb and store heat and is protected from the environment by the collector glazing. Unfinished concrete masonry of a dark color performs best in these conditions. In Direct Gain systems with exterior wall storage, however, the outside face of a solid CMU wall is not protected by the collector glazing. If this face is exposed directly to the environment and is not protected by some form of insulation and finish, stored heat will travel through the wall and be lost to the outside, severely reducing the thermal performance of the overall system.

There are a variety of methods for finishing the exterior of a concrete masonry wall being used for Direct Gain storage. Because it is recommended that insulation be placed on the *outside* of exterior walls being used for Direct Gain storage, such finishes will need to be applied *over* this insulation. In a standard masonry bearing wall configuration, stucco or siding can be applied over insulation that has been adhered or fastened to the CMU wall. In typical exterior CMU veneer wall construction, the configuration of the materials can be inverted, moving the masonry to the inside for heat storage and the insulated bearing frame wall to the outside. Siding is then applied in the conventional manner.

Such external finishes will affect the thermal performance of a Direct Gain storage component only insofar as they increase the effective insulation value of the wall. If a wall that is absorbing heat at its interior face is well insulated on its exterior, it will retain more of the heat energy imparted to it and will thus serve as a more effective solar energy storage medium.

It is recommended that interior finishes be kept to a minimum to preserve the inherent thermal performance capabilities of the CMUs. For example, if a finish such as wood paneling or gypsum board is applied to the interior face of a wall intended to absorb and store solar energy, it will reduce both the amount of heat that reaches the wall and the amount that is reradiated from it to the space. This is especially true when covering a CMU paver floor or CMU radiant floor system with carpet, which is emphatically *not* recommended. Although the negative effects of such interior finishes have not yet been quantified, a good rule of thumb is to assume that, relative to other finish materials, the higher the insulating value of a given finish, the more it will inhibit the passive thermal performance of the component to which it is applied. Several acceptable finish materials, such as tile products, can be effectively combined with concrete masonry construction (for further information see the discussion of finishes for concrete storage components later in this chapter).

It should also be noted that a variety of concrete masonry products, such as architectural block, have been designed specifically as finish materials and are suitable for residential construction.

Carrying this concept one step further, it is obvious that putting insulation on the interior surface of the CMU used for energy storage is extremely counterproductive and will essentially nullify any passive benefits that might have been derived from the component. It is for this reason that it is recommended that no passive storage component (wall, floor, or roof/ceiling) be insulated on its interior face.

Color

Where concrete masonry is used both to absorb and to store solar energy, the surface exposed to solar radiation should be as dark a color as possible to increase absorption of light and heat. Other surfaces in a space that are not being used for storage, such as standard wood frame walls, should be light in color in order to reflect more light to the darker storage material (for a listing of the solar absorptance characteristics of various colors/materials, see Table 7.9).

In the case of concrete masonry, natural color can vary from light to dark grey and also to tints of buff, red, or brown, depending on the color of aggregate, cement, and other mix ingredients. Again, it is recommended that the darker tones be specified when the CMUs are being used unfinished for heat storage. In customized units, specific dark colors can be requested from the manufacturer.

The concrete masonry can also be painted or stained. Care must be taken to specify paints or stains that do not degrade when subjected to high temperatures in applications where such temperatures are expected (e.g., Thermal Storage Walls).

Another possible finish is the so-called selective surface, a coating that has a very high absorptivity of incoming solar radiation but a very low emissivity of heat. Such coatings can be very effective at increasing the thermal efficiency of passive solar storage components, and manufacturers' data should be referenced for specific performance characteristics (for further information, see Absorber Components earlier in this chapter).

Texture

One final note should be added concerning customized CMUs that have architecturally finished faces. Such units present a greater amount of exposed surface area for heat exchange to the surrounding air than do standard flat-surfaced CMUs. It might be assumed that such increased exposure might influence the performance of installations using these materials, but this has not, as yet, been proven to be true. Until more accurate data are compiled in this regard, it is recommended that customized CMUs be used interchangeably with standard units in passive solar storage applications.

Thermal Properties

From the point of view of thermal performance in a passive solar application, the most important material characteristics of any particular concrete masonry unit are its conductivity, density, specific heat, and thickness.

A material's conductivity (k) is a measure of how readily the material will transfer or "conduct" heat: the higher the conductivity, the faster the heat will travel through a material. In general, thermal performance increases as the conductivity increases.

The remaining three basic material characteristics (density, specific heat, and thickness) can also be used as indicators of anticipated thermal performance. When multiplied together, they yield the so-called "thermal storage capacity" of a specific unit of material. Essentially, this characteristic is a measure of the ability of a given unit of material to store heat. The higher the thermal storage capacity, the greater the amount of heat the unit can absorb and store.

In the case of concrete masonry, the specific heat for the majority of available units remains basically constant. Thus, the major variables that will determine the thermal storage capacity of a CMU are its density (measured in pounds per cubic foot) and its thickness.

In general, concrete masonry units can be divided into two broad categories of material density:

- Dense or Normal Weight—Employing sand, gravel, crushed stone, or air-cooled blast-furnace slag as aggregate.

- Lightweight—Using expanded shale, clay, and slate, expanded blast-furnace slag, sintered fly ash, coal cinders, or natural lightweight materials such as pumice or scoria.

By weight, concrete masonry is considered lightweight if its density is 105 lb/ft^3 or less, medium weight if 105-125 lb/ft^3, and normal weight if over 125 lb/ft^3. It is recommended that the heavier-weight CMUs be used in passive solar storage applications.

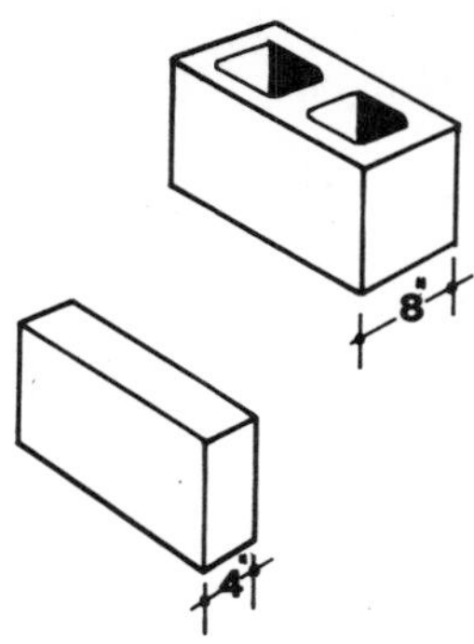

Figure 7.13

The weight of any specific concrete masonry unit will be a function of the density of the concrete used and the volume of this concrete in the particular unit. In the case of hollow units, the volume of concrete is equal to the overall volume of the unit minus the volume of the cores. The concept of "equivalent solid thickness," used for rating a unit's fire-resistance capability, comes to mind in this regard as a method for visualizing the "real" shape of a hollow-core block (see Figure 7.13).

In terms of thermal performance, however, such a model is misleading. In passive solar applications where concrete masonry is being used as a storage medium, an 8'' hollow block with a 50% core area does not perform like a 4'' solid unit. Heat transfer through the webs and across the airspaces formed by the cores is not uniform, and the amount of heat reaching the far side of the block is considerably less in an 8'' hollow unit than it is in a 4'' solid one.

It is for this reason that filling all hollow concrete masonry cores when they are used as passive solar storage media is recommended (grout, small-size normal-weight aggregate, or other materials are suitable). For the purpose of analysis, when the grout or other material in the cores is of different density from the CMU itself, a weighted average of the densities involved should be used. The density of the joint mortar employed should be considered to be the same as that of the CMU itself.

Table 7.11 presents thermal performance data for a variety of CMUs with different thicknesses, densities, and grout types. The table can be used for general performance comparisons between the various material combinations presented, or specific information can be extracted for use with the Analysis Methods discussed in Appendix C. It should be noted that the conductivity (k) values presented are for the CMU material only.

In general, the higher the thermal storage capacity of a given unit, the better its performance from a passive solar point of view. This should be kept in mind when selecting a specific CMU for a specific installation. For example, if a nominal 8'' CMU wall is desired for a particular passive installation, the combination of unit density and grout type that results in the highest possible thermal storage capacity should generally be specified (subject to local cost considerations).

For grout types and block densities not covered in the tables, a good rule of

MINIMUM REQUIREMENTS FOR NON-STRUCTURAL GROUT*

Where concrete masonry walls are grouted for the purpose of increasing thermal storage capacity, i.e., non-structural purposes, the grout may consist of a mixture of Portland Cement, aggregate and water.

Portland Cement shall conform to the requirements of ASTM C-150, Standard Specification for Portland Cement, or ASTM C-595, Specification for Blended Hydraulic Cements.

Aggregate shall conform to the requirements for fine aggregate of ASTM C-33, Standard Specification for Concrete Aggregates (except that the grading Requirements need not apply), or ASTM C-404, Standard Specification for Aggregates for Masonry Grout.

Grout shall consist of a mixture of Portland Cement and aggregate proportioned such that the aggregate to cement ratio does not exceed 12:1; and sufficient water to permit the mixture to flow readily, without segregation, into the cores of the masonry.

*Courtesy of the National Concrete Masonry Association
2302 Horse Pen Road
P.O.Box 781
Herndon, Va. 22070

Concrete Masonry Units: Physical Properties — **Table 7.11**

Unit Thickness*	Unit Density∞	Specific Heatθ	Thermal Storage CapacityΣ			R-ValueΔ			ConductivityΩ
3"	80	.22	3.5			0.95			2.5
	95	.22	4.1			0.72			3.3
	105	.22	4.5			0.66			3.6
	125	.21	5.2			0.39			6.1
	135	.21	5.6			0.29			8.1
			A	B	C	A	B	C	
4"	80	.22	3.5	5.9	6.6	n/a	n/a	n/a	2.5
	95	.22	4.0	6.5	7.2	n/a	n/a	n/a	3.3
	105	.22	4.5	6.9	7.6	n/a	n/a	n/a	3.6
	125	.21	5.0	7.3	8.1	n/a	n/a	n/a	6.1
	135	.21	5.5	7.8	8.5	n/a	n/a	n/a	8.1
6"	80	.22	5.3	9.1	10.2	1.79	1.84	1.37	2.5
	95	.22	6.3	10.0	11.2	1.58	1.61	1.17	3.3
	105	.22	6.9	10.7	11.8	1.45	1.48	1.06	3.6
	125	.21	7.9	11.5	12.5	1.21	1.24	.85	6.1
	135	.21	8.5	12.1	13.1	1.08	1.11	.76	8.1
8"	80	.22	6.3	12.4	14.1	2.07	2.48	1.83	2.5
	95	.22	7.6	13.7	15.4	1.79	2.18	1.56	3.3
	105	.22	8.4	14.4	16.1	1.64	2.01	1.42	3.6
	125	.21	9.5	15.3	17.0	1.35	1.68	1.15	6.1
	135	.21	10.2	16.0	17.7	1.21	1.49	1.02	8.1
10"	80	.22	7.2	15.8	18.4	2.22	3.10	2.22	2.5
	95	.22	8.5	17.1	19.7	1.92	2.75	1.92	3.3
	105	.22	9.5	18.1	20.7	1.76	2.55	1.76	3.6
	125	.21	10.8	19.0	21.5	1.45	2.15	1.45	6.1
	135	.21	11.7	19.9	22.3	1.30	1.92	1.30	8.1
12"	80	.22	8.4	19.2	22.5	2.39	3.73	2.64	2.5
	95	.22	9.9	20.7	24.1	2.06	3.32	2.29	3.3
	105	.22	10.9	21.7	25.1	1.88	3.08	2.10	3.6
	125	.21	12.5	22.9	26.1	1.54	2.59	1.75	6.1
	135	.21	13.5	23.9	27.1	1.38	2.31	1.57	8.1

* 3" units solid CMU pavers.

A Cores hollow.
B Cores filled with 100 lb/ft^2 grout.
C Cores filled with 130 lb/ft^2 grout.
n/a data not available

∞ – Units of lb/ft^3
θ – Units of btu/lb/°F
Σ – Units of $btu/ft^2/°F$
Δ – Units of hr $°F/ft^2/btu$
Ω – Units of btu $in/hr/°F/ft^2$

thumb is to assume that, for a given unit thickness, the higher the density of the unit and/or of the grout, the greater the overall thermal storage capacity. It can also be assumed that the thermal storage capacities for thicknesses greater than 12'' will roughly equal the sum of the capacities of smaller units which, when added together, will equal the total overall thickness desired. For example, the thermal storage capacity per square foot for a 20''-thick wall of 80-lb-density block and 100-lb grout will roughly equal the sum of the capacities of one 8'' and one 12'' block wall (12.4 + 19.2 = 31.6) or of two 10'' walls (15.8 + 15.8 = 31.6).

It should be noted that some concrete masonry manufacturers in the United States produce CMUs with higher densities than those listed in Table 7.11. Should higher densities be required, local manufacturers of concrete masonry units should be consulted in advance.

Concrete

Concrete is essentially a mixture of two elements: aggregate and paste. The paste is made of portland cement and water, and it binds the aggregate (sand plus gravel or crushed stone) into a solid mass as it hardens due to a chemical reaction between the cement and water. The paste is usually from 25-40% of the total volume of the concrete, with aggregates comprising the remainder.

Concrete can be separated into the following four general categories based on density:

- Insulating Lightweight Concrete (15-90 lb/ft^3)—Using pumice, scoria, perlite, vermiculite, or diatomite as aggregate.

- Structural Lightweight Concrete (85-115 lb/ft^3).

- Normal-Weight Concrete (135-160 lb/ft^3)—Using sand, gravel, crushed stone, or air-cooled blast-furnace slag as aggregate.

- Heavyweight Concrete (160-400 lb/ft^3)—Using barlite, limonite, magnetite, ilmenite, hematite, iron, or steel slugs as aggregate.

The most commonly used concrete is normal weight, and this general type is recommended for most passive solar applications. Structural lightweight is appropriate for use as floor fill over structural wood floors.

Concrete can be used in a wide variety of wall, floor, and roof/ceiling applications that make use of either cast-in-place or precast elements. In the single-family, detached residential sector, the use of precast concrete elements is quite uncommon. Cast-in-place concrete, on the other hand, is commonly used for footings, foundation/basement walls, and slabs-on-grade. Although the possible thicknesses of cast-in-place elements in such applications are basically infinite, dimensions are generally as follows:

- Walls—6, 8, 10, or 12''.

- Structural Floors—4-8''.

- Slabs-on-Grade—3-6''.

- Floor Fill—1½-2½''.

- Roofs/Ceilings—Not applicable.

Precast elements also come in a wide variety of possible dimensions. The most common thicknesses are listed below.

- Solid Slabs and Tees—2, 3, 4, 5, 6, and 8″ (tee thickness is for slab portion only, exclusive of stem).
- Hollow-Core Slabs—6, 8, 10, and 12″.

Lengths can vary, as can widths, for solid slabs. Widths for tees and hollow-core slabs are generally 2, 4, or 8′.

Construction

In most passive installations, construction with concrete will follow standard practice. The only major departure from typical construction is placing insulation on the outside of concrete used for Direct Gain storage (e.g., in a solid concrete exterior wall). This change requires only minor modifications to the building process and is fully covered in the construction details in Chapters 3 through 6.

Finishes

In Thermal Storage Wall and Attached Sunspace construction, the exterior face of a concrete wall is used to absorb and store heat and is protected from the environment by the collector glazing. Unfinished concrete of a dark color performs best in these conditions. In Direct Gain systems, however, the exterior face of a solid concrete wall is not protected by the collector glazing. If the concrete is exposed directly to the environment and is not protected by some form of insulation and finish, stored heat will travel through the wall and be lost to the outside, severely reducing the thermal performance of the overall system.

There are a variety of methods for finishing the exterior of a concrete wall, including paint, stain, stucco, and siding. Placing insulation on the *outside* of exterior walls being used for passive solar energy storage requires that such finishes be applied *over* this insulation. Methods of applying these finishes are discussed in Chapter 3.

Exterior finishes will affect the thermal performance of a Direct Gain storage component only insofar as they increase the effective insulation of that component. If a wall that is absorbing heat at its interior face is well insulated on its exterior, it will retain more of the heat energy imparted to it and will thus serve as a more efficient solar energy storage medium.

It is recommended that interior finishes be kept to a minimum in order to preserve the inherent thermal performance capabilities of the concrete itself. For example, covering the interior face of a concrete storage component, such as a Direct Gain wall, with gypsum board or wood paneling will reduce both the amount of solar energy that reaches the wall and the amount that is reradiated from it to the space. This is especially true of covering a concrete-slab floor with wall-to-wall carpet, which is emphatically *not* recommended.

An acceptable range of choices for concrete wall and floor finishes can be found in various ceramic tile materials. Details for the installation of tile over concrete and masonry walls and floors may be found in the *Handbook for Ceramic Tile Installation*, published annually by, and available from, the Tile Council of America, Inc., P.O. Box 326, Princeton, NJ 08540.

A general rule of thumb in assessing the effects that coverings have on thermal performance is that the higher the insulating value of a given finish, the more it will inhibit the passive thermal performance of the component to which it is applied. Relative to the finish materials suggested in the accompanying discussions, gypsum board and paneling have high insulative properties.

Carrying this concept one step further, it is obvious that putting insulation on the interior surface of a concrete component (e.g., a Direct Gain wall) used for energy storage is extremely counterproductive and will essentially nullify any passive benefits that might have been derived from the installation. It is for this reason that it is recommended that no passive solar storage component (wall, floor, or roof/ceiling) be insulated on its interior face.

Special mention must be made concerning finishes for slab-on-grade floors that are used for thermal storage. As previously mentioned, such floors should never be covered with carpet. One alternative to carpet is to stamp the slab, while wet, with any one of a variety of surface textures. If done with proper care, this technique will result in a very handsome floor surface that can be waxed or polished and that will retain its inherent thermal performance capability. Imaginative layouts of the control joints can also help make standard wood-float or steel-trowel finished slabs more attractive. Alternate finished surfaces, such as exposed aggregate, brick or CMU pavers, and tile, mentioned earlier, can also be used effectively.

Color

Where concrete is used both to absorb and to store solar energy, the surface exposed to solar radiation should be as dark a color as possible to increase absorption of light and heat. Other surfaces in a space, such as wood frame walls, which are not being used for storage, should be light in color in order to reflect more light to the darker storage material (for a list of the solar absorptance characteristics of various colors/materials, see Table 7.9).

Colored concrete can be produced by utilizing colored aggregate, by adding color pigments, or both. The concrete can also be stained or painted. It is recommended that paints or stains that do not degrade when subjected to high temperatures be specified in applications where such temperatures are expected (e.g., Thermal Storage Walls).

Another possible finish is the so-called selective surface, a coating that has a very high absorptivity of incoming solar radiation but a very low emissivity of heat. Such coatings can be very effective at increasing the thermal efficiency of passive solar storage components, and manufacturers' data should be referenced for specific performance characteristics (for further information, see Absorber Components earlier in this chapter).

Texture

Exposed concrete can also be textured in a number of different fashions. Patterns can be created by the particular mold employed, or the aggregate can be exposed by a number of different methods, including washing, brushing, using retarders, scrubbing, abrasive blasting, tooling, or grinding. The net result of such techiniques is to expose more surface area for heat exchange with surrounding air.

It might be assumed that such increased exposure would influence the heat storage performance of components using these textured concretes, but this has not, as yet, been proven to be true. It is therefore recommended that textured and unfinished concrete be used on the basis of aesthetic preference in passive solar storage applications.

Thermal Properties

From the point of view of thermal performance in a passive solar application, the most important material characteristics of a particular concrete mixture are its conductivity, density, specific heat, and thickness.

A material's conductivity (k) is a measure of how readily the material will transfer or "conduct" heat: the higher the conductivity, the faster the heat will travel through a material. In general, thermal performance increases as the conductivity increases.

The remaining three basic material characteristics (density, specific heat, and thickness) can also be used as indicators of anticipated thermal performance. When multiplied together, they yield the so-called "thermal storage capacity" of a specific unit of material. Essentially, this characteristic is a measure of the ability of a given unit of material to store heat. The higher the thermal storage capacity, the greater the amount of heat the unit can absorb and store.

In the case of concrete, the specific heat for the majority of mixtures and densities commonly used in residential construction varies only slightly. Thus, the major variables that will determine the thermal storage capacity of a concrete component are its density (measured in pounds per cubic foot) and its thickness.

As discussed above, concrete can be divided into four broad density classifications:

- Insulating Lightweight—15-90 lb/ft^3.
- Structural Lightweight—85-115 lb/ft^3.
- Normal Weight—135-160 lb/ft^3.
- Heavyweight—160-400 lb/ft^3.

Tables 7.12 and 7.13 present thermal performance data, including thermal storage capacities, for a variety of concrete densities and thicknesses. The tables can be used for general performance comparisons between various thickness/density combinations, or specific information can be extracted for use with the Analysis Methods discussed in detail in Appendix C.

The higher densities for heavyweight concrete are not shown, since this is not a very common building material and is principally used for radiation shielding. Its heavier density will improve its thermal performance in a passive solar application, but the exact extent of this improvement is not known. If use of heavyweight concrete is anticipated, manufacturers' specifications should be referenced, where possible, for specific thermal performance data.

In general, the higher the thermal storage capacity of a given component, the better its performance from a passive solar point of view. This should be kept in mind when selecting a concrete mixture for a specific installation. For example, if an 8" concrete wall is desired for a particular passive installation, the density that results in the highest possible thermal storage

(continued on page 226)

Concrete: Physical Properties Table 7.12

Unit Thickness	Unit Density∞	Specific Heatθ	Thermal Storage CapacityΣ	R-ValueΔ	ConductivityΩ
2"	20	0.24	0.9	2.86	0.70
	50	0.23	1.9	1.34	1.50
	70	0.22	2.6	0.90	2.20
	90	0.22	3.3	0.60	3.30
	100	0.22	3.6	0.48	4.20
	110	0.21	3.9	0.38	5.30
	130	0.21	4.6	0.22	9.10
	145	0.21	5.0	0.15	13.30
3"	20	0.24	1.2	4.29	0.70
	50	0.23	2.9	2.01	1.50
	70	0.22	3.9	1.35	2.20
	90	0.22	4.9	0.90	3.30
	100	0.22	5.5	0.72	4.20
	110	0.21	5.8	0.57	5.30
	130	0.21	6.8	0.33	9.10
	145	0.21	7.6	0.23	13.30
4"	20	0.24	1.6	5.72	0.70
	50	0.23	3.9	2.68	1.50
	70	0.22	5.2	1.80	2.20
	90	0.22	6.6	1.20	3.30
	100	0.22	7.3	0.96	4.20
	110	0.21	7.6	0.76	5.30
	130	0.21	9.1	0.44	9.10
	145	0.21	10.2	0.30	13.30
5"	20	0.24	2.0	7.15	0.70
	50	0.23	4.8	3.35	1.50
	70	0.22	6.5	2.25	2.20
	90	0.22	8.2	1.50	3.30
	100	0.22	9.2	1.20	4.20
	110	0.21	9.7	0.95	5.30
	130	0.21	11.4	0.55	9.10
	145	0.21	12.7	0.38	13.30

∞ - Units of lb/ft^3
θ - Units of btu/lb/°F
Σ - Units of btu/ft^2/°F
Δ - Units of hr °F/ft^2/btu
Ω - Units of btu in/hr/°F/ft^2

Concrete: Physical Properties — Table 7.13

Unit Thickness	Unit Density∞	Specific Heatθ	Thermal Storage CapacityΣ	R-ValueΔ	ConductivityΩ
6"	20	0.24	2.5	8.58	0.70
	50	0.23	5.8	4.02	1.50
	70	0.22	7.6	2.70	2.20
	90	0.22	9.9	1.80	3.30
	100	0.22	10.9	1.44	4.20
	110	0 21	11.5	1.14	5.30
	130	0.21	13.7	0.66	9.10
	145	0.21	15.3	0.45	13.30
8"	20	0.24	3.2	11.44	0.70
	50	0.23	7.6	5.36	1.50
	70	0.22	10.2	3.60	2.20
	90	0.22	13.3	2.40	3.30
	100	0.22	14.7	1.92	4.20
	110	0.21	15.4	1.52	5.30
	130	0.21	18.1	0.88	9.10
	145	0.21	20.3	0.60	13.30
10"	20	0.24	4.0	14.30	0.70
	50	0.23	9.7	6.70	1.50
	70	0.22	13.4	4.50	2.20
	90	0.22	16.6	3.00	3.30
	100	0.22	18.3	2.40	4.20
	110	0.21	19.8	1.90	5.30
	130	0.21	22.8	1.10	9.10
	145	0.21	25.3	0.75	13.30
12"	20	0.24	4.8	17.16	0.70
	50	0.23	11.5	8.04	1.50
	70	0.22	15.4	5.40	2.20
	90	0.22	19.9	3.60	3.30
	100	0.22	22.0	2.88	4.20
	110	0.21	23.0	2.28	5.30
	130	0.21	27.4	1.32	9.10
	145	0.21	30.4	0.90	13.30

∞ - Units of lb/ft^3
θ - Units of btu/lb/°F
Σ - Units of $btu/ft^2/°F$
Δ - Units of $hr\ °F/ft^2/btu$
Ω - Units of $btu\ in/hr/°F/ft^2$

capacity should generally be specified (subject to cost considerations).

It can also be assumed that the thermal storage capacities for thicknesses greater than 12'' will roughly equal the sum of capacities of smaller thicknesses, which, when added together, will equal the total overall dimension desired. For example, the thermal storage capacity per square foot of a 20''-thick wall will roughly equal the sum of the capacities of one 8'' and one 12'' wall (14.7 + 22.0 = 36.7) or of two 10'' walls (18.3+ 18.3 = 36.6).

Brick

A brick is generally defined as a small building unit, solid or cored, formed from clay or shale into a rectangle, and hardened by heat. A brick is referred to as "solid" when the net cross-sectional area is 75% or more of its gross cross-sectional area. Most "cored" units are in this category. A brick is referred to as "hollow" when the net cross-sectional area is less than 75% but more than 60% of its gross cross-sectional area. It is recommended that all hollow bricks used for passive solar storage be fully grouted in addition to regular mortaring.

These general categorizations comprise a number of specific brick types that differ from each other in terms of texture, color, degree of water absorption, strength, and durability. These differences, however, do not significantly affect the thermal performance characteristics of the majority of brick types.

Brick should be specified by the appropriate American Society for Testing and Materials (ASTM) Standards:

- **ASTM C-216**—Standard specification for facing brick (solid masonry units made from clay or shale).
- **ASTM C-62**—Standard specification for building brick (solid masonry units made from clay or shale).
- **ASTM C-652**—Standard specification for hollow brick (hollow masonry units made from clay or shale).
- **ASTM C-902**—Standard specification for pedestrian and light traffic paving brick.
- **ASTM C-126**—Standard specification for ceramic glazed structural clay facing tile, facing brick, and solid masonry units.

Brick construction is based on a modular system of dimensions for all standard products. Table 7.14 lists some of the typical industry sizes. The dimensions of the units listed in the table are "nominal." Actual dimensions are equal to the nominal dimension minus the thickness of the mortar joint with which the unit is designed to be laid. In general, such mortar joints are either ⅜" or ½" thick. Thus, a brick that is nominally 4" × 2¾" × 8" can be 3½" × 2¼" × 7½" or 3⅝" × 2¼" × 7⅝", depending on the specifications for a particular application. Note that the vertical dimension does not always change, and the difference in mortar joint size is made up in the coursing of the brick as it is constructed.

Although some solid brick units are manufactured without cores, most bricks are cored to some degree to reduce unit weight and to improve firing, resulting in a denser and more durable product.

Brick: Typical Sizes — Table 7.14

Nominal Dimensions t	h	l	Joint Thickness	Manufactured Dimensions t	h	l	Modular Coursing*
4"	4"	12"	3/8"	3-5/8"	3-5/8"	11-5/8"	1C=4"
4"	4"	12"	1/2"	3-1/2"	3-1/2"	11-1/2"	1C=4"
6"	4"	12"	3/8"	5-5/8"	3-5/8"	11-5/8"	1C=4"
6"	4"	12"	1/2"	5-1/2"	3-1/2"	11-1/2"	1C=4"
8"	4"	12"	3/8"	7-5/8"	3-5/8"	11-5/8"	1C=4"
8"	4"	12"	1/2"	7-1/2"	3-1/2"	11-1/2"	1C=4"
12"	4"	12"	3/8"	11-5/8"	3-5/8"	11-5/8"	1C=4"
12"	4"	12"	1/2"	11-1/2"	3-1/2"	11-1/2"	1C=4"
6"	5-1/3"	12"	3/8"	5-5/8"	4-7/8"	11-5/8"	3C=16"
6"	5-1/3"	12"	1/2"	5-1/2"	4-3/4"	11-1/2"	3C=16"
8"	5-1/3"	12"	3/8"	7-5/8"	4-7/8"	11-5/8"	3C=16"
8"	5-1/3"	12"	1/2"	7-1/2"	4-3/4"	11-1/2"	3C=16"
4"	8"	8"	3/8"	3-5/8"	7-7/8"	7-7/8"	1C=8"
4"	8"	8"	1/2"	3-1/2"	7-1/2"	7-1/2"	1C=8"

t thickness
h height
l length

* Dimensions are for the number of courses indicated (e.g. 1C = 1 course).

The cores in solid brick units will not significantly affect the thermal performance of brick masonry used in passive solar storage applications. The process of laying brick usually results in the cores being filled with mortar, and thus the thermal storage is not significantly different from comparable solid units. However, when using hollow brick units as the thermal storage material, large cores should be grouted.

Illustrated in Figures 7.14 and 7.15 are some of the representative sizes and shapes of bricks currently available.

Construction

In most passive installations, construction with brick will follow standard practice. The only major departure from typical construction is placing insulation on the *outside* of a brick being used for Direct Gain storage (e.g., in a solid brick exterior wall). This and other changes require only minor modifications to the building process and are fully covered in the details illustrated in Chapters 3 through 5.

Finishes

In Thermal Storage Wall and Attached Sunspace construction, the exterior face of a brick wall is used to absorb and store heat and is protected from the environment by the collector glazing. Unfinished brick of a dark color performs best in these conditions. In Direct Gain systems, however, the exterior face of a solid brick

(continued on page 230)

Brick: Typical Shapes **Figure 7.14**

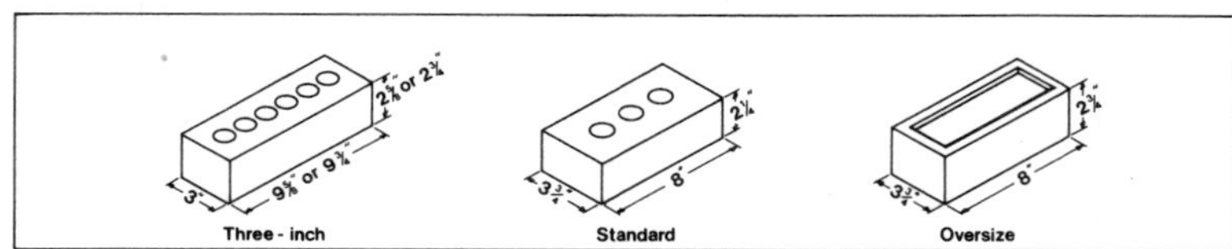

• NON-MODULAR SOLID BRICK • (Actual Dimensions)

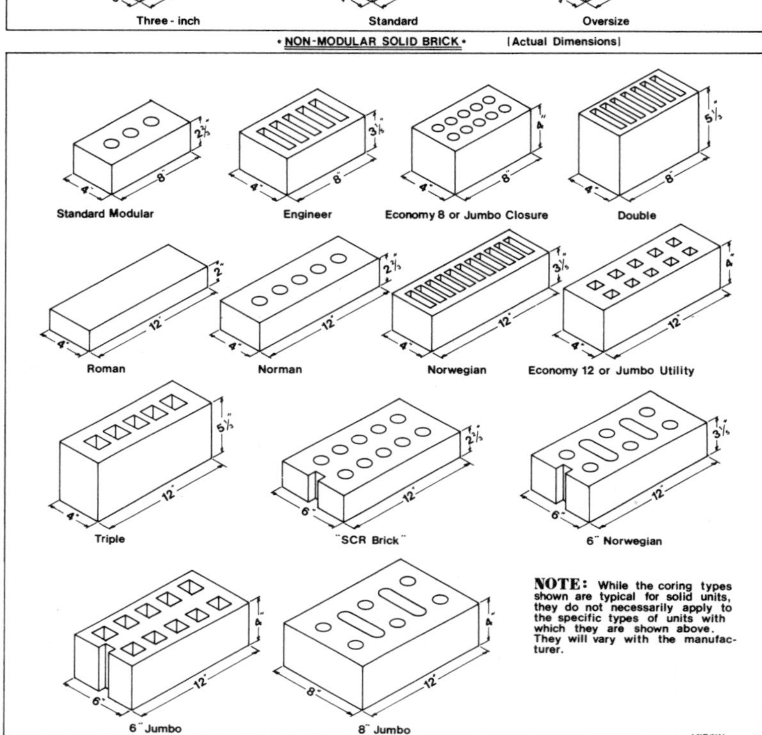

• MODULAR SOLID BRICK • (Nominal Dimensions)

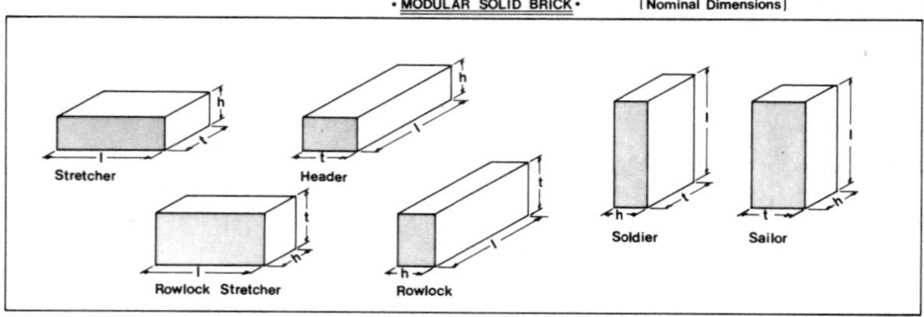

• BRICK POSITIONS IN WALL •

Brick: Typical Shapes **Figure 7.15**

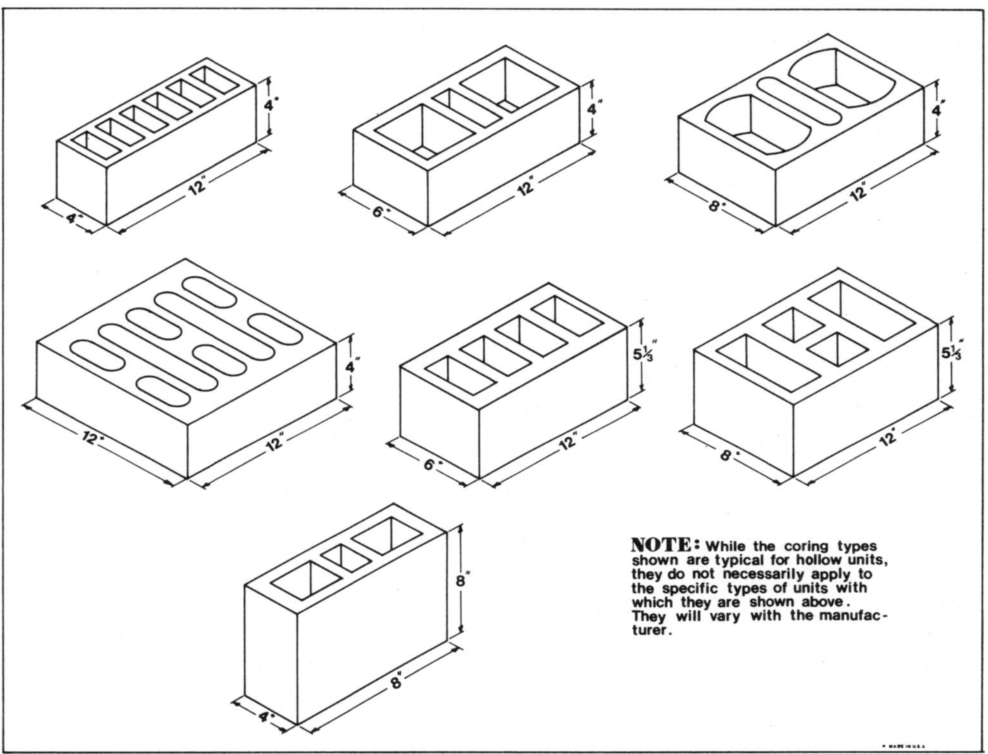

* MODULAR HOLLOW BRICK * (Nominal Dimensions)

* MODULAR PAVING BRICK * (Nominal Dimensions)

wall is not protected by the collector glazing. If this exterior face is exposed directly to the environment and is not protected by some form of insulation and finish, stored heat will travel through it and be lost to the outside, severely reducing the thermal performance of the overall system.

There are a variety of methods for finishing the exterior of a brick wall being used for Direct Gain storage. Because it is recommended that insulation be placed on the *outside* of exterior walls being used for Direct Gain storage, such finishes will need to be applied *over* this insulation. In a standard masonry bearing wall configuration, stucco or siding can be applied over insulation that has been adhered to the brick wall. In typical brick veneer wall construction, the configuration of the materials can be inverted, moving the masonry to the inside for heat storage and the insulated bearing frame wall to the outside. Siding can then be applied in the conventional manner. Neither of these configurations, however, provides a brick appearance on the outside of the house. In order to obtain this appearance while maintaining the passive solar performance characteristics of interior brick masonry, a cavity wall can be specified. If well insulated, such a wall will provide both the aesthetic appeal of brick on the exterior (exterior wythe) and the heat storage capacity of brick on the interior (interior wythe). The cavity insulation will ensure that stored heat will not be lost, while still providing a brick appearance on the exterior.

It is recommended that interior finishes be avoided in order to preserve the inherent thermal performance capabilities of the brick itself. For example, if a finish such as wood paneling or gypsum board is applied to the interior face of a brick wall intended to absorb and store solar energy, it will not only hide the brick but will reduce both the amount of heat reaching the wall and the amount reradiated from the wall back to the space. This is especially true of covering a brick paver floor with wall-to-wall carpet, which is emphatically *not* recommended. Although the exact negative effects of such coverings have not been quantified, a good rule of thumb is to assume that, relative to other finish materials, the higher the insulating value of a given finish, the more it will inhibit the passive thermal performance of the material to which it is applied.

Carrying this concept one step further, it is obvious that putting insulation on the interior surface of a brick component (e.g., Direct Gain wall) used for energy storage is extremely counterproductive and essentially nullifies any passive benefits that might have been derived from the component. It is for this reason that it is recommended that no passive solar storage component (wall, floor, or roof/ceiling) be insulated on its interior face.

Color

Where brick is used both to absorb and to store solar energy, the surface exposed to solar radiation should be as dark a color as possible to increase absorption of light and heat. Other surfaces, such as standard frame walls, that are not being used for storage should be light in color in order to reflect as much sunlight as possible onto the darker storage material.

Although black is the most desirable storage material color from a thermal performance point of view, certain of the darker natural brick colors (browns, blues, and reds) will perform almost as effectively. Using the exposed natural brick will avoid maintenance problems that might result from the use of paint (for a listing of the solar absorptance characteristics of various colors/materials, see Table 7.9).

Brick with a glazed ceramic coating should be avoided, since the finish will reflect some of the solar radiation striking it. Many brick manufacturers, however, can supply brick with dull black ceramic glazed faces that perform as well as the dull finishes on natural colored brick.

Another possible finish is the so-called selective surface, a coating that has a very high absorptivity of incoming solar radiation but a very low emissivity of heat. Such coatings can be very effective at increasing the thermal efficiency of passive solar components, and manufacturers' data should be referenced for specific performance characteristics (for further information, see Absorber Components earlier in this chapter).

Texture

Although it would seem at first glance that rough-textured brick, by providing more surface area for the collection of energy, would be more effective than smooth brick for energy storage, this is not the case. It appears that brick texture does not have a major impact on the performance of passive solar installations, and any desired texture can be used without significant loss or gain in effectiveness.

Thermal Properties

In terms of thermal performance, the most important material characteristics of a brick unit are its conductivity, density, specific heat, and thickness.

A material's conductivity (k) is a measure of how readily the material will transfer or "conduct" heat. The higher the conductivity, the faster the heat will travel through a material. In general, thermal performance increases as the conductivity increases.

The remaining three basic material characteristics (density, specific heat, and thickness) can also be used as indicators of anticipated thermal performance. When multiplied together, they yield the so-called "thermal storage capacity" of a specific unit of material. Essentially, this characteristic is a measure of the ability of a given unit of material to store heat. The higher the thermal storage capacity, the greater the amount of heat the unit can absorb and store.

In the case of brick, the specific heat for the majority of units commonly used in residential construction remains basically constant. Thus, the major variables that will determine the thermal storage capacity of a brick unit are its density (measured in pounds per cubic foot), which varies only slightly, and its thickness.

Table 7.15 presents thermal performance data, including thermal storage capacities, for a variety of bricks with different thicknesses and two different densities. The table can be used for general performance comparisons between the various brick types presented, or specific information can be extracted for use with the Analysis Methods discussed in Appendix C.

In general, the higher the thermal storage capacity of a given unit, the better its performance in a passive solar application. This should be kept in mind when selecting a specific brick for use as a storage component. It is evident from Table 7.15 that thicker bricks will have higher thermal storage capacities and, within the rule of thumb guidelines set out in Chapters 3 through 6, will generally perform better than thinner units.

It can also be assumed that the thermal storage capacities for thicknesses greater than 16" will roughly equal the sum of the capacities of smaller units which, when added together, will equal the overall thickness desired. For example, the thermal capacity per square foot of a 20"-thick wall of facing brick will roughly equal the sum of the capacities of one 8" and one 12" wall (18.3 + 27.9 = 46.2) or of two 10" walls (23.1 + 23.1 = 46.2).

Mention should be made of the potential use of magnesia brick as a passive storage medium. Magnesia brick exhibits much higher thermal conductivities than clay and shale brick. It is not, however, markedly superior in its capacity to store heat. Rather, it is capable of delivering more heat to the interior of a space and insulating against losses from the space. Such a

Brick: Physical Properties — Table 7.15

Brick Type[1]	Unit Thickness	Unit Density∞	Specific Heat[2] θ	Thermal Storage Capacity[3] Σ	R-ValueΔ	ConductivityΩ
Paver	1-5/8"	120	0.24	3.90	0.36	5.00
Paver		130	0.24	4.23	0.20	9.00
Paver	2-1/4"	120	0.24	5.40	0.50	5.00
Paver		130	0.24	5.85	0.27	9.00
Common	4"	120	0.24	8.70	0.80	5.00
Facing		130	0.24	9.43	0.44	9.00
Common	6"	120	0.24	13.50	1.20	5.00
Facing		130	0.24	14.63	0.66	9.00
Common	8"	120	0.24	18.30	1.60	5.00
Facing		130	0.24	19.83	0.88	9.00
Common	10"	120	0.24	23.10	2.00	5.00
Facing		130	0.24	25.03	1.10	9.00
Common	12"	120	0.24	27.90	2.40	5.00
Facing		130	0.24	30.23	1.32	9.00
Common	16"	120	0.24	37.50	3.20	5.00
Facing		130	0.24	40.63	1.76	9.00

1 When using hollow units grouted, use values for Common Brick. When using paving units, use values for Facing Brick.
2 Specific heat may vary from 0.20 to 0.26; 0.24 is typical.
3 Thermal Storage Capacities are for actual dimensions of units listed.

∞ - Units of lb/ft^3
θ - Units of btu/lb/°F
Σ - Units of $btu/ft^2/°F$
Δ - Units of hr °F ft^2/btu
Ω - Units of btu in/hr/°F/ft^2

capability can be useful in situations where relatively rapid heat transfer is desired, but it can also have the drawback of causing a space to overheat during sunny days. Because magnesia brick is also significantly more expensive than common brick, its advantages and disadvantages should be carefully weighed before it is substituted for common brick in passive solar applications.

It must also be noted that little documentation is available regarding the thermal performance of the various types of mortar that can be used in brick construction. Therefore, when conducting a thermal performance analysis, it is recommended that a brick installation be considered solid brick without any mortar joints at all. This is equivalent to assuming that the mortar will exhibit the same thermal performance characteristics as the brick itself.

Water

Within the same volume, water can store more than twice as much heat as masonry. Thus, to store a given amount of heat, less than one-half the volume and approximately one-fifth the weight of water will be required compared to a storage system utilizing masonry materials.

When evaluating water as a storage material for a passive solar system, however, two aspects of its use must be considered and evaluated. First, the use of water requires that it must be contained, and the cost of the container, although generally inexpensive, will need to be considered. Second, unlikely masonry materials, which can also provide structural support, water storage containers are rarely designed as structural elements of the home and frequently require framing for bracing.

Although water containers, due to their high storage capacity, can be narrower than masonry, the specification, design, and placement of the containers must be carefully planned and constructed.

Site assembly, however, is almost always faster and easier than for masonry, since the containers are prefabricated and are quickly filled on site.

Construction

As with all storage materials, water storage for passive solar systems should be located in direct sunlight. To function efficiently, the water storage container should present a large amount of surface area both to the sun, for absorption of heat, and to the living space, for release of the heat.

In general, a wide range of off-the-shelf water container types specifically designed to be used in passive systems are available. Of course, appropriate products are not limited to those advertised for use as thermal storage, although a few basic characteristics, discussed below, will need to be considered before any product is specified.

Containers

A variety of container types and designs are commercially available. Some containers are designed to be left uncovered, their surface providing the interior finish. Most containers, however, are not intended to be left exposed and may be enclosed in walls or incorporated into features such as planters or window seats. The less expensive the container, the more likely it is to be of this concealed type. Care must be taken, when containers are concealed, that they not be inadvertently punctured (for example, by nails in hanging artwork).

Unfortunately, covering the containers will degrade the system's performance. One of the most common applications in which the containers are concealed is in a Thermal Storage Wall. Where a frame wall is constructed to cover the containers, the wall sheathing, typically gypsum board, will need to contain vents at both the bottom and the top of the wall to allow heat flow into the adjacent living space.

Regardless of the type of container chosen for water storage, the water itself will need to be chemically treated to prohibit the growth of algae or other organisms within the container. In addition, all metal containers should be coated either by the manufacturer or at the site to prevent rust or corrosion.

The most widely used containers are tubes, culverts, and drums, which can be used alone or in combination (see Figure 7.16). All have inherent advantages and disadvantages. The following provides a brief description of each of the major container types.

Tubes—Fiberglass tubes are manufactured for the specific purpose of water thermal storage. They are often left exposed and freestanding for functional and aesthetic reasons. Before filling, the tubes are lightweight and easy to install but must be handled carefully to prevent cracking or puncturing. Fiberglass tubes have the advantage of being noncorrosive and are available in sizes ranging from 7-18" in diameter, and 4-10' in height.

Metal Culverts—Culvert pipe, available from pipe supply houses, can be used for water storage. Culverts can be cut to any height, providing more design flexibility than is possible with fiberglass tubes. Further, culverts are sturdier than fiberglass and less subject to damage, although their heavier weight will need to be considered in sizing the supporting foundation.

Culverts can be painted but should first be treated with two coats of zinc chromate primer. Typically, a plate is welded to the bottom and a covering cap is fabricated from sheet steel. Culverts must be treated to resist corrosion, and it is generally recommended that a plastic liner bag be installed to further protect against leaks. Culverts are generally available in diameters of 12" and larger.

Metal Drums—Thirty- and 50-gal metal drums can be used as water storage containers. However, their use is recommended in applications where their industrial appearance is not objectionable. Drums are generally low in cost, but they are difficult to clean and may need to be treated to prevent corrosion.

If drums are to be specified, be sure to check on the former contents to ensure against toxic reactions. Drums are available from chemical supply companies and/or drum manufacturers.

Plastic Drums—Five-gal high-density polyethylene drums are available for use as water storage containers. These are self-supporting and will stack safely. Most plastic drums are constructed with a finish increasing their solar heat absorptivity and are not affected by corrosive action. The drums are easy to handle, but because of their small size, a large number may be required to provide sufficient storage area.

Tanks—Metal tanks are available off-the-shelf, or they can be custom-built. Such tanks may be incorporated into the exte-

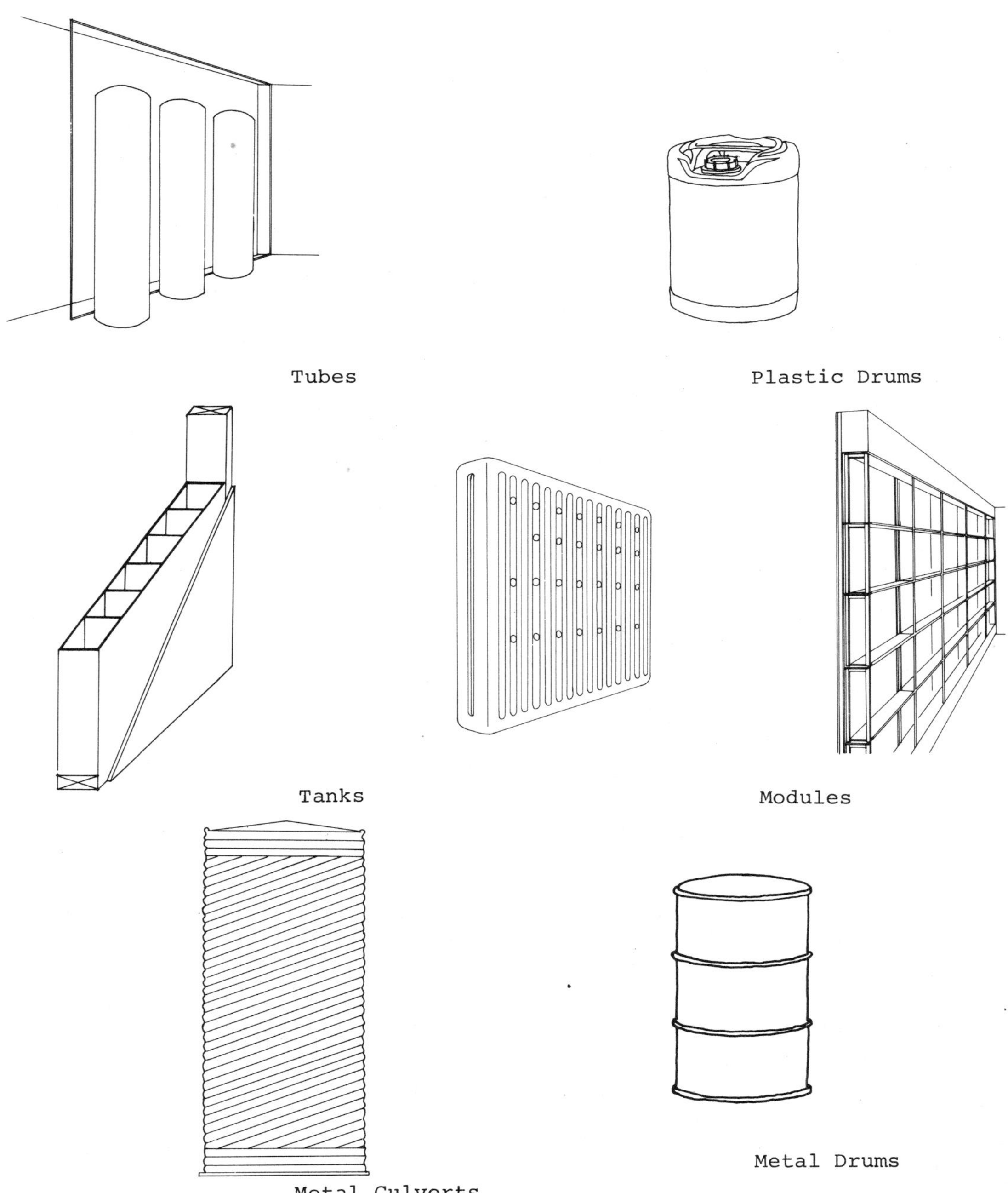
Tubes
Plastic Drums
Tanks
Modules
Metal Culverts
Metal Drums

Figure 7.16

rior wall framing system and covered with standard finish materials. Metal tanks must be treated to prevent corrosion, and it is generally recommended that a plastic bag be used to line the tank. If custom-made, be certain that all welds are continuous to prevent leaks. Prefabricated tank systems should be installed per manufacturers' instructions.

Modules—A variety of modular water storage systems are available that employ several different materials: plastic, metal, and glass. Such systems are typically manufactured to fit 2″ × 4″, 2″ × 6″, or 2″ × 8″ conventional framing systems. Each system design should be examined on its own merits, with particular attention to design flexibility, aesthetics, cost, and thermal performance.

Color

When water is used both to absorb and to store solar energy, the surface exposed to solar radiation should be as dark a color as possible to increase absorption of light and heat. Other surfaces in the space, such as wood frame walls, that are not being used for storage should be light in color in order to reflect more light to the darker storage material (for a listing of the solar absorptance characteristics of various colors and materials, see Table 7.9).

Color can be applied to water storage systems in two ways. In the first, and most common, the container surface can be painted, or the container can be ordered in a dark color. Plastic containers can be ordered in dark colors, while metal is often painted in the field with a suitable primer applied first. An effective alternative involves specifying a translucent or transparent container material and treating the water with a dark dye.

Water containers can also be finished with a selective surface, a coating that has a very high solar absorptivity. Such coatings can be very effective in increasing the thermal efficiency of the passive solar storage component (for further information, see Absorber Components earlier in this chapter). In the case of water storage systems, such a surface should be placed on the side of the container that faces the sun.

Texture

The texture of the water storage container can vary, depending on the material specified. Most metal or plastic containers are smooth. In general, however, the surface texture will not have a significant impact on performance.

Thermal Properties

From the point of view of thermal performance in a passive solar application, the most important material characteristics of a particular water storage system are its conductivity, density, specific heat, and thickness.

A material's conductivity (k) is a measure of how readily the material will transfer or "conduct" heat. The higher the conductivity, the faster the heat will travel through a material. In general, thermal performance increases as the conductivity increases.

The remaining three basic material characteristics (density, specific heat, and thickness) can also be used as an indication of anticipated thermal performance. When multiplied together, they yield the so-called "thermal storage capacity" of a specific unit of material. Essentially, this characteristic is a measure of the ability of a given volume of water to store heat.

The higher the thermal storage capacity, the greater the amount of heat that can be absorbed and stored.

In the case of water, the specific heat and the density are constant. Therefore, the only variable that will influence thermal storage capacity is the thickness. The performance of a water storage system will increase and temperature fluctuations within the living space will decrease as thickness increases.

Table 7.16 presents thermal performance data for a variety of water container thicknesses. The thicknesses listed are for rectangular containers, and the values will need to be adjusted for cylindrical containers.

Comparing the thermal storage capacity values in Table 7.16 with corresponding values for masonry materials will confirm the use of water as a suitable storage material. As stated earlier, the storage capacity per unit of thickness far exceeds that of masonry materials, a factor that should be seriously considered in designing the storage component.

Phase-Change Materials

A new and potentially highly effective, lightweight storage material appropriate for use in passive solar systems is a group of chemical compounds known as phase-change materials (PCMs).

The storage materials previously discussed in this section, including masonry and water, store and release heat by changing their temperature. Phase-change materials, on the other hand, store and release most of their collected heat by changing form or "phase," usually from solid to liquid. In effect, when heated by sunlight, the PCM, which is solid, melts without significantly changing in temperature. Later, when heat is required in the living spaces, the now-liquid PCM begins to solidify, giving off heat to the space.

Water: Physical Properties — **Table 7.16**

Unit Thickness	Unit Density[1]	Specific Heat[2]	Thermal Storage Capacity[3]
4″	62.32	1.0	20.8
5″	62.32	1.0	26.0
6″	62.32	1.0	31.2
7″	62.32	1.0	36.4
8″	62.32	1.0	41.6
9″	62.32	1.0	46.8
10″	62.32	1.0	52.0
11″	62.32	1.0	57.2
12″	62.32	1.0	62.4
13″	62.32	1.0	67.6
14″	62.32	1.0	72.8
15″	62.32	1.0	78.0
16″	62.32	1.0	83.2
17″	62.32	1.0	88.4
18″	62.32	1.0	93.6

1-Units of lb/ft^3
2-Units of $Btu/lb/°F$
3-Units of $Btu/ft^2/°F$

Various PCM products require different amounts of heat to cause this change of state and, depending on the precise chemical makeup, will change phase at different temperatures. Ideally, a phase-change material will have a capacity to store a great deal of heat as it changes in phase, resulting in a storage component more compact and lighter than that possible with other storage materials.

One of the major concerns with the use of phase-change materials is the question of degradation of performance over time. At present, few nationally recognized standards for evaluating PCM products and materials exist, and long-term performance, in actual residences, is not well known. Therefore, specific products should be carefully evaluated before being specified, and builders who have experience with the use of PCMs should be consulted.

Construction

Since PCMs are available in a variety of shapes and forms, some designed for specific types of applications, it is difficult to generalize about their construction. However, they are all lightweight, which means that they can be installed in a wide range of locations. For example, products are available that require little or no increase in conventional framing requirements. In addition, most products are thin, minimizing the amount of space in the home required for the storage component.

Containers

PCMs are packaged in a variety of container types, often designed for particular and unique applications. All containers, however, must be leakproof, easily transportable, and easy to install.

While many products are designed to be left exposed as the finish material, some will need to be covered for aesthetic reasons. Where PCMs are concealed, air is circulated over the containers to distribute heat to the living spaces. Concealing PCMs degrades the system performance and should be avoided if possible.

The following are brief descriptions of the most common container types. Illustrations appear in Figure 7.17.

Pods—Pods can be mounted in a conventional window frame, hung with clips from framing members, or mounted in a frame provided by a PCM manufacturer. Pods are available in 24″ and 48″ widths. They are typically translucent but can be ordered with a black coating to increase absorptivity. A typical 16″ × 24″ × 2″ pod weighs 15 lb.

Cans—Cans are available in diameters of about 4″, enabling them to be used within the wall, ceiling, and floor cavity using, at maximum, 2″ × 6″ framing. Care must be taken to allow for the small added load in ceiling and wall applications. A 4″-diameter can 7″ high weighs approximately 4½ lb.

Packets—Packets are usually made of aluminum foil and are placed in concealed locations, either underfloor or in ceiling spaces. They can also be stacked in racks in a specially designed storage compartment or room. A typical-sized packet is 12″ × 24″ × ¾″ and weighs 10 lb.

Rods—Rods may be installed vertically or horizontally, although for most products, vertical installations are not recommended. Installation may be concealed or exposed, and rods can be opaque or translucent. Rods are available in varying lengths from 2-6′ and from 3½-4¾″ in diameter. A typical 3½″-diameter, 6′-long rod weighs 35 lb.

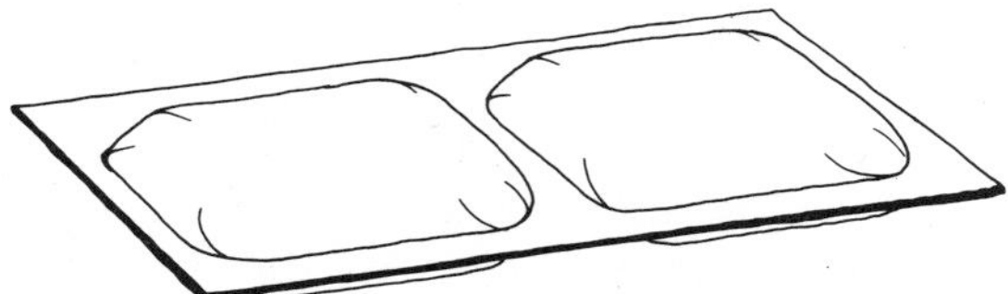

Packets

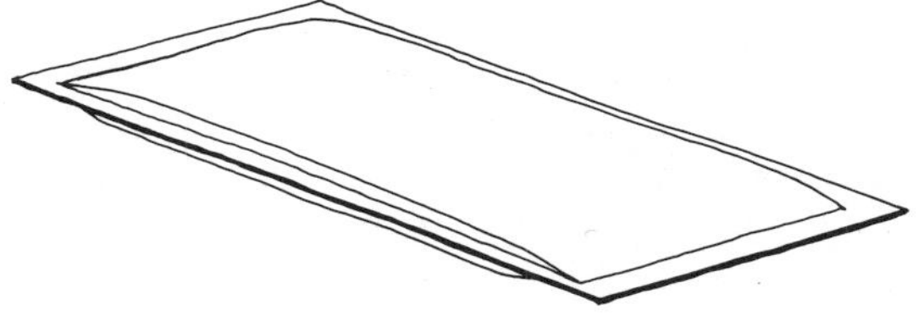

Pods

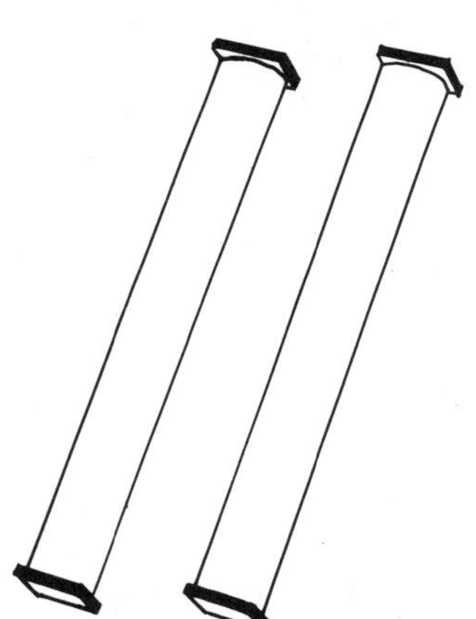

Rods

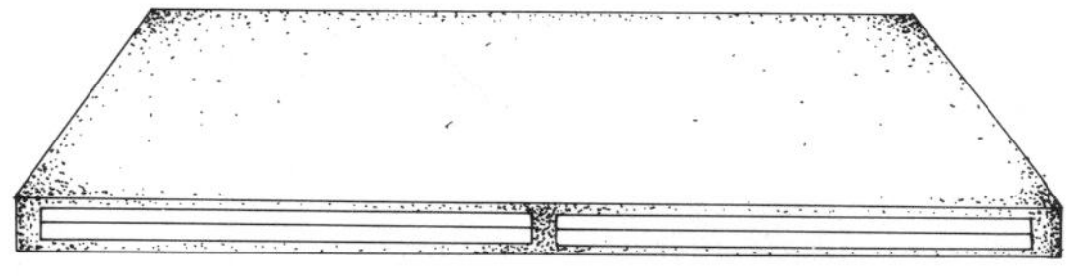

Tiles

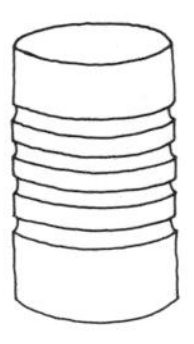

Cans

Figure 7.17

Tiles—PCMs are manufactured in fiberglass or reinforced polymer concrete, which is suitable as a floor material or ceiling tile. Tiles are available in a variety of finishes. Care must be taken to provide sufficient support in both ceiling and floor applications, as each tile, typically 2′×2′×1¼″, weighs approximately 43 lb.

Thermal Properties

If compared on an equal volume basis, PCMs are significantly more thermally effective than either masonry or water.

In addition, the thermal performance of PCMs will depend upon the type of chemical compound used. The most common PCM compounds are listed in Table 7.17. Each compound has a characteristic phase-change temperature defined as the point at which the compound changes from solid to liquid and has a latent heat value indicating the amount of heat that can be stored in 1 lb of material. The type of compound used should not be the only consideration in evaluating the relative merits of alternate products. To assist in product selection, a few of the major PCM manufacturers and characteristics of their products are listed in Table 7.18.

Properties of Phase-Change Materials **Table 7.17**

Chemical Compound	Phase-Change Temperature (°F)	Latent Heat (Btu/lb)
Sodium Thisulfate	116	90
Calcium Chloride Hexahydrate	81	75
Sodium Sulfate Decahydrate (Glauber's Salt)	89	129

Rock Storage

In many passive designs, it may be preferable to locate the storage component out of direct sunlight, remote from both collector and absorber. Remote storage has the advantage of added design flexibility, since it can be placed virtually anywhere in the home. The major disadvantages with remote storage are the necessity for a more elaborate distribution system and a slightly lower thermal efficiency than storage placed in direct sunlight.

Rocks, the most popular remote storage material, are laid in a bed usually under, or adjacent to, the space to be heated. Rock-bed storage is frequently used with Convective Loop systems that require a storage component and occasionally with Attached Sunspaces (particularly with air exchange subsystem designs).

There are two basic types of rock-bed storage systems. In the first type, the rocks are heated by air circulated between absorber and storage. When heat is required, air is circulated between storage and living spaces with the rocks releasing their heat into the airstream. Since storage is completely isolated from the living spaces, heat delivery in this type of rock storage, commonly called active discharge rock-bed storage, is by warm air only.

In the second type of rock-bed storage, referred to as radiant release, the rocks are heated in the same manner as in active discharge storage. However, delivery of heat to the living spaces is by radiation from the rock bed. In other words, the rock bed is *not* thermally isolated from the living spaces.

In most rock-bed designs, the flow of air is in one direction, with inlet and outlet plenums provided to distribute air equally throughout the bed (see Figure 7.18).

In addition, airflow through the rock bed can be in either the vertical or the horizontal direction, depending upon the relative location of rock bed and living space. Radiant release rock beds are almost always of the horizontal feed type (see Figures 7.19 and 7.20). It is, therefore, necessary for the rock bed to be located next to a slab floor or wall for this radiant release to work properly.

Construction

No matter what type of rock-bed design is selected, it should be well insulated. Depending on the passive system type, heat delivered to the rock bed may be at a low temperature (80-90°F), and care must be taken to ensure that heat losses are minimized. Also, the rock-bed air inlets and outlets should be designed to expose as much of the rock-bed area to moving air as possible.

The length of travel of the air across the bed will influence its performance and will require careful designing. If the travel path is too short, available heat in the moving air will not have enough opportunity to be transferred to the rocks during the charging cycle. If the path of travel is too long, the power requirements for the fans may be excessive.

The rock bed is, in fact, simply a well-insulated container that must be structurally secure to resist the outward pressure of the rocks. The structural considerations usually dictate that the container be placed below grade with the soil providing sufficient back pressure on the container.

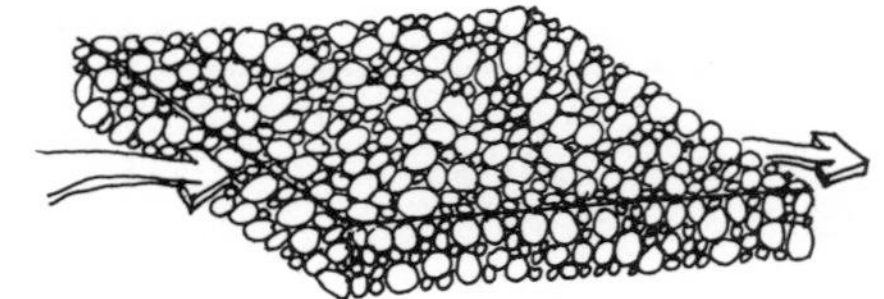

Figure 7.18

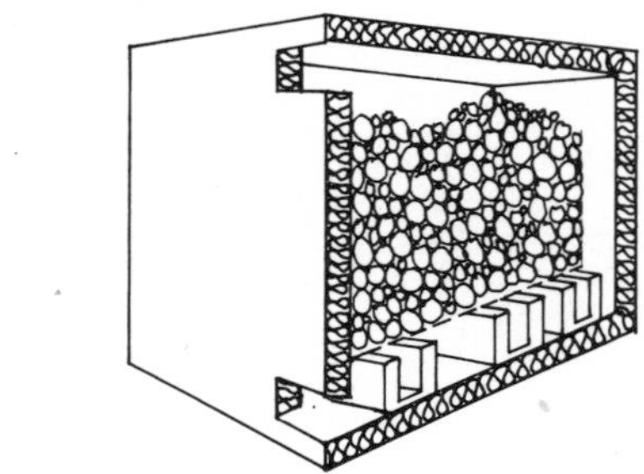

Figure 7.19

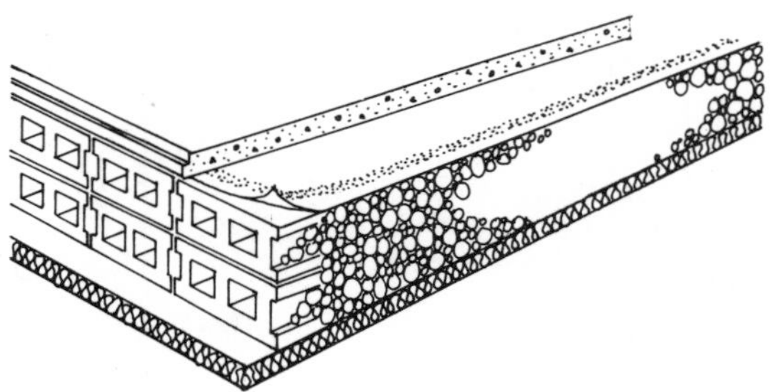

Figure 7.20

In some instances, however, vertical rock beds have been employed that double as an interior partition (see Chapter 6: Convective Loop). Such designs require careful structural design and may need to be reinforced.

(continued on page 244)

Commercial Suppliers of Phase-Change Materials*

Manufacturer	Product Name	Phase-Change Material	Container
Addison Products Co. Addison, MI 49220 (517) 547-6131	Solar Therm	Paraffin	Steel, round, 1 gal
Architectural Research Corp. 40 Water St. New York, NY 10004 (212) 943-3160	Sol-Ar Tile™	Glauber's salt	Polymer resinous concrete tile, 2′ × 2′
Blue Lakes Engineering Pace Corp. P.O. Box 1033 Appleton, WI 43912 (414) 733-0941 (a distributor for PSI Energy Systems Inc.)	Thermol-81	Calcium chloride hexahydrate with additives	Black polyethylene tube 6′ long, 3½″ diameter
Boardman Energy Systems, Inc. 5720 Kenneth Pike P.O. Box 4198 Wilmington, DE 19807 (617) 475-3276	Boardman Tube	Sodium Sulfate	Plated steel tubes 30″ long and 4″ diameter, built-in spaces, selective coating available
Colloidal Materials, Inc. P.O. Box 696 Andover, MA 01810 (617) 475-3276 (a licensee of Cabot Corp.)	Heat Pac™	Sodium Sulfate	3-ply aluminum foil laminate pouch, ¾″ × 2′ × 2′
Energy Materials Inc. 2622 South Zuni Englewood, CO 80110 (303) 934-2444 (under OEM agreement with Dow Chemical Co)	Thermairod-27	Calcium chloride hexahydrate	Black polyethylene pipe, 6′ long, 3½″ diameter
PSI Energy System Inc. 1533 Fenpark Drive St. Louis, MO 63026 (314) 343-7666 (under OEM agreement with Dow Chemical Co)	Thermol-81	Calcium chloride hexahydrate, with additives	Black polyethylene tube, 6′ long, 3½″ diameter
Texxor Corporation 9910 North 48th St. Omaha, NE 68152 (402) 453-7558	Texxor Heat Cell	Calcium chloride hexahydrate (Bisol II)	Steel cylinder, 7″ long, 4.26″ diameter

*This chart appeared in an article in the May 1980 issue of *Solar Age*,

Table 7.18

Total Weight (lb)	Phase-Change Temperature (°F)	Heat Storage Capacity (at phase-change temperature) (Btu)	Warranty (yr)
7.25	115	80	1
44	73	1,000	2
35	81	2,460	10
22	45, 64, 74, 78, 81, 89	1,444-2,000 varies with phase-change temperature	
10	73 (may be varied between 65 and 89)	2,542	10-year limited
35	81	2,542	10-year limited
35	81	2,460	10
35	81	345	5

The bed should be well insulated. Generally, the insulation provided for the container is placed inside the structure to trap heat in the rocks, not in the container itself.

Rocks

Rocks used in the bed will need to be clean and dry. Pressure-washed rocks should be specified. Rocks can be washed after installation in the rock storage container, but only if there is a floor drain or other provision for removing the wash water. If this is done, make sure the rocks are adequately dry before enclosing the container.

The storage system and the duct system should be airtight to minimize heat losses, and all duct joints should be sealed. The inside of the storage container should be lined with polyethylene, or if rigid insulation is used, the joints should be taped. Clean rocks, which can be purchased from the local ready-mix concrete supplier, can be placed in the rock storage area using a standard concrete chute.

Thermal Properties

Unlike the other storage materials, which generally store solar heat directly, the performance of rocks in a rock bed will depend not only on their material properties but also on the design of the bed and distribution system. The rocks have properties similar to those of concrete. However, the extent to which the storage capabilities, specifically the thermal storage capacity, of the rocks is utilized will depend upon such aspects as airflow through the rocks, delivery temperature, and rock-bed configuration.

Appendix A contains a simple and straightforward procedure for designing an effective rock bed. This information supplements the rules of thumb contained in Chapter 5: Attached Sunspace, and Chapter 6: Convective Loop.

Distribution Components

The distribution components move heat or "coolth" throughout the passive system, including into the living space where it is finally utilized. In most system designs, the distribution component receives little consideration. In these instances, such as Direct Gain design, the distribution of heat is by natural means (e.g., movement of air from space to space), and particular products or materials do not need to be specified.

In Thermal Storage Wall design, the only instance in which distribution is considered is in the design of a vented wall. The vents in this case are considered the distribution component. These are normally standard vents used in conventional heating design.

The use of distribution with Attached Sunspaces will vary depending upon the subsystem selected. The open wall and direct gain subsystems distribute heated air by natural means, and a specific distribution component is not specified. In both the air exchange and thermal storage wall subsystems, air movement is through vents, and, as for Thermal Storage Walls, the vent alone constitutes the distribution component.

Methods of heat distribution in the Attached Sunspace remote storage subsystem and in Convective Loop systems are similar and can be discussed together. In these two types of designs, heat is collected at a point removed from storage. If the heat is to be delivered directly into the living spaces, this can typically be easily accomplished, again, simply by providing vents into the space.

Where heat is to be delivered into storage, which is typically not adjacent to the collector and absorber, a system of ducts will need to be provided. These ducts should be tightly taped and otherwise sealed and insulated to reduce system heat losses. In most cases, a low-velocity fan should be installed in the ducts to assist in the movement of air.

The operation of the fan is generally thermostatically controlled. The thermostat senses the temperatures in the living space and at the collector/absorber to determine whether the heat should be diverted into storage.

Selection, operation, and control of the fan(s) will vary, depending upon the design of the system and the products specified. Manufacturers' literature should be consulted in this regard. Information on specific products is currently available from mechanical equipment manufacturers and suppliers.

Control Components

Introduction

Without the provision of control components, glazing areas, often associated with a passive home, can be a mixed blessing. During the evening hours in the heating season, the large expanse of collector can become a liability, often losing a substantial amount of heat. In some areas of the country, the nighttime heat losses can exceed the daytime solar heat gains. In the cooling season, the same glass can allow solar heat to increase the cooling energy requirements, often offsetting the energy savings from the heating season. In both seasons, control components substantially reduce many of these liabilities, while assisting in the proper functioning of the other passive components.

The three basic types of control components are shading, reflecting, and insulating. Shading is a cooling-season measure that reduces the amount of solar heat that reaches the living spaces. Reflecting, conversely, is a heating-season control device whose purpose is to increase the amount of collected solar heat. Insulating devices slow the movement of heat into, and out of, the living spaces.

Insulating components are divided into two categories: fixed and movable. Fixed insulation generally remains in place throughout the year. Movable insulation is, as its name indicates, moved into position, typically on a daily basis.

Many of the products used in one type of control application can also be used in another, such as a reflector, which, if shut in the evening, provides insulating value. Products that can serve a dual role often provide the highest dollar return in energy savings for dollar invested.

Each of the three control components, shading, reflecting, and insulating, is covered in a separate discussion below. It may be evident that the relative importance of the components will vary depending upon home design and geographical location. However, for most well-designed passive homes in most areas, movable insulation will have the highest energy savings associated with its use, with shading of particular importance in the South and reflecting in the North.

Shading

In most passive systems, some form of shading is required to avoid overheating during the cooling season. Shading devices can be placed on either the inside or the outside of the collector glazing.

If placed on the exterior, the shade will intercept direct solar radiation before it strikes the collector. Shades can be fixed or operable. Fixed exterior shading is not as effective as movable exterior shading, since it will block some of the sunlight from reaching the collector during the heating season. However, fixed shading is generally less expensive and less prone to maintenance problems than movable shading. Examples of common shading devices on the exterior include simple overhangs, awnings (fixed or adjustable), trellises, and shutters.

Shading devices placed inside the collector are not as thermally efficient as those placed on the exterior because they allow sunlight to penetrate into the space, where it can cause some heat gain. However, this heat buildup is usually restricted to the window surface and can be vented. To reduce the amount of heat reaching the living space, the interior shade should have edges sealed at the top, bottom, and both sides.

Interior shades are typically light in color to reflect sunlight back out the window before it can be converted to heat. The major advantage of interior shades is that they are typically easier to operate and maintain than most operable exterior elements.

Interior fixed shades are rarely specified. Some examples of interior shading components include roller shades, drapes, and movable panels. The components are especially effective if they can also be designed to provide insulation during the day in the cooling periods and at night in the heating periods (see Insulating later in this chapter).

Types of Shading Devices

There are a vast variety of shading products and configurations effective in passive solar applications. Selection will depend largely on ease of construction and operation, aesthetics, and cost. It is important to remember that, regardless of the system selected, the shading device will not be effective if it is not operated appropriately. Therefore, convenience and ease of operation are of foremost importance in the selection process of this control component.

The following are brief descriptions of the most commonly specified shading devices. Examples are shown in Figure 7.21.

Trellises and Overhangs—Simple trellises and overhangs are popular and inexpensive exterior shading options. Since they are fixed, they must be sized to reduce the amount of sun reaching the collector in the cooling season, while allowing as much sunlight as possible to strike the collector in the heating season.

Trellises and overhangs are usually simple extensions of the roofline and typically do not require any maintenance. By encouraging air movement across the collector glazing, trellises are slightly more effective than overhangs in reducing the cooling requirements.

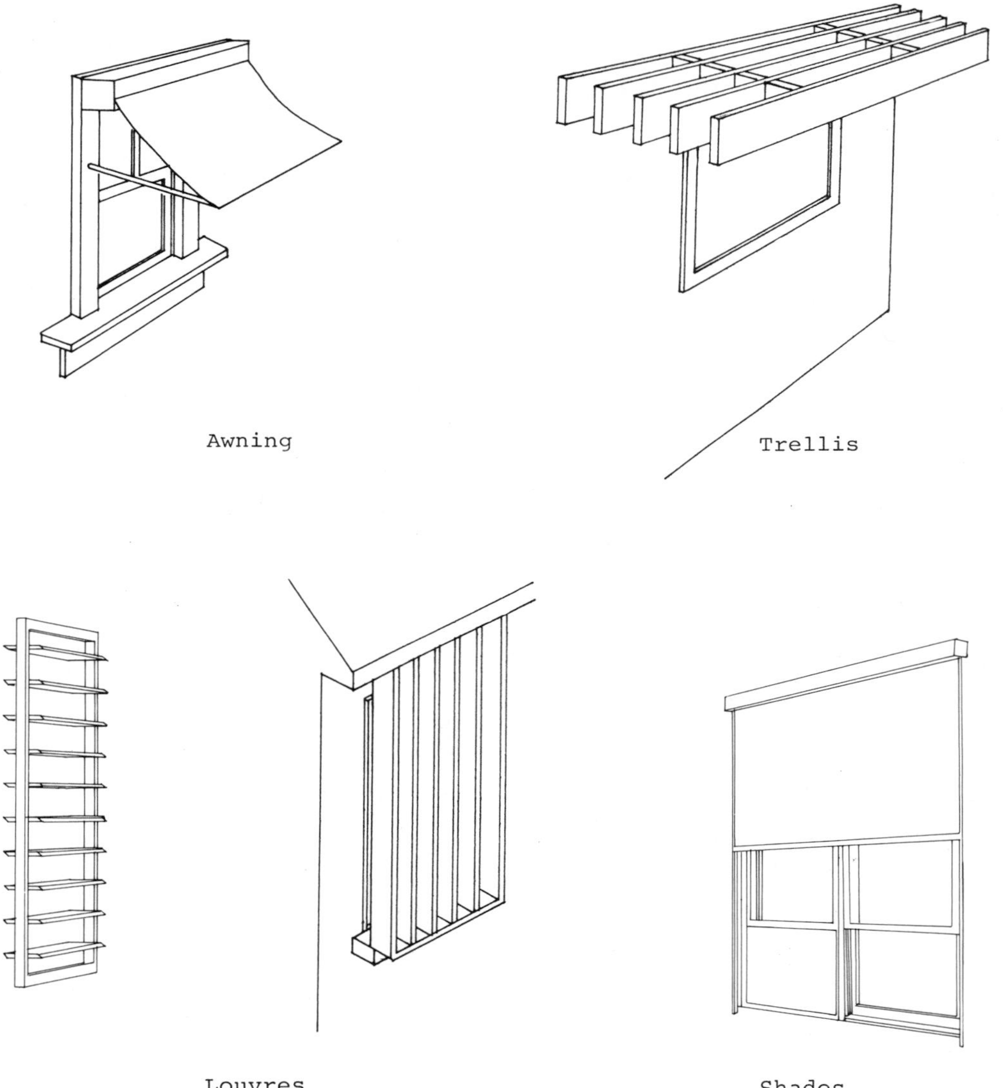

Figure 7.21

Roll Shades—Roll shades can be installed on the exterior or interior. They can be solid, or translucent, allowing some light into the living spaces. All fabrics used should be resistant to ultraviolet degradation, and exterior applications must be weather-resistant.

Interior applications may be designed with insulative materials to combine shading and insulating benefits. In addition, interior shades should be provided with seals at the edges to further restrict the flow of heat into the living space.

Awnings—Awnings may be fixed or movable. As mentioned earlier, fixed awnings should be sized to reduce heat gain during the cooling season yet allow for solar gain during the heating season. In order to reduce possible maintenance problems, operable awnings should be removed and stored during the heating months. Movable awnings provide the flexibility required to provide shade only when needed, adjusting to the angle of the sun. Operation of movable awnings can be either manual or motorized.

Louvers—Louvers can be installed on the exterior or interior. In either location, movable, adjustable louvers can reduce solar gain and glare and also provide insulating and/or reflecting qualities.

Louvers or slats can be horizontal or vertical. The latter configuration is particularly effective for southeast and southwest exposures. Both horizontal-mounted and vertical-mounted louvers are available. Operation may be manual or motorized.

Reflecting

Unlike a shading device, whose function is to limit the amount of solar heat in the passive system, a reflector is designed to enhance the collection of solar energy.

Reflectors increase the effective collector area by increasing the amount of solar heat passing through the glazing. The reflective surface acts like a mirror, reflecting much of the sunlight that strikes it. If properly oriented, the sunlight can be reflected onto the glazing, increasing the amount of solar heat available to the passive system. This increase in collected solar heat can, in effect, be considered to be the equivalent of adding the additional glazing area that would have been required to provide the same amount of heat.

Some significant performance increases can result from reflectors used in areas where local site obstructions, such as large trees or other homes, shade the collector at certain times of the day or year. A well-designed reflector can increase solar heat collection as much as 30-40% in certain situations.

Reflectors can be movable or fixed. A movable reflector in a passive solar system is most commonly constructed or purchased as a rigid panel with a highly reflective surface such as aluminum or foil on one side. A typical installation might include either vertical or horizontal hinges to permit the panel to close over the collector during periods of heat loss. The reflector is nearly always designed to double as movable nighttime insulation and, for ease of use, can be motorized to operate automatically.

Even if such panels are not feasible or not desired for certain applications, the performance of a passive solar system can be improved, although not as significantly as with a movable reflective surface, through the use of fixed reflectors.

Reflectivity Values of Various Surfaces Table 7.19

Surface	Reflectivity (%)
Aluminum Foil, bright	92-97
Aluminum Sheet	80-95
Aluminum-Coated Paper (polished)	75-84
Steel, galvanized, bright	70-80
Aluminum Paint	30-70
Brick	30-48
Paint (white)	75
Marble (white)	45
Concrete	40
Cement	27
Gravel	13
Grass	6

Concrete patios or white gravel planting beds adjacent to the collector, light colors on exterior walls, and decks or balconies with light-colored finishes are all examples of how reflective benefits can be obtained from standard fixed building components designed as reflectors.

Construction

Movable reflectors are typically fabricated specifically for the job. This "one-off" approach is generally necessary, since the requirements for different applications tend to vary.

Panels can be readily fabricated, incorporating insulation characteristics as well, by finishing the face of a rigid insulation board with a reflective material. All panels exposed to the outside should be weather-resistant, and operation should be made as simple as possible.

Table 7.19 contains the reflectivities of various materials. These values should be used for comparative purposes. The actual amount of sunlight reflected onto and through the collector will vary with time of year, location, angle of reflector, weather conditions, collector design, and material selection.

Insulating

Passive solar insulating components can be separated into two basic categories: fixed insulation that is permanently installed to reduce heat loss from a system's storage components; and movable insulation that is usually located at the collector and that is operated periodically to reduce nighttime and cloudy-day heat loss (during the heating season) and daytime heat gain (during the cooling season).

Fixed Insulation

In order to be effective, fixed insulation must be placed on the exterior surface of storage components that would otherwise be directly exposed to the outside. This type of installation will prevent the heat absorbed and stored at the interior surface of, for example, a Direct Gain floor or wall from being lost to the outside environment. If standard practice is followed and insulation is placed on the *inside* face of a storage component, the passive system is rendered ineffective.

Any of the materials commonly used for residential insulation can be employed as fixed exterior insulation for passive solar storage elements. The methods for attaching this insulation to the storage elements may differ slightly from standard practice, but those differences require only minor modifications to standard construction practices (see Chapters 3 through 6).

Movable Insulation

Passive solar collectors, in addition to admitting and trapping large amounts of solar energy, can also lose large amounts of heat during the night and during cloudy periods unless properly insulated. Provision for some form of movable insulation to prevent these losses will have a significant impact on the overall thermal performance of any passive solar system. Such insulation, if in place during the daytime hours, can also be used to prevent unwanted heat gain during the cooling season.

Movable insulation comes in a variety of forms, including rigid panels, insulating shutters, insulating drapes and shades, and even double glazing with insulation integrated into the airspace.Depending on their physical characteristics, these devices can be used on the exterior or the interior of the collector. Interior applications are generally easier to operate and maintain, and they avoid degradation due to weathering. Exterior applications have the advantage, during the cooling season, of also acting as exterior shades.

If interior movable insulation is used, the side facing the glass should be light colored to reflect sunlight and thereby avoid heat buildup in the airspace between the insulation and glass. This buildup can occur even on sunny winter days if the insulation is left in place. Therefore, all interior movable insulation should be light colored on the side facing the glass, whether it is used to prevent summer heat gain or not.

The most distinctive characteristic of movable insulation is that it is operable and, in general, must be managed on a day-to-day basis. Although some varieties can be thermostatically controlled, the majority require some operation, either manual or automatic switch controlled. The choice of product will depend, of course, on cost and ease of operation, as well as on thermal effectiveness.

There are a variety of movable insulation product types that can be installed on either the interior or the exterior of the windows. All will function well if properly installed and operated, and a mix of types will probably be appropriate for the typical home. Each product type is briefly discussed below. Some typical products of these types are shown in Figures 7.22 and 7.23.

Interior Movable Insulation Products

Drapes—Insulating drapes commonly employ quilted fabrics, reflective liners, or multiple layers of materials to help reduce heat flow. Manufacturers' specifications should be consulted for rated R-values, but performance in the R-1 to R-4 range can be expected.

Very careful attention must be paid to sealing the edges, tops, and bottoms of drapes. Velcro or magnetic strips can be used at the edges, and weights in the bottom hem are appropriate if the drapes reach the floor or rest on the windowsill. If they fall past the sill but do not reach the floor, special detailing will be required to ensure a continuous seal at the bottom.

It should be noted that standard drapes, not specifically designed to have a significant R-value, can nonetheless provide some insulating benefit if they are properly sealed at the edges and bottom. The R-value for sealed, standard drapes is approximately R-1.

Shades—Insulating shades can be regular, quilted, or multi-layered and can incorporate fiberfill, insulating airspaces, or heat-reflective materials. R-values range roughly from R-3 to R-12, and, again, manufacturers' specifications should be consulted for precise values.

As in the case of insulating curtains, edge and bottom seals are critical to the effec-

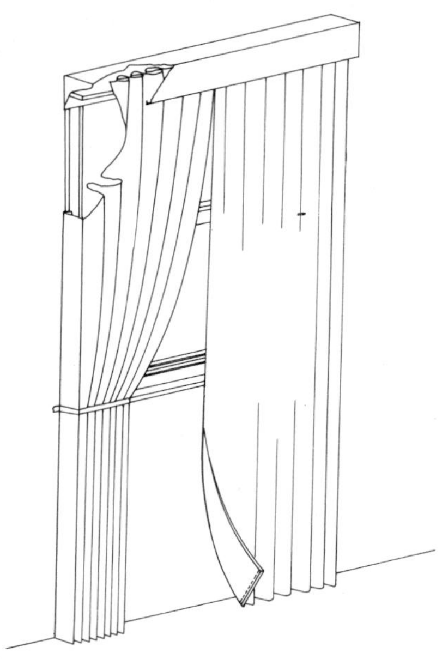

Drapes

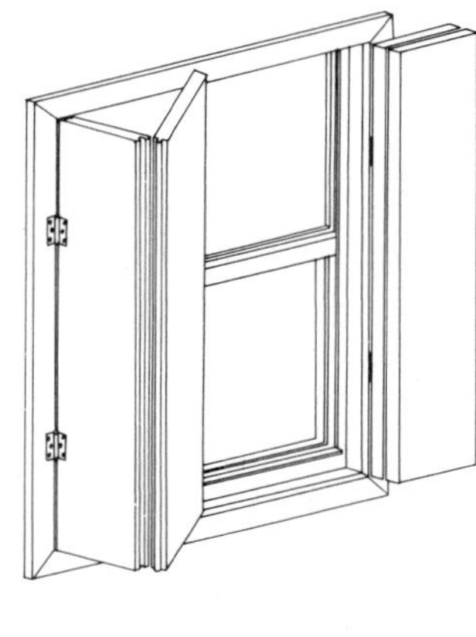

Shutters

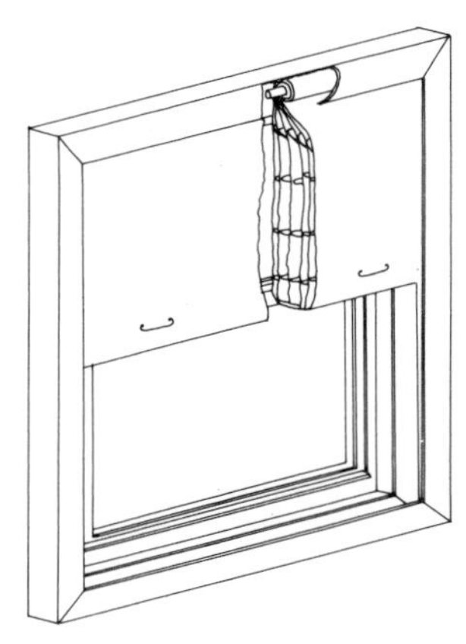

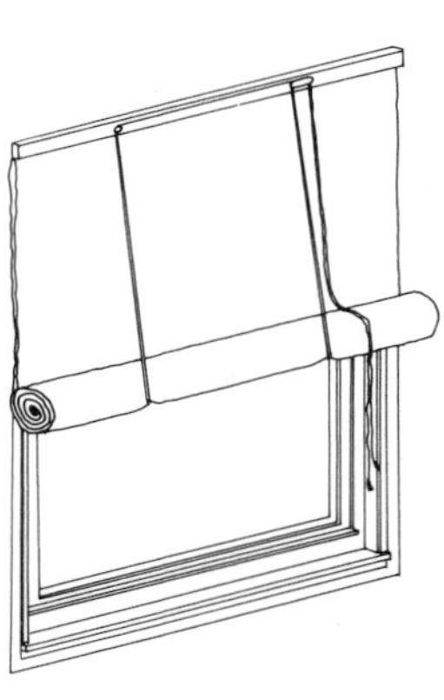

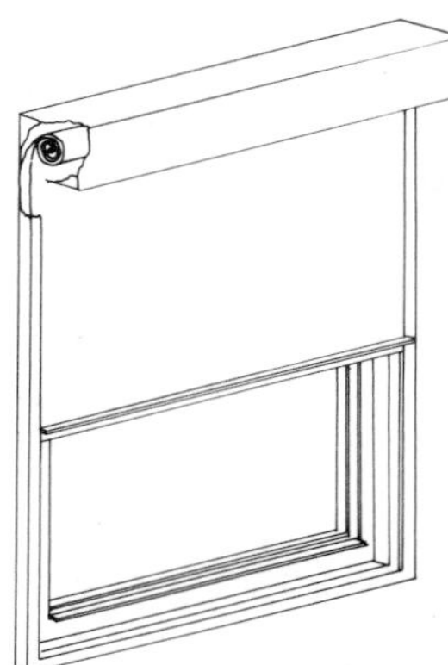

Shades

Figure 7.22

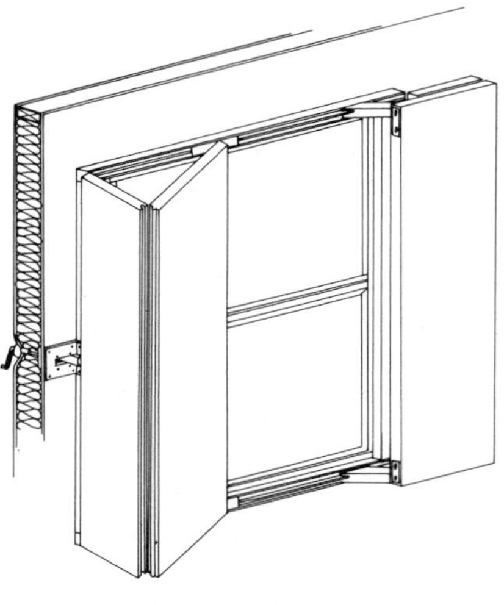

Shutters

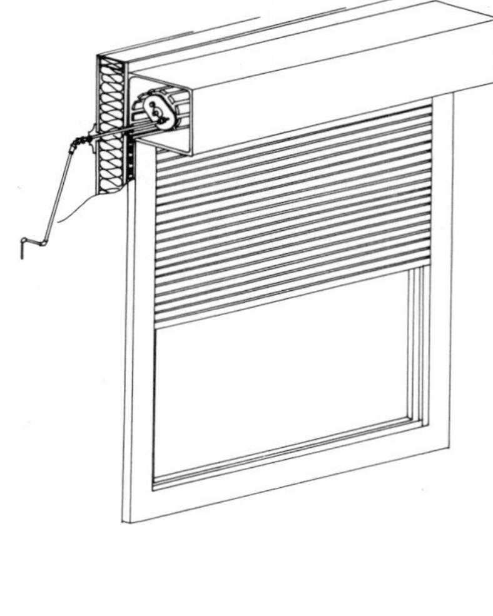

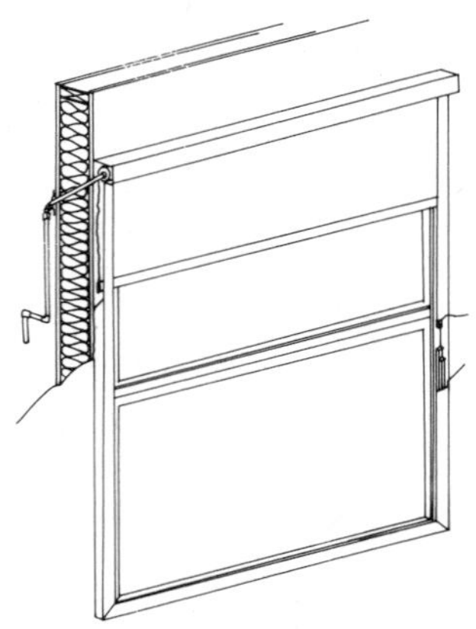

Rolldown Shutters

Figure 7.23

tive performance of insulating shades. Velcro or magnetic strips can be used (sides and bottom) to seal shades to the window frame, or side tracks can be mounted on the frame with the bottom of the shade fitting into an additional track or simply resting on the sill. In the latter case, a small piece of foam plastic material attached to the bottom of the shade may be desirable to ensure a tight seal.

Insulating shades may be bulky relative to more standard varieties and may require larger valances.

Standard shades, not specifically designed to have an insulating capability, can provide some insulating benefit if properly sealed at edges and bottom. Again, an estimated R-value of R-1 should be assumed for well-sealed shades.

Shutters—Insulating shutters are typically fabricated from rigid insulating boards (polystyrene, polyurethane, or polyisocyanurate) sandwiched between some form of finish material. Rated R-values are available from manufacturers or can be determined from the R-value of the insulation used. Performance will generally range from R-4 to R-8 per inch of insulation, depending on the material used.

Typical shutter configurations are hinged, bifold, or track mounted. In any configuration, there should be a tight fit between the shutters and frame, and all edges should be well sealed. Manufacturers' recommendations should be followed for sealing, or common weatherstripping details can be applied.

Panels—Removable insulating panels are similar to insulating shutters in configuration, except that they are not mechanically attached to the window frame but are removed and stored after each use. Such panels should be detailed so that edges are continuously sealed with materials such as Velcro, magnetic strips, or tight-fitting buttons that hold the panel firmly against the frame.

Exterior Movable Insulation Products

Shutters and Panels—Insulating exterior shutters and panels are constructed and perform similarly to those placed on the interior, except that they require operation from the outside or special detailing to allow operation from the inside of the home. As in the case of interior insulation, all edges should be well sealed.

Rolldown Shutters—Exterior shutters are typically made from interlocking wood, aluminum, or PVC slats. A range of product R-values are available. Side tracks on rolldown shutters should be detailed to provide a tight seal, as should the bottom of the shutter where it meets the sill.

Construction

In many passive systems, the installation and operation of movable insulation may be difficult. For this reason, it is strongly recommended that consideration be given to movable insulation components at the earliest stages of design. This will ensure that the passive system as a whole operates at optimum performance levels. It is also advisable to consult manufacturers' recommendations when specifying movable insulation components to ensure that proper allowances and clearances are provided.

BIBLIOGRAPHY

The following is a selected bibliography of books, articles, and papers that will provide additional information on topics covered in the *Handbook*. Included at the end of the bibliography are several trade periodicals that provide up-to-date information on new technologies, products, and ideas related to passive and active solar energy. Entries are listed alphabetically by title.

The number(s) following each entry will classify the overall subject matter into the following groups:

1. Design Primer
2. Case Studies
3. Conference Proceedings
4. Economics
5. Energy Analysis
6. Technical Literature
7. Construction Guide
8. Bibliography
9. Product Guide
10. Marketing

1. Sabine, Hale J. and Myron B. Lacher, *Acoustical and Thermal Performance of Exterior Residential Walls, Doors, and Windows,* U.S. Government Printing Office, Washington, DC (1975). [6]

2. The Solar Energy Research Institute, *Analysis Methods for Solar Heating and Cooling Applications,* 3rd ed., U.S. Department of Energy, Golden, CO (1980). [5]

3. Elmiger, A., *Architectural and Engineering Concrete Masonry Details for Building Construction,* National Concrete Masonry Association, Herndon, VA (1976). [7]

4. Ramsey, Charles G. and Harold R. Sleeper, *Architectural Graphic Standards,* ed. by Joseph N. Boaz, American Institute of Architects, John Wiley and Sons, Inc., New York, NY (1970). [7]

5. Prestressed Concrete Institute, *Architectural Precast Concrete,* PCI, Chicago, IL (1973). [7]

6. *1977 ASHRAE Handbook & Product Directory, Fundamentals,* American Society of Heating, Refrigerating and Air-Conditioning Engineers, Inc., New York, NY (1977). [6]

7. Holtz, Michael J., Wayne Place and Ronald Kammerud, *A Classification Scheme for the Common Passive and Hybrid Heating and Cooling Systems,* Solar Energy Research Institute, Golden, CO (1979). [1]

8. Steven Winter Associates, Inc., *Code Manual for Passive Solar Design—Single Family Residential Construction,* Southern Solar Energy Center, Atlanta, GA (1979). [6,7]

9. *The Concrete Approach to Energy Conservation,* Portland Cement Association, Skokie, IL (1974). [1]

10. *Concrete Energy Conservation Guidelines,* Portland Cement Association, Skokie, IL (1980). [1]

11. Spears, Ralph E., *Concrete Floors on Ground,* Portland Cement Association, Skokie, IL (1978). [7]

12. Randall, Frank A. and William C. Panarese, *Concrete Masonry Handbook,* Portland Cement Association, Skokie, IL (1976). [7]

13. Cook, Jeffrey, AIA, *Cool Houses for Desert Suburbs,* Arizona Solar Energy Commission, Phoenix, AZ (1979). [1,2]

14. Small Homes Council—Building Research Council, *Current House Construction Practices,* University of Illinois at Urbana-Champaign, Champaign, IL (1974). [7]

15. *Design and Control of Concrete Mixtures,* 12th ed., Portland Cement Association, Skokie, IL (1979). [6,7]

16. Watson, Donald, *Designing and Building a Solar House,* Garden Way Publishing, Charlotte, VT (1977). [1,2]

17. Balcomb, J. D., *Designing Passive Solar Buildings to Reduce Temperature Swings,* Los Alamos Scientific Laborabory, Los Alamos, NM (1975). [6]

18. Lewis, Dan, Charles Michal and Paul Pietz, Total Environmental Action, *Design of Residential Buildings Utilizing Natural Thermal Storage,* National Technical Information Service, U.S. Department of Commerce, Springfield, VA (1979). [1,2]

19. Olgyay, Victor, *Design with Climate,* Princeton University Press, Princeton, NJ (1973). [1]

20. Professional Builder, *Energy and the Builder,* Cahners Publishing Company, Chicago, IL (1977). [1,9,10]

21. Marshall, Harold E. and Rosalie T. Ruegg, *Energy Conservation in Buildings: An Economics Guide for Investment Decisions,* NBS Handbook 132, U.S. Government Printing Office, Washington, DC (1980). [4]

22. *Energy Efficient Building and Rebuilding: The Profit Opportunities,* Proceedings of the Builders and Remodelers '80 Conference, Northeast Solar Energy Center, Boston, MA (1980). [3,6]

23. Balcomb, J. D., *Energy Savings Obtainable Through Passive Solar Techniques,* Los Alamos Scientific Laboratory, Los Alamos, NM (1978). [6]

24. Yanda, Bill and Rick Fisher, *The Food and Heat Producing Solar Greenhouse,* John Muir Publications, Inc., Santa Fe, NM (1976). [1,2,7]

25. Brown, Robert J. and Rudolph R. Yanuck, *Life Cycle Costing: A Practical Guide for Energy Managers,* Fairmont Press, Inc., Atlanta, GA (1980). [4]

26. Reynolds, Smith & Hills, Inc., *Life Cycle Costing Emphasizing Energy Conservation,* National Technical Information Service, U.S. Department of Commerce, Springfield, VA (1976). [4]

27. Givoni, B., *Man, Climate and Architecture,* Applied Science Publishers Ltd., London, England (1976). [1,6]

28. *A Manual of Facts on Concrete Masonry,* National Concrete Masonry Association, Herndon, VA (1980). [7]

29. Robson, William M., Scott A. Noll and Fred J. Roach, *Market Potential Indexing in Support of Solar Action Week,* Los Alamos Scientific Laboratory, Los Alamos, NM (1980). [10]

30. Wass, Alonzo, *Methods and Materials of Residential Construction,* Reston Publishing Company, Inc., Reston, VA (1977). [7]

31. Wright, D., *Natural Solar Architecture: A Passive Primer,* Van Nostrand Reinhold Co., New York, NY (1978). [1]

32. *1979 NYSERDA Passive Solar Design Awards,* New York State Energy Research and Development Authority, Albany, NY (1979). [2]

33. *Passive Cooling Handbook,* Lawrence Berkeley Laboratory, Berkeley, CA (1980). [3]

34. *Passive Design Ideas for the Energy Conscious Builder,* National Solar Heating and Cooling Information Center, Philadelphia, PA. [1]

35. Berkeley Solar Group, *Passive Design Saves Energy and Money,* Concrete Masonry Association of California and Nevada, Sacramento, CA (1979). [1]

36. Southern Solar Energy Center, *Passive Retrofit Handbook,* SSEC, Atlanta, GA (1981). [1,7]

37. *Passive Solar Design: An Extensive Bibliography,* National Technical Information Service, U.S. Department of Commerce, Springfield, VA (1978). [8]

38. Balcomb, J. D., et al., *Passive Solar Design Handbook, Volume 2: Passive Solar Design Analysis,* National Technical Information Service, U.S. Department of Commerce, Springfield, VA (1980). [1,5,6]

39. Total Environment Action, Inc., *Passive Solar Design Handbook, Volume 1: Passive Solar Design Concepts,* National Technical Information Service, U.S. Department of Commerce, Springfield, VA (1980). [1,5,6]

40. Mazria, Edward, *The Passive Solar Energy Book,* Rodale Press, Emmaus, PA (1979). [1]

41. Gibson, Stephanie, *Passive Solar Products Directory,* Northeast Solar Energy Center, Boston, MA (1980). [9]

42. *Passive Solar Workshop Handbook,* Passive Solar Associates, Santa Fe, NM (1978). [1,2,7]

43. *PCI Design Handbook, Precast Prestressed Concrete,* Prestressed Concrete Institute, Chicago, IL (1978). [7]

44. Prowler, Don, ed., *Proceedings of the 2nd National Passive Solar Conference,* Vols. 1-3, American Section of the International Solar Energy Society, Inc., Newark, DE (1978). [3]

45. Miller, Harry, Michael Riordan, and David Richards, eds., *Proceedings of the 3rd National Passive Solar Conference,* American Section of the International Solar Energy Society, Inc., Newark, DE (1979). [3]

46. Franta, Gregory, ed., *Proceedings of the 4th National Passive Solar Conference,* American Section of the International Solar Energy Society, Inc., Newark, DE (1979). [3]

47. Hayes, John and Rachel Snyder, eds., *Proceedings of the 5th National Passive Solar Conference,* Vols. 1-2, American Section of the International Solar Energy Society, Newark, DE (1980). [3]

48. Hayes, John and William Kolar, eds., *Proceedings of the 6th National Passive Solar Conference,* Vol. 6, American Section of the International Solar Energy Society, Inc., Newark, DE (1981). [3]

49. *Proceedings of the Solar Glazing 1979 Topical Conference,* MidAtlantic Solar Energy Association, Philadelphia, PA (1979). [3]

50. Northeast Solar Energy Center, *Proceedings of U.S. Department of Energy Passive and Hybrid Solar Energy Program Update Meeting,* National Technical Information Service, U.S. Department of Commerce, Springfield, VA (1980). [3]

51. *Recommended Requirements to Code Officials for Solar Heating, Cooling and Hot Water Systems,* Council of American Building Officials, Washington, DC (1980). [7]

52. The AIA Research Corporation, *Regional Guidelines for Building Passive and Energy Conserving Homes,* U.S. Department of Housing and Urban Development, Washington, DC (1978). [1]

53. The AIA Research Corporation, "Passive Technology," *Research & Design Quarterly,* Vol. 11, No. 1, Washington, DC (January, 1979). [1]

54. The AIA Research Corporation, "Passive Cooling," *Research & Design Quarterly,* Vol. 11, No. 3, Washington, DC (Fall, 1979). [1]

55. *Selling the Solar Home '80: Market Findings for the Housing Industry,* Real Estate Research Corporation, Chicago, IL (1980). [10]

56. Wray, W. O. and J. D. Balcomb, *Sensitivity of Direct Gain Space Heating Performance to Fundamental Parameter Variations,* Los Alamos Scientific Laboratory, Los Alamos, NM (1978). [10]

57. Marshall, Harold E. and Rosalie T. Ruegg, *Simplified Energy Design Economics,* National Bureau of Standard Publication 544, Washington, DC (1980). [4]

58. The AIA Research Corporation, *Solar Dwelling Design Concepts,* U.S. Government Printing Office, Washington, DC (1976). [1]

59. Mid-American Solar Energy Complex, *Solar 80 Home Designs,* MASEC, Bloomington, MN (1980). [2]

60. California Energy Commission, *Solar Gain: Passive Solar Design Competition,* Rodale Press, Emmaus, PA (1980). [2]

61. McCullagh, James C., ed., *The Solar Greenhouse Book,* Rodale Press, Emmaus, PA (1978). [1,2]

62. The AIA Research Corporation, *Solar Heating and Cooling Demonstration Program: A Descriptive Summary of HUD Cycle 2 Solar Residential Projects,* U.S. Government Printing Office, Washington, DC (1976). [2]

63. The AIA Research Corporation, *Solar Heating and Cooling Demonstration Program: A Descriptive Summary of HUD Cycle 3 Solar Residential Projects,* U.S. Government Printing Office, Washington, DC (1977). [2]

64. The AIA Research Corporation, *Solar Heating and Cooling Demonstration Program: A Descriptive Summary of HUD Cycle 4 and 4A Solar Residential Projects,* U.S. Government Printing Office, Washington, DC (1979). [2]

65. The AIA Research Corporation, *Solar Heating and Cooling Demonstration Program: A Descriptive Summary of HUD Cycle 5 Solar Residential Projects,* U.S. Government Printing Office, Washington, DC (1980). [2]

66. Anderson, B. and M. Riordan, *Solar Home Book,* Brick House Publishing Co., Andover, MA (1976). [1]

67. Tennessee Valley Authority, *Solar Homes for the Valley: 1979 Design Portfolio,* TVA, Knoxville, TN (1979). [2]

68. Solar Age Magazine, *Solar Products Specifications Guide,* SolarVision, Inc., Harrisville, NH (1979). [9]

69. Solar Energy Applications Laboratory, Colorado State University, and Granklin Research Center, *Solar Workshops of the Financial Community: Technical, Economic, and Legal Considerations for Evaluating Solar-Heated Buildings,* U.S. Department of Energy, Washington, DC (1979). [4]

70. *South Carolina Passive Solar Home Design: Low Cost Energy Efficient Opportunities,* Governor's Office, Division of Energy Resources, Columbia, SC (1980). [1,2]

71. Newman, Morton, *Standard Structural Details for Building Construction,* McGraw-Hill, Inc., New York, NY (1968). [7]

72. Steven Winter Associates, Inc., *Suntempering in the Northeast: A Selection of Builders' Designs,* Northeast Solar Energy Center, Boston, MA (1980). [2]

73. The AIA Research Corporation, *A Survey of Passive Solar Buildings,* U.S. Department of Energy, Washington, DC (1980). [2]

74. The AIA Research Corporation, *A Survey of Passive Solar Homes,* U.S. Department of Energy, Washington, DC (1980). [2]

75. *Technical Notes on Brick Construction,* Brick Institute of America, McLean, VA (1980). [6,7]

76. Shurcliff, William A., *Thermal Shutters and Shades,* Brick House Publishing Company, Inc., Andover, MA (1980). [1,9]

77. Wilson, Alex, *Thermal Storage Wall Design Manual,* Modern Press, Albuquerque, NM (1979). [1]

78. Bainbridge, David, Judy Corbett and John Hofacre, *Village Homes' Solar House Designs,* Rodale Press, Emmaus, PA (1979). [2]

79. Hastings, Robert S. and Richard W. Crenshaw, *Window Design Strategies,* NBS Building Science Series 104, U.S. Government Printing Office, Washington, DC (1977). [1]

80. Selkowitz, Steven, *Windows for Energy Efficient Buildings,* Vol. 1, Nos. 1-2, Lawrence Berkeley Laboratory, Berkeley, CA (1980). [1,9]

81. Beckett, H. E. and J. A. Godfrey, *Windows: Performance, Design and Installation,* Van Nostrand Reinhold Company, New York, NY (1974). [7]

82. Anderson, L. O., *Wood Frame House Construction,* U.S. Government Printing Office, Washington, DC (1970). [7]

Periodicals

1. *Energy Conservation Digest,* Washington, DC (bi-weekly).
2. *Solar Age,* SolarVision, Inc., Harrisville, NH (monthly).
3. *Solar Bibliography,* National Solar Heating and Cooling Information Center, Rockville, MD (irregular monthly).
4. *Solar Energy,* Pergamon Press, Inc., Elmsford, NY (monthly).
5. *Solar Engineering,* Solar Engineering Publishers, Inc., Dallas, TX (monthly).
6. *Solar Heating and Cooling,* Gordon Publications, Morristown, NJ (monthly).
7. *Solar Source Sheet,* National Solar Heating and Cooling Information Center, Rockville, MD (irregular monthly).

GLOSSARY

ABSORBER—A solid surface, usually dark colored, which is exposed to, and struck by, sunlight, and which transforms radiant solar energy to heat energy.

ABSORPTANCE—The ratio of the radiation absorbed by a surface to the total energy falling on that surface, measured as a percentage.

ABSORPTIVITY—See Absorptance.

ACTIVE SOLAR SYSTEM—A system that uses mechanical devices and an external energy source, in addition to solar energy, to collect, store, and distribute thermal energy (heat).

AIR CHANGE—The replacement of a quantity of air in a volume within a given period of time. This is expressed in number of changes per hour. If a house has 1 air change per hour, all the air in the house will be replaced in a 1-hour period.

AIR-LOCK ENTRY—A vestibule enclosed with two airtight doors for permitting entrance while limiting air or heat exchange.

AMBIENT TEMPERATURE—The prevailing temperature outside a building.

ANGLE OF INCIDENCE—The angle that the sun's rays make with a line perpendicular to a surface.

ANNUAL FUEL REQUIREMENT—Annual heating load divided by the product of the Mechanical Plant Efficiency (MPE) (usually obtainable from the equipment manufacturer) and the fuel conversion factor.

APERTURE—That part of the south-facing glazing that contributes to solar heating; literally, an opening.

ASHRAE—Abbreviation for the American Society of Heating, Refrigerating and Air-Conditioning Engineers.

AUXILIARY ENERGY SYSTEM—Equipment utilizing energy other than solar, both to supplement the output provided by the solar energy system (as required by the design conditions) and to provide full energy backup requirements during periods when the solar heating or domestic hot water systems are inoperable.

AUXILIARY HEATING FRACTION (AHF)—That part of the total building heating requirements supplied by the auxiliary heating system.

AZIMUTH—The angular distance between true south and the point on the horizon directly below the sun.

BACKUP SYSTEM—See Auxiliary Energy System.

BERM—A man-made mound or small hill of earth.

BLACK BODY—A theoretically perfect absorber.

Btu (BRITISH THERMAL UNIT)—Basic heat measurement, equivalent to the amount of heat needed to raise 1 lb of water 1°F.

Btu/DD/FT^2 (HEATED AREA)—A unit commonly used to express the inherent ability of the building shell to resist heat loss.

BUILDING ENVELOPE—The elements of a building that enclose conditioned spaces through which thermal energy may be transferred to or from the exterior.

BUILDING LOAD COEFFICIENT (BLC)—A rough measure of the insulating quality of the home, exclusive of the south facade, used in estimating the Solar Savings Fraction (SSF) (which is the percentage of heating needs supplied by the passive solar system).

BUILDING SKIN CONDUCTANCE—The weighted average conductance of all the components of the building skin.

CALORIE (cal)—The quantity of heat needed to raise the temperature of 1 gram of water 1°C. One calorie is approximately 0.004 Btu.

CLERESTORY—A vertical window placed high in a wall near the eaves or between a lower roof and a higher roof, used for light, heat gain, and ventilation.

COLLECTOR—An area of transparent or translucent glazing commonly located on the south-facing side of the home.

COLLECTOR EFFICIENCY—The amount of energy gathered by a collector compared to the amount striking it.

COMFORT RANGE—The range of climatic conditions within which people feel comfortable.

COMPONENT—An individually distinguishable product that forms part of a more complex product (i.e., subsystem of a system).

CONDUCTANCE (C)—The quantity of heat (Btu's) that will flow through 1 sq ft of material in 1 hour, when there is a 1°F temperature difference between both surfaces. Conductance values are given for a specific thickness of materials, not per inch of thickness.

CONDUCTIVITY (k)—The quantity of heat (Btu's) that will flow through 1 sq ft of material, 1'' thick, in 1 hour, when there is a temperature difference of 1°F between both surfaces.

CONTROL—A device or devices that regulates heat flow between the building and the exterior.

CONVECTION—The transfer of heat by movement of a fluid (liquid or gas).

CONVECTION, FORCED—Heat transfer through a medium such as air or water by currents caused by a device powered by an external energy source.

CONVECTION, NATURAL—Heat transfer of a fluid such as air or water that results from the natural rising of the lighter, warm fluid and the sinking of the heavier, cool fluid.

CONVECTIVE LOOP—A passive system in which a fluid, typically air, is circulated in a closed path induced by rising hot air and falling cold air.

COOLING SEASON—The period(s) during the year when the outside temperature and humidity conditions require that the living spaces must be naturally or mechanically cooled to be comfortable.

COOLTH—A term used to describe the quality of a material at a lower than ambient temperature. Coolth is to cooling as warmth is to heating.

DECLINATION—A deviation, as from a specific direction or standard. Used primarily in relation to magnetic declination (magnetic variation), which is the angle between true north and magnetic north. The declination varies with different geographical areas.

DEGREE-DAY (DD), COOLING—See Degree-Day, Heating, except that cooling degree-days are measured as the variation of degrees above, not below, the standard temperature (usually 65°F).

DEGREE-DAY (DD), HEATING—A unit of heat measurement equal to 1° variation below a standard temperature (usually 65°F) in the average temperature of one day. If the standard is 65°F and the average outside temperature for one day is 50°F, then the number of degree-days recorded for that day would be 15.

DELTA T (ΔT)—A difference in temperature.

DENSITY—The mass of a substance, which is expressed in pounds per cubic foot.

DESIGN LIFE—The period of time during which a heating, cooling, or domestic hot water system is expected to perform its intended function without requiring maintenance or replacement.

DESIGN TEMPERATURE—A designated temperature close to the most severe winter or summer temperature extremes of an area, used in estimating heating and/or cooling demand.

DIFFUSE RADIATION—Sunlight that is scattered by air molecules, dust, water vapor, and translucent materials.

DIRECT GAIN (DG)—A passive system in which solar radiation is admitted directly into the conditioned (or living) space, where it is converted to heat and stored.

DIRECT RADIATION—Light that has traveled a straight path from the sun, as opposed to diffuse radiation.

DISTRIBUTION—The method by which heat is delivered to the living areas.

DRY BULB TEMPERATURE—A measure of the sensible temperature of the air.

DHW—Domestic hot water.

ECONOMIC EFFICIENCY—Maximizing net benefits or minimizing costs for a given level of benefits.

EFFICIENCY—In solar applications, the amount of useful solar energy collected, divided by the amount of solar energy available to the collector.

EMISSIVITY—The ability to radiate heat in the form of long-wave radiation.

EMITTANCE—The ratio of the amount of heat radiated by a surface to the amount that would be radiated by a black body at the same temperature. Emittance values range from 0.05 for brightly polished metals to 0.96 for flat black paint.

ENERGY—The capacity for doing work. It takes a number of forms, which may be transformed from one into another, such as thermal (heat), mechanical (work), electrical, and chemical; customarily measured in kilowatt-hours (kwh) or British thermal units (Btu).

ENERGY TRANSPORT SYSTEM—The portion of the heating and domestic hot water systems that contains heat transfer fluids and transports energy throughout the system.

EQUINOX—Either of the two times during the year when the sun crosses the celestial equator and when the length of day and night are approximately equal. These are the autumnal equinox on or about September 22 and the vernal equinox on or about March 22.

FIRST COST—A measure of the initial cost of a component or system.

FLAT PLATE COLLECTOR—A panel of metal or other suitable material that converts sunlight into heat. The solar radiative absorbing surface is essentially flat, and the aperture and absorber are similar in area and geometry.

GLAZING—A covering of transparent or translucent material (glass or plastic) used for admitting light. Glazing reduces heat losses from reradiation and convection. Examples: windows, skylights, greenhouses, and clerestories.

GREENHOUSE EFFECT—The ability of a glazing material to both transmit short-wave solar radiation into a space and trap long-wave heat generated by the conversion of the short-wave radiation into heat.

GROSS FLOOR AREA—The sum of the areas of all floors of a building, including basements, cellars, mezzanine, and intermediate floored tiers, and penthouses of headroom height, measured from the exterior faces of exterior walls or from the centerline of the walls separating buildings.

GROSS WALL AREA—The gross area of exterior walls consists of all opaque wall areas (including foundation walls, areas between floor spandrels, peripheral edges of floors, window areas including sash, and door areas) where such surfaces are exposed to outdoor air and enclose a heated or mechanically cooled space, including areas between two such spaces.

HEAT—The form of energy that is transferred by virtue of a temperature difference.

HEAT CAPACITY—The property of a material defined as the quantity of heat needed to raise 1 cu ft of the material 1°F. Numerically, the density multiplied by the specific heat.

HEAT EXCHANGER—A device specifically designed to transfer heat between two physically separated fluids.

HEAT GAIN—An increase in the amount of heat contained in a space, resulting from solar radiation and the flow of heat through the building envelope, plus internal heat gain.

HEAT LOSS—A decrease in the amount of heat contained in a space, resulting from heat flow through walls, windows, roof, and other building envelope components.

HEAT LOSS COEFFICIENT (UA)—The rate of energy transfer through the walls, roof, and floor of a house, calculated in Btu's per hour per degree F (Btu/hr/°F).

HEAT SINK—A substance that is capable of accepting and storing heat, and therefore may also act as a heat source.

HEAT TRANSFER MEDIUM—A medium—liquid, air, or solid—that is used to transport thermal energy.

HEATED SPACE—Space within a building that is provided with a positive heat supply to maintain the air temperature at 50°F or higher.

HVAC—Heating, ventilating, and air conditioning.

HVAC SYSTEM—A system that provides, either collectively or individually, the processes of comfort control, including heating, ventilating, and/or air conditioning within or associated with a building.

HYBRID SYSTEM—A solar heating system that combines active and passive techniques.

INCIDENT ANGLE—The angle between the incident ray from the sun and a line drawn perpendicular to the solar collector surface.

INDIRECT GAIN PASSIVE SYSTEM—A solar heating system in which sunlight first strikes a thermal mass located between the sun and a space. The sunlight absorbed by the mass is converted to heat and then transferred into the living space.

INFILTRATION—The uncontrolled inward air leakage through cracks and interstices in any building element and around windows and doors of a building, caused by the pressure effects of wind and/or the effect of differences in the indoor and outdoor air density.

INSOLATION—The total amount of solar radiation incident upon an exposed surface measured in Btu's per hour per square foot (Btu/hr/ft^2) or in Langleys.

INSULATION—A material having a relatively high resistance to heat flow and used principally to reduce heat flow.

INTERNAL HEAT GAIN—Heat generated by equipment, appliances, lights, and people.

ISOLATED GAIN PASSIVE SYSTEM—A system in which solar collection and heat storage are isolated from the living spaces.

LIFE-CYCLE COST—A measure of total system cost, including initial, maintenance, and operating costs over its life span. The accumulation generally includes a discounting of future costs to reflect the relative value of money over time.

LOAD COLLECTOR RATIO (LCR)—The Building Load Coefficient (BLC) divided by the total passive solar collector area. The LCR is used to determine the Solar Savings Fraction.

MAGNETIC SOUTH—South as indicated by a compass; it changes markedly with latitude.

MECHANICAL PLANT EFFICIENCY—The efficiency of the mechanical system, usually obtainable from the equipment manufacturer.

MICROCLIMATE—The climate of a defined local area, such as a house or building site, formed by a unique combination of factors such as wind, topography, solar exposure, soil, and vegetation.

MMBtu—Million (10^6) Btu's. The predominant unit of measure for energy in the United States.

MOVABLE INSULATION—Insulation placed over windows, clerestories, skylights, and other glazing when needed to prevent heat loss or gain, and removed for light, view, venting, or heat.

NATURAL VENTILATION—See Ventilation, Natural.

NIGHT INSULATION (NI)—See Movable Insulation.

NOCTURNAL COOLING—Cooling by nighttime radiation, convection, and evaporation.

OPAQUE—Impenetrable by light.

OUTSIDE AIR—Air taken from the outdoors and, therefore, not previously circulated through the system.

PASSIVE SOLAR SYSTEM—An assembly of natural and architectural components, including collectors, thermal storage device(s), and transfer fluid, which converts solar energy into thermal energy in a controlled manner and in which no fans or pumps are used to accomplish the transfer of thermal energy.

PAYBACK—A traditional measure of economic viability of investment projects. A payback period is defined in several ways, one of which is the number of years required to accumulate fuel savings that exactly equals the initial capital cost of the system.

PEAK LOAD—The design heating and cooling load used in mechanical system sizing. Usually set to meet human comfort requirements 93-97% of the time.

PERCENT SOLAR—A crude measure of the amount of heating or cooling provided by a solar system compared to the total demand.

PHASE-CHANGE MATERIAL—A material such as salt or wax that stores thermal energy when the material melts, and releases heat when it solidifies.

PLENUM—A chamber used for the even distribution of air entering into and exiting from remote storage systems.

POWER—In connection with machines, power is the time rate of doing work. In connection with the transmission of energy of all types, power refers to the rate at which energy is transmitted. In customary units, it is measured in watts (W) or British thermal units per hour (Btu/hr).

RADIATION—The direct transport of energy through a space by means of electromagnetic waves.

RADIATION, SOLAR—See Solar Radiation.

RECOVERED ENERGY—Energy utilized that would otherwise be wasted from an energy utilization system.

REFLECTANCE—The ratio of the light reflected by a surface to the light falling upon it.

REFLECTED RADIATION—Solar radiation reflected by light-colored or polished surfaces. It can be used to increase solar gain.

REFRACTION—The change in direction of light rays as they enter a transparent medium such as water, air, or glass.

RESISTANCE (R)—The tendency of a material to reduce the flow of heat (see R-Value).

REVERSE THERMOCIRCULATION—The convective circulation that occurs when a warm fluid is cooled, causing it to drop. (see Thermocirculation).

ROCK STORAGE SYSTEM—A solar energy system in which the collected heat is stored in a rock bin for later use. This type of storage can be used in an active, hybrid, or passive system. However, rock storage is primarily used with a system that circulates air as the transfer medium between the collector and storage, and from the storage to the heated space.

R-VALUE—A unit of thermal resistance used for comparing insulating values of different materials. The reciprocal of the conductivity. The higher the R-value of a material, the greater its insulating capabilities.

SELECTIVE SURFACE—A coating with high solar radiation absorptance and low thermal emittance, used on the surface of an absorber to increase system efficiency.

SENSIBLE HEAT—Heat that results in a temperature change.

SHADING COEFFICIENT (SC)—The ratio of the solar heat gain through a specific glazing system under a given set of conditions to the total solar heat gain through a single layer of clear, double-strength glass under the same conditions.

SOLAR ALTITUDE—The angle of the sun above the horizon, measured in a vertical plane.

SOLAR CONSTANT—The amount of radiation or heat energy that reaches the outside of the earth's atmosphere.

SOLAR HEATING FRACTION (SHF)—The percentage of heating needs supplied by the passive solar system.

SOLAR RADIATION—Electromagnetic radiation emitted by the sun.

SOLAR RETROFIT—The application of a solar heating or cooling system to an existing building.

SOLAR SAVING FRACTION (SSF)—The difference in auxiliary heat required with and without the solar collection aperture (solar wall). The "nonsolar" building is simply the same building without a solar wall.

$$\text{SSF} = \frac{\text{solar savings}}{\text{reference net thermal load}}$$

or

$$\text{SSF} = \frac{1 - \text{auxiliary heat required by solar building}}{\text{auxiliary heat required by nonsolar building}}$$

SOLAR TIME—The hours of the day reckoned by the apparent position of the sun. Solar noon is that instant on any day at which time the sun reaches its maximum altitude for that day. Solar time is very rarely the same as local standard time in any locality.

SOLAR WINDOW—An opening that is designed or placed primarily to admit solar energy into a space.

SPECIFIC HEAT—The number of Btu's required to raise the temperature of 1 lb of a material 1°F in temperature.

SPECULAR—Resembling, or produced by, a mirror, polished metal plate, or other reflector device.

STORAGE—The assembly used for storing thermal energy so that it can be used when required.

STRATIFICATION—In the context of solar heating, the formation of layers in a substance in which the top layer is warmer than the bottom.

SUNTEMPERED—A structure that is designed or oriented to take into account climatic conditions but that does not possess strict passive features such as thermal mass.

THERMAL BREAK—An element of low heat conductivity placed in such a way as to reduce or prevent the flow of heat.

THERMAL CAPACITY—See Heat Capacity.

THERMAL CONDUCTANCE—See Conductance.

THERMAL ENERGY—Heat possessed by a material resulting from the motion of molecules that can do work.

THERMAL MASS—A thermally absorptive component used to store heat energy. In a passive solar system, the mass absorbs the sun's heat during the day and radiates it at night as the temperatures drop. Thermal mass can also refer to the overall amount of heat storage capacity available in a given system or assembly.

THERMAL STORAGE CAPACITY—The ability of a material, per square foot of exposed surface area, to absorb and store heat. Numerically, the density times the specific heat times the thickness.

THERMAL STORAGE ROOF—A passive system in which the storage mass is located on the roof. Mass can be water or masonry and usually has movable insulation.

THERMAL STORAGE WALL—A passive system in which the storage mass is a wall located between the collector and the living space(s) to be heated. The mass can be a variety of materials including water or masonry.

THERMAL TRANSMITTANCE (U)—Overall coefficient of heat transmission (air-to-air) expressed in units of Btu per hour per square foot per degree F. It is the time rate of heat flow. The U-value applies to combinations of different materials used in a series along the heat flow path: single materials that comprise a building section, cavity air spaces, and surface air films on both sides of a building element.

THERMOCIRCULATION—The convective circulation of fluid, such as water or air, that occurs when warm fluid rises and is displaced by denser, cooler fluid in the same system.

THERMOSIPHON—See Thermocirculation.

TILT ANGLE—The angle of a collector relative to the ground. A rule of thumb collector angle for winter heating is the latitude + 10°, while the rule for year-round heating is the latitude − 10°.

TIME LAG—The period of time between the absorption of solar radiation by a material and its release into a space. Time lag is an important consideration in sizing a thermal storage wall.

TRANSLUCENT—Having the characteristic of transmitting light but causing sufficient diffusion to eliminate perception of distinct images.

TRANSMITTANCE—The ratio of radiant energy transmitted through a transparent or translucent substance to the total radiant energy incident on its surface.

TRANSPARENT—Having the characteristic of transmitting light so that objects or images can be seen as if there were no intervening material.

TROMBE WALL—Another name for a thermal storage wall, named after its inventor, Dr. Felix Trombe.

U-VALUE—The number of Btu's that flow through 1 sq ft of roof, wall, or floor, in 1 hour, when there is a 1°F difference in temperature between the inside and the outside air, under steady-state conditions. The U-value is the reciprocal of the resistance or R-value (see Thermal Transmittance).

VAPOR BARRIER—A layer of material, resistant to the flow of water in the gaseous state, used to prevent condensation of water within insulation or dead-air spaces.

VENTILATION, FORCED—The mechanically assisted movement of fresh air through a building using some sort of fan or blower.

VENTILATION, INDUCED—The thermally assisted movement of fresh air through a building, such as by thermocirculation.

VENTILATION, NATURAL—The unassisted movement of fresh air through a building.

WATER WALL—A passive technique for collecting solar energy. Water walls are usually black, water-filled containers exposed to the sun. These collect and store heat, which is then used to warm a living space.

WET BULB TEMPERATURE—The lowest temperature attainable by evaporating water into the air.

WYTHE—Each continuous vertical section of a wall one masonry unit in thickness and tied to an adjacent vertical section or sections.

ZONE—A space or group of spaces within a building with heating and/or cooling requirements sufficiently similar so that comfort conditions can be maintained throughout by a single controlling device.

Appendix A
SIMPLIFIED ROCK-BED DESIGN

The following is a step-by-step procedure for sizing rock beds. The fan-forced rock bed described is appropriate for remote storage in Attached Sunspaces, although the values can be modified for use with Convective Loop systems. This design procedure has been extracted from the *Passive Solar Design Handbook, Volume 2: Passive Solar Design Analysis* by J. Douglas Balcomb, et al., Los Alamos Scientific Laboratory, 1980, prepared for the U.S. Department of Energy (see Bibliography, reference no. 38).

Terminology and Formulas (see Figure A.1)

Volume = Face Area × Bed Length
Face Velocity = Airflow/Face Area
For most rock (independent of rock size):
Density = 165 lb/ ft^3 (solid)
Void = 42% (solid = 58%)
Heat Capacity = 0.21 Btu/lb − °F
Therefore, the volumetric heat capacity is:
Density × % Solid × Heat Capacity (rock)
(165) × (0.58) × (0.21) = 20 Btu/ft^3 − °F
Heat Storage = Heat Capacity (Working Bed $\triangle$T) × (Volume)

Figure A.1

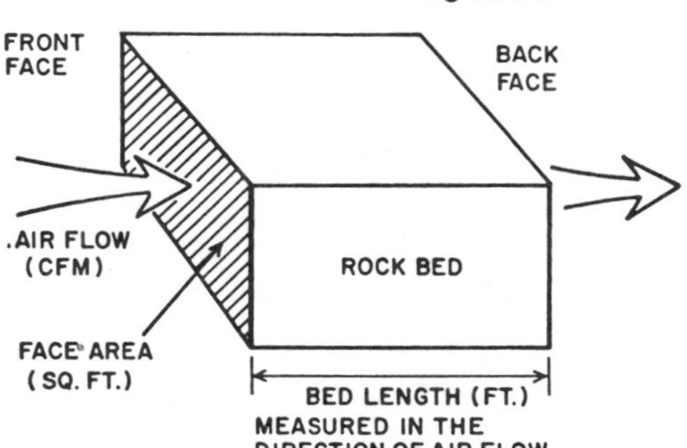

Design Procedure

As a rule of thumb, for Sunspaces, it is not advisable to design for transfer of more than approximately one-third of the net heat out of a space to a rock bed. "Net heat" means the solar energy input minus the daytime losses. To exceed one-third will result in either excessive airflow rates or excessive temperatures. The bulk of the net heat should be passively stored within the enclosing surfaces of the Sunspace. The active part of the system then becomes a minor element and the solar heating is not totally dependent on external power or the reliable functioning of mechanical equipment.

The design procedure described below consists of finding a good match between the pressure rise characteristics of the fan and the pressure drop characteristics of the rock bed and associated ducting.

In order to begin the procedure it is necessary to determine two values from the building: the working-air ΔT and working-bed ΔT.

The working-air ΔT is the air temperature into the rock bed minus the return-air temperature from the rock bed. As a rule of thumb, the working-bed ΔT can then be taken equal to about half of the working-air ΔT.

The working-air ΔT is determined by the air temperature available in the Sunspace and the average temperature at which the rock bed will tend to stabilize. In most applications, the Sunspace temperature is typically from 80-90°F and the rock-bed temperature is typically from 65-70°F. Since the air exits the rock bed at something less than the average temperature of the rock bed, a slightly lower value can be used for the return-air temperature. Thus a typical working-air ΔT is 15-20°F.

With these two values in hand, one can begin a sizing procedure as follows:

Step 1:

Establish rock-bed size and airflow requirements

$$\text{Rock-Bed Volume (cu ft)} = \frac{\text{Desired Heat Storage (Btu)}}{(20\ \text{Btu °F} - \text{cu ft}) \times (\text{Working-Bed } \Delta T,\ \text{°F})}$$

$$\text{Airflow (CFM)} = \frac{\text{Heat Transport Rate (Btuh)}}{(\text{Working-Air } \Delta T) \times (1.08\ \text{Btuh/cfm} - \text{°F})^{*}}$$

Step 2:

Select rock bed length (ft)

Calculate Face Area (sq ft)

$$\text{Face Area} = \frac{\text{Bed Volume (cu ft)}}{\text{Bed Length (ft)}}$$

*See Table A.1 for values other than those at sea level.

Step 3:

Calculate Face Velocity (ft/min)

$$\text{Face Velocity} = \frac{\text{Airflow (from Step 1)}}{\text{Face Area (from Step 2)}}$$

Step 4:

Select rock size (generally 2-4" in diameter)
Determine pressure drop per foot across bed (Figure A.2)
Calculate pressure drop (ΔP)

$$P = (\text{Pressure Drop per Foot}) \times (\text{Bed Length from Step 2})$$

Check if selected rock-bed length is greater than length recommended in Figure A.2 for good heat transfer (return to Step 2 if necessary)

Step 5:

Calculate duct area

$$\text{Duct Area} = [3.14 \times (\text{Diameter})^2 \times 0.25]$$

Calculate duct velocity (not over 700 fpm for low noise)

$$\text{Duct Velocity} = \frac{\text{Airflow (cfm) (from Step 1)}}{\text{Duct Area (sq ft)}}$$

Find Duct Friction Loss (see Figure A.3 using duct velocity and diameter)
Determine total pressure drop in ducts ($\triangle P$)

$$\triangle P = \frac{\text{[Duct Friction Loss] [Length of Duct]} + (\text{Number of Turns} \times \text{Equivalent Diameters per Turn})^{**} \times \text{Diameter of Duct (ft)}}{100}$$

Step 6:

Consult a catalog of fan sizes and select a fan that matches the required cfm and total $\triangle P$.

One can return to Step 2 or Step 4, or both, to obtain a good match.

**A value of 15 is suggested for other values. Consult *ASHRAE Handbook of Fundamentals* for more details

Example

It is desired to remove excess heat from a greenhouse Sunspace to an underfloor rock bed. The Sunspace temperature is to be limited to 85° by removing 15,000 Btuh. The rock bed is normally at about 70°F, so the working-air $\triangle T$ is 15°F. The ducting has a diameter of 18″, a length of 60′, and has 5 turns.

Given = Heat Transport Rate = 15,000 Btuh (peak) at 15°F working-air $\triangle T$. Desired Heat Storage = 2 days' accumulation (or 8 hours at peak rate).

Step 1: Desired Heat Storage = 15,000 × 8 = 120,000 Btu
Rock-Bed Volume = 120,000/(20 × 15 × 0.5) = 800 cu ft (38.3 tons)
Airflow = (15,000)/15 × 0.84*) = 1,190 cfm

Step 2: Desire a rock-bed length of 10′
Face Area = 800/10 = 80 sq ft

Step 3: Face Velocity = 1,190/80 = 14.9 ft/min

Step 4: Use 2″ rock
Pressure drop = 0.0048 water/ft (from Figure A.2)
$\triangle P$ = (0.0048) (10) = 0.048″ water
(Minimum length for good heat transfer is 12″, which is less than 120″ used; from Figure A.2)

Step 5: Duct Area = [3.14 × (1.5 ft)^2 × 0.25] = 1.77 sq ft
Duct Velocity = 1,190/1.77 = 672 fpm
Friction Loss = 0.04″ water/100′ (from Figure A.3)
$\triangle P$ – [Friction Loss] [Duct Length (ft)] + [Number of Turns] × [Diameter per Turn] × [Diameter of Duct]
$\triangle P$ = 0.04 (60 + 5 × 15 × 18/12)/100 = 0.07″ water

Step 6: One might choose a fan that will deliver 1,200 cfm @ 0.12″ water (no need to change length or rock size)

*See Table A.1 for heat capacities of air at various altitudes.

Heat Capacities of Air at Various Altitudes	Table A.1
Altitude (ft)	Heat Capacity (Btuh/F – cfm)
0	1.08
1,000	1.04
2,000	1.01
3,000	0.97
4,000	0.93
5,000	0.90
6,000	0.86
8,000	0.81
10,000	0.74

Figure A.2

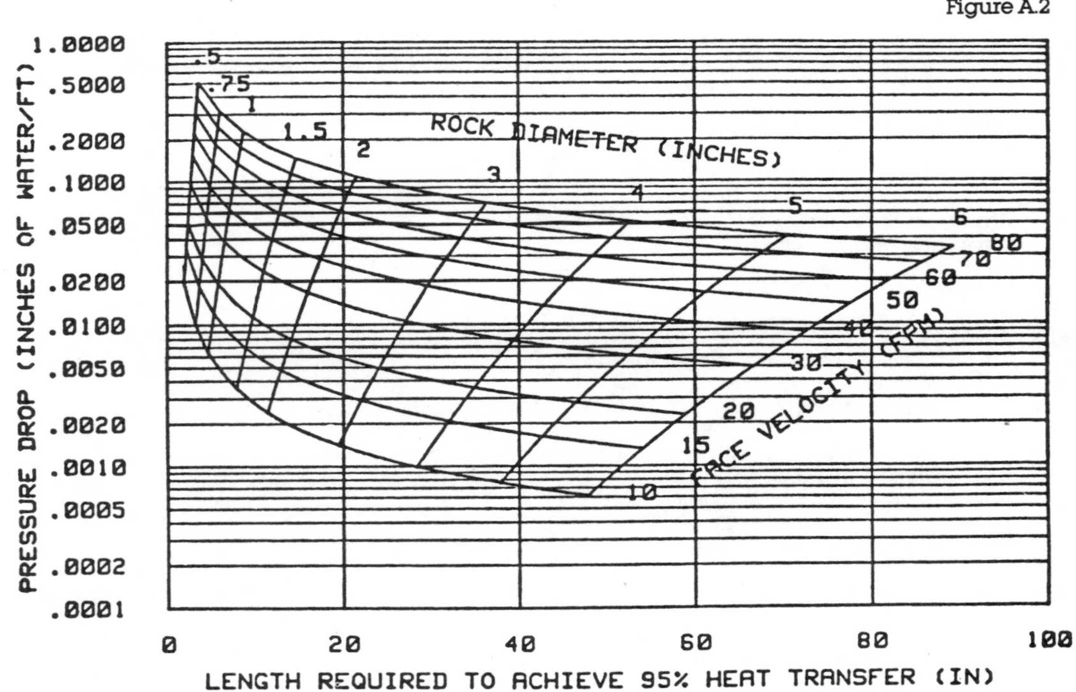

Figure A.3

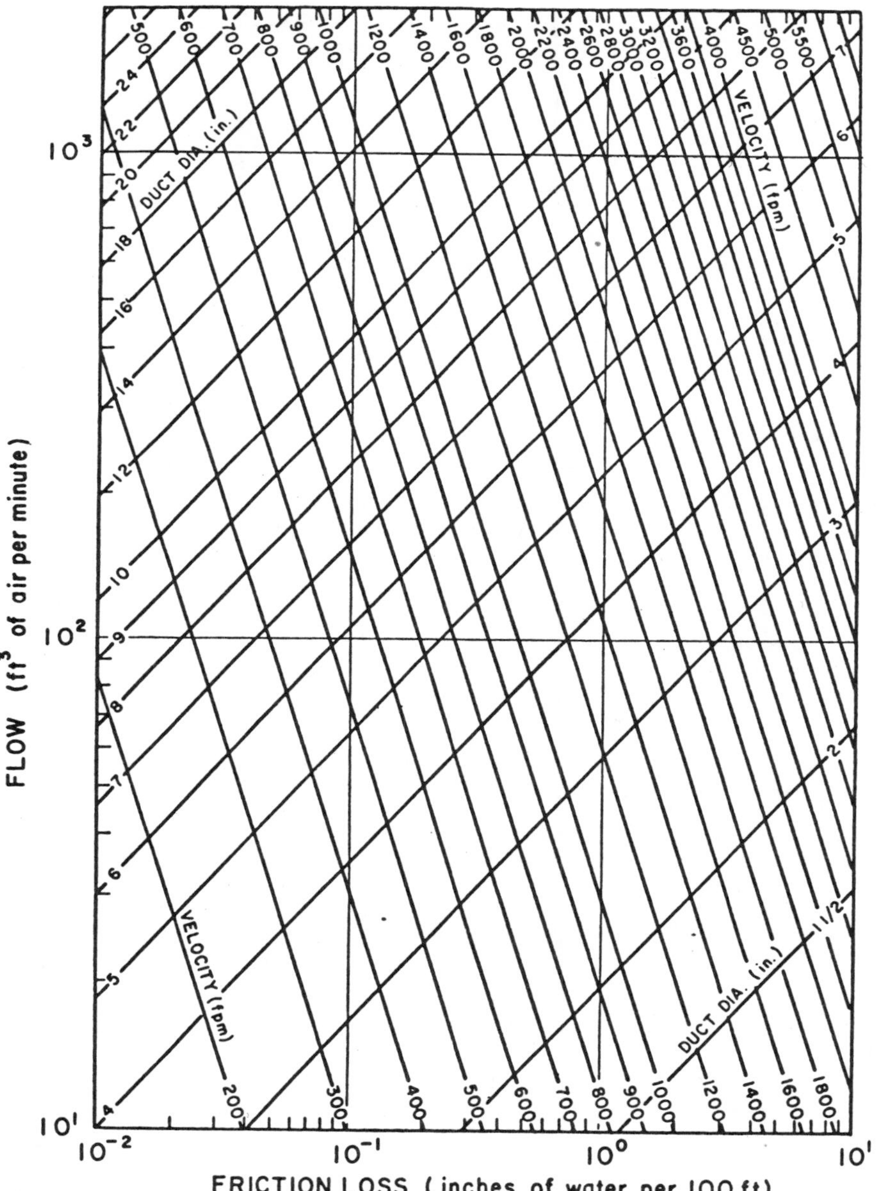
FLOW (ft³ of air per minute)
FRICTION LOSS (inches of water per 100 ft)
DUCT DIA. (in.)
VELOCITY (fpm)
DUCT DIA. (in.)
VELOCITY (fpm)

Appendix B
RECOMMENDED ANALYSIS TOOL

Throughout the design/development stages it is useful to periodically estimate the projected energy consumption of the home and the energy savings attributable to the selected passive system. Basic design guidelines and rules of thumb offer some guidance but typically do not provide estimates of the projected energy savings attributable to solar. A variety of methods is available which aid in the selection and sizing of the appropriate passive system(s). However, most of these methods tend to be cumbersome, requiring a costly piece of hardware (e.g., a microcomputer) or too much time to be practical as a design tool.

Described below is a simple and fast technique for estimating the performance of a passive design. The technique, referred to as the Load Collector Ratio or "LCR" method, is particularly useful during the schematic design stages when only the general characteristics of the home have been established and relatively major modifications can be considered.

Of course, the simplicity of the LCR method also imposes certain limitations. For example, certain design situations, such as underground structures, cannot be accurately evaluated. Furthermore, the method is based on conservative assumptions as to how the passive system is constructed (e.g., the thickness of the Thermal Storage Wall storage material is assumed to be 18"). It should be emphasized, however, that the LCR method is simply intended to provide "quick and dirty" estimates of energy savings as a tool for making design decisions. At later stages in the design process it may be appropriate to use a more sophisticated and flexible analysis method to fine-tune the estimates of energy savings derived from the LCR method (see Appendix C).

The Procedure

Although the LCR method is appropriate for the early stages of design, a few characteristics of the home need to be established prior to performing the analysis. These include: the gross floor area and length of the perimeter, the number of stories, the foundation type (slab, crawl space or basement), and the R-Values of the walls and roof.

The method consists of four steps. In Step 1 the Building Load Coefficient (BLC), a rough measure reflecting the insulating quality of the home, is estimated In Step 2 the solar collector area anticipated for the passive system, is determined. In Step 3 the BLC is divided by the collector area to yield the Load Collector Ratio (LCR). Finally, the LCR is used, in conjunction with a series of data tables, to determine the amount of auxiliary heat saved by utilizing the passive system. This savings is expressed as a fraction conventionally referred to as the Solar Savings Fraction (SSF).

Step 1: Determining the Building Load Coefficient (BLC)

This step consists of estimating and totaling the individual heat losses of each of the major components of the home.

a) Calculate Areas

Calculate the gross floor area (sq.ft.), the length of perimeter (ft.), and the area of the non-south (east, west, and north) glazing (sq.ft.). If the projected amount of glazing is not known use the following formula to derive a rough estimate:

- Non-south glazing area = [2/3] x [gross east, west and north wall area] x [non-south window fraction (see note)]

Note: The non-south window fraction is a factor specifically derived for use with this formula. It will vary between 0.05, for a home with a minimum amount of non-south window area, and 0.10 for a home with a substantial amount of non-south glazing.

b) Compute R-Values

For each of the components including walls, floor(s), and roof determine the overall R-Value. Also derive the R-Value for the perimeter insulation. These values can be derived from the American Society of Heating, Refrigerating and Air-Conditioning Engineers (ASHRAE), Handbook of Fundamentals (see BIBLIOGRAPHY reference No. 6).

c) Compute Component Heat Loss Factors

For each component (e.g., wall, roof) a factor is computed which represents the insulating quality for that component.

WALL FACTOR

- $L_w = 24 \times \frac{\text{wall area (sq.ft.)}}{\text{R-Value of walls}}$

Note: Wall area = [perimeter (ft.)] x [ceiling height (ft.)] - [area of non-south glazing (sq.ft.)] - [south window area (sq.ft.)]

NON-SOUTH GLAZING FACTOR

- $L_g = 26 \times \frac{\text{non-south glazing area}}{\text{number of glazings}}$

Note: For number of glazings use 1 for single glazed, 2 for double glazed, etc.

FLOOR FACTOR

Slab-on-Grade:

- $L_f = 100 \times \frac{\text{[length of foundation perimeter (ft.)]}}{\text{[R-Value of perimeter insulation + 5]}}$

Crawl Space (vented):

- $L_f = 24 \times \frac{\text{[area of ground floor (sq.ft.)]}}{\text{[R-Value of floor]}}$

Note: If unheated basement is specified, use crawl space calculation.

Basement (heated):

o $L_b = 256 \times \frac{\text{[perimeter length of basement wall (ft)]}}{\text{[R-Value of wall insulation + 8]}}$

Note: If more than one floor/foundation type is used, individual values for each type should be combined.

ROOF FACTOR

o $L_r = 24 \times \frac{\text{[area of insulated roof (sq.ft.)]}}{\text{[R-Value of roof]}}$

INFILTRATION FACTOR

o L_i = [0.432] x [ADR (see note)] x [ceiling height (ft.)] x [combined area of all floors (sq.ft.)] x [air change rate]

Note: The value of Air Density Ratio (ADR) can be determined from the graph below:

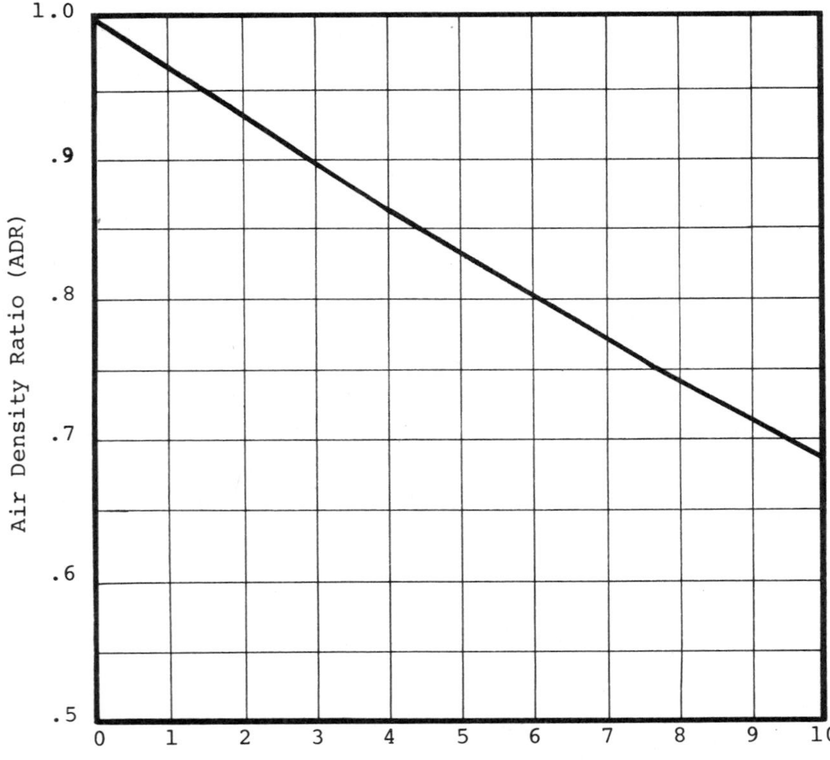

d) Determine the Building Load Coefficient (BLC)

The BLC is simply the sum of the component heat loss factors

$$BLC = L_w + L_g + L_r + L_p \text{ (and/or } L_f\text{; and/or } L_b\text{)} + L_i$$

Step 2: Determine the South Facing Glazed Area

Determine the area of south facing glazing (A_G). This area is the net glazed area of the collector excluding framing.

Step 3: Determine the Load Collector Ratio (LCR)

The LCR is a simple ratio of the BLC and the south facing glazed area (A_G):

$$LCR = BLC/A_G$$

Step 4: Identify the Solar Savings Fraction (SSF)

Armed with the LCR, identify on the tables on pages 282 through 309 the location closest to the anticipated site for the passive home and the type of passive system selected. The values calculated are for water wall, Trombe Wall (Thermal Storage Wall), and Direct Gain systems. Attached Sunspaces should be modeled as Trombe Walls. Simply locate the system type by the following code:

WW - Water Wall
WWNI - Water Wall with Night Insulation
TW - Trombe Wall
TWNI - Trombe Wall with Night Insulation
DG - Direct Gain
DGNI - Direct Gain with Night Insulation

Read across the table to the closest LCR, and read up the table to determine the SSF. To find the SSF if more than one passive system is employed:

a) Determine the SSF for each system using the BLC and total south facing glazing area in each case;
b) Average the SSF in proportion to the fraction of the total collector area represented by each system.

The SSF will is a rough estimate of the amount of auxiliary heat that will be saved by using the passive system.

Example

The following example illustrates the use of the LCR method:

A 1,800 sq. ft. two story home (25' x 36' plan) is constructed on a slab with an R-19 wall, R-30 roof and R-10 perimeter insulation. Seven and one-half percent of the east, west and north wall area is glazed (double glazing is used throughout). The air change rate is 1.0 air changes per hour. The home is to be built in Dodge City, Kansas (elev. 2580 ft.). The ceiling height is 8 ft. The south glazing is 220 sq. ft. of Direct Gain with night insulation and 280 sq. ft. of Thermal Storage Wall (Trombe Wall) without night insulation.

Step 1: BLC Calculations

a) Areas:

Plan Area = 900 sq. ft.
Gross Floor Area = 1800 sq. ft.
Perimeter Length = 122 ft.
Wall Height (2 story) = 16 ft.
Gross (non-south facing) Wall Area: (25 ft. + 25 ft. + 36 ft.) x (16 ft.)
= 1376 sq.ft.
Non-South* Glazing Area: (2/3) x (1376 sq.ft.) x (0.075)
= 69 sq.ft.
Note: Assumes average amount of glazing.

b) R-Values:

Wall	R-19
Roof	R-30
Perimeter	R-10

c) Heat Loss Factors:

Wall Area = (122 ft. x 16 ft.) - (69 sq.ft.) - (500 sq.ft.)
= 1383 sq.ft.

$$L_w = 24 \times \frac{1383}{19} = 1746$$

$$L_g = 26 \times \frac{69}{2} = 897$$

$$L_f = 100 \times \frac{122}{10 + 5} = 813$$

$$L_r = 24 \times \frac{900}{30} = 720$$

$$L_i = (0.432) \times (1) \times (0.92) \times (8 \text{ ft.}) \times (1800 \text{ sq.ft.}) = 5723$$

d) BLC = 1746 + 897 + 813 + 720 + 5723
= 9899

Step 2: South Facing Glazed Area

A_G = [20 sq.ft. (Direct Gain)] + [280 sq.ft. (Thermal Storage Wall)]
= 500 sq.ft.

Step 3: Load Collector Ratio (LCR)

$LCR = \frac{9899}{500} = 19.8$

Step 4: Solar Savings Fraction (SSF)

Find the LCR values for Dodge City, Kansas on the tables on pages 282 through 309. For the first system, Direct Gain with night insulation (DGNI), read across the table to the value closest to a LCR of 19.8 (20 in this case). Reading up the table we see that a LCR of 20 corresponds to a SSF of 0.70 or 70%. Repeating the procedure for the Thermal Storage Wall without night insulation (interpolating between an LCR of 17 and 25) yields a SSF of approximately 0.47. Proportioning the SSF by the area of each system yields:

o Direct Gain:

$\frac{220 \text{ sq.ft.}}{500 \text{ sq.ft. (total)}}$ = 0.44 x 0.7 (SSF) = 0.31

o Thermal Storage:

$\frac{280 \text{ sq.ft.}}{500 \text{ sq.ft. (total)}}$ = 0.65 x 0.47 (SSF) = 0.26

o SSF (total) = (0.31) + (0.26) = 0.57.

Approximate energy consumption for heating is then:

o AUX = (1-SSF)(DD)(BLC) = (1-.57)(5046)(9899) = 21.47 MMBTU/YR

NANAIMO BRITISH COLUMBIA		SSF =	.1	.2	.3	.4	.5	.6	.7	.8	.9
		WW	140	61	35	21	13	7	-	-	-
49.2 N	L	WWNI	229	105	65	44	32	23	17	12	8
5515 DD	C	TW	135	58	32	20	12	7	-	-	-
T(JAN)=37	R	TWNI	214	98	60	40	29	21	15	10	6
		DG	126	43	-	-	-	-	-	-	-
		DGNI	233	106	63	42	29	20	14	9	4
		WW	12.2	9.5	6.9	4.4	2.6	1.1	-	-	-
		WWNI	21.5	18.2	15.4	12.9	9.9	7.4	5.4	3.4	1.3
	D	TW	12.0	8.8	6.1	4.1	2.3	1.0	-	-	-
		TWNI	19.8	16.7	14.1	11.2	8.7	6.5	4.5	2.5	.9
		DG	9.7	4.1	-	-	-	-	-	-	-
		DGNI	21.3	17.4	13.9	10.7	8.0	5.5	3.3	1.6	.3

VANCOUVER BRITISH COLUMBIA		SSF =	.1	.2	.3	.4	.5	.6	.7	.8	.9
		WW	111	47	26	15	9	4	-	-	-
49.3 N	L	WWNI	193	89	54	37	27	20	15	11	7
5515 DD	C	TW	109	46	25	15	8	4	-	-	-
T(JAN)=37	R	TWNI	181	83	50	34	24	18	13	9	6
		DG	86	-	-	-	-	-	-	-	-
		DGNI	195	88	52	34	24	17	11	7	3
		WW	9.5	6.9	4.7	2.8	1.4	.1	-	-	-
		WWNI	18.3	15.2	12.7	10.8	8.4	6.6	4.9	3.2	1.2
	D	TW	9.5	6.7	4.5	2.8	1.4	.4	-	-	-
		TWNI	16.9	14.1	11.7	9.5	7.6	5.7	4.0	2.3	.8
		DG	5.9	-	-	-	-	-	-	-	-
		DGNI	17.8	14.3	11.4	8.8	6.6	4.6	2.7	1.3	.2

WINNIPEG MANITOBA		SSF =	.1	.2	.3	.4	.5	.6	.7	.8	.9
		WW	53	21	9	-	-	-	-	-	-
49.9 N	L	WWNI	125	58	36	25	18	13	10	7	5
10679 DD	C	TW	59	23	11	4	-	-	-	-	-
T(JAN)= -	R	TWNI	119	55	34	23	17	12	9	6	4
		DG	-	-	-	-	-	-	-	-	-
		DGNI	121	54	32	21	15	10	6	3	-
		WW	4.3	2.7	1.0	-	-	-	-	-	-
		WWNI	11.4	10.1	8.8	7.4	5.9	4.7	3.4	2.2	.8
	D	TW	4.8	3.1	1.6	.2	-	-	-	-	-
		TWNI	10.7	9.4	8.1	6.6	5.3	4.1	2.8	1.6	.5
		DG	-	-	-	-	-	-	-	-	-
		DGNI	10.8	9.0	7.1	5.4	3.9	2.5	1.3	.4	-

DARTMOUTH NOVA SCOTIA		SSF =	.1	.2	.3	.4	.5	.6	.7	.8	.9
		WW	73	31	18	11	7	4	-	-	-
44.6 N	L	WWNI	148	68	43	30	23	18	14	11	7
7361 DD	C	TW	76	33	18	11	6	3	-	-	-
T(JAN)=26	R	TWNI	140	64	40	28	21	16	12	9	6
		DG	34	-	-	-	-	-	-	-	-
		DGNI	146	66	41	28	20	15	11	7	3
		WW	6.0	4.8	3.7	2.7	1.5	.3	-	-	-
		WWNI	13.5	11.9	11.1	9.8	8.8	7.5	5.7	3.7	1.4
	D	TW	6.6	5.0	3.5	2.2	1.1	.3	-	-	-
		TWNI	12.9	11.4	10.2	9.1	7.7	6.1	4.3	2.5	1.0
		DG	1.5	-	-	-	-	-	-	-	-
		DGNI	13.1	11.3	9.8	8.3	6.5	4.6	2.9	1.4	.3

MOOSONEE ONTARIO		SSF =	.1	.2	.3	.4	.5	.6	.7	.8	.9
		WW	34	11	-	-	-	-	-	-	-
51.3 N	L	WWNI	102	48	30	21	15	11	8	6	4
11572 DD	C	TW	42	15	-	-	-	-	-	-	-
T(JAN)=-1	R	TWNI	98	45	28	19	14	10	7	5	3
		DG	-	-	-	-	-	-	-	-	-
		DGNI	96	43	26	17	11	7	4	-	-
		WW	2.5	.9	-	-	-	-	-	-	-
		WWNI	9.7	8.3	7.3	6.4	4.9	3.6	2.5	1.6	.5
	D	TW	3.4	1.7	-	-	-	-	-	-	-
		TWNI	9.2	7.8	6.9	5.6	4.3	3.2	2.2	1.2	.4
		DG	-	-	-	-	-	-	-	-	-
		DGNI	8.7	7.1	5.6	4.0	2.7	1.6	.6	-	-

OTTAWA ONTARIO		SSF =	.1	.2	.3	.4	.5	.6	.7	.8	.9
		WW	65	26	14	8	-	-	-	-	-
45.5 N	L	WWNI	140	63	39	27	20	16	12	9	6
8735 DD	C	TW	69	28	15	8	3	-	-	-	-
T(JAN)=13	R	TWNI	133	60	37	26	19	14	11	8	5
		DG	-	-	-	-	-	-	-	-	-
		DGNI	137	60	36	25	17	12	9	5	2
		WW	5.1	3.6	2.4	1.2	-	-	-	-	-
		WWNI	12.6	10.7	9.7	8.6	7.4	6.0	4.6	3.0	1.1
	D	TW	5.7	4.0	2.5	1.3	.2	-	-	-	-
		TWNI	12.0	10.2	9.0	7.8	6.5	5.1	3.5	2.1	.8
		DG	-	-	-	-	-	-	-	-	-
		DGNI	12.0	10.0	8.4	6.8	5.2	3.6	2.1	.9	.1

TORONTO ONTARIO		SSF =	.1	.2	.3	.4	.5	.6	.7	.8	.9
		WW	73	31	17	11	7	-	-	-	-
43.7 N	L	WWNI	148	68	43	30	22	17	14	10	7
6827 DD	C	TW	76	32	18	11	6	2	-	-	-
T(JAN)=25	R	TWNI	140	65	40	28	21	16	12	9	6
		DG	32	-	-	-	-	-	-	-	-
		DGNI	146	66	40	28	20	14	10	7	3
		WW	6.0	4.6	3.4	2.4	1.2	-	-	-	-
		WWNI	13.7	11.9	10.9	9.5	8.3	7.1	5.5	3.6	1.4
	D	TW	6.5	4.9	3.3	2.0	1.0	.1	-	-	-
		TWNI	12.8	11.3	10.0	8.8	7.5	5.9	4.1	2.4	.9
		DG	1.2	-	-	-	-	-	-	-	-
		DGNI	13.0	11.2	9.5	8.0	6.3	4.4	2.7	1.3	.3

NORMANDIN QUEBEC		SSF =	.1	.2	.3	.4	.5	.6	.7	.8	.9
		WW	49	19	9	-	-	-	-	-	-
48.8 N	L	WWNI	118	56	35	24	18	13	10	7	5
10528 DD	C	TW	54	22	11	4	-	-	-	-	-
T(JAN)= 4	R	TWNI	112	52	33	23	16	12	9	6	4
		DG	-	-	-	-	-	-	-	-	-
		DGNI	113	52	31	21	14	10	6	4	-
		WW	4.1	2.5	1.0	-	-	-	-	-	-
		WWNI	11.4	9.9	8.6	7.7	6.2	4.8	3.4	2.2	.8
	D	TW	4.6	2.9	1.5	.2	-	-	-	-	-
		TWNI	10.6	9.2	8.1	6.7	5.4	4.1	2.8	1.6	.5
		DG	-	-	-	-	-	-	-	-	-
		DGNI	10.6	8.7	7.1	5.5	3.9	2.6	1.4	.4	-

BIRMINGHAM ALABAMA		SSF =	.1	.2	.3	.4	.5	.6	.7	.8	.9
		WW	186	87	53	37	27	21	15	10	5
33.6 N	L	WWNI	281	134	85	60	45	36	29	23	16
2844 DD	C	TW	179	82	50	33	23	16	11	7	4
T(JAN)=44	R	TWNI	265	126	79	56	42	32	25	18	12
		DG	192	83	47	28	15	-	-	-	-
		DGNI	296	139	86	61	45	34	25	18	11
		WW	17.1	14.8	12.6	11.4	9.5	7.4	4.6	2.2	.4
		WWNI	26.1	25.0	21.2	19.4	18.3	15.6	12.9	8.3	3.4
	D	TW	16.3	13.9	11.6	8.9	6.6	4.5	2.9	1.6	.6
		TWNI	24.9	22.5	20.0	18.2	15.8	12.7	9.0	5.5	2.2
		DG	16.6	12.9	8.9	5.1	2.0	-	-	-	-
		DGNI	27.7	24.3	21.8	19.0	15.9	11.9	8.2	4.8	1.8

MOBILE ALABAMA		SSF =	.1	.2	.3	.4	.5	.6	.7	.8	.9
		WW	318	144	89	64	48	37	29	21	13
30.7 N	L	WWNI	441	205	127	92	71	56	45	36	25
1684 DD	C	TW	293	135	83	57	40	30	22	15	10
T(JAN)=51	R	TWNI	412	191	120	86	65	51	39	29	19
		DG	354	159	96	63	43	29	18	10	-
		DGNI	469	216	135	96	72	55	42	30	20
		WW	29.8	24.2	23.0	21.1	18.3	15.1	10.5	5.9	2.0
		WWNI	42.6	34.9	32.9	32.4	29.0	25.3	20.8	13.4	5.6
	D	TW	27.0	23.4	19.7	16.1	12.6	9.5	6.7	4.2	1.8
		TWNI	39.1	33.2	31.4	28.4	25.2	20.3	14.5	8.9	3.7
		DG	31.7	26.4	21.5	15.9	11.0	7.0	3.6	1.2	-
		DGNI	43.8	37.6	34.9	31.0	26.5	20.2	14.3	8.6	3.5

MONTGOMERY ALABAMA		SSF =	.1	.2	.3	.4	.5	.6	.7	.8	.9
		WW	236	111	68	48	36	27	21	15	8
32.3 N	L	WWNI	339	165	103	73	56	44	35	28	20
2269 DD	C	TW	224	103	63	43	30	22	15	11	7
T(JAN)=48	R	TWNI	320	153	96	68	51	40	30	23	15
		DG	257	115	67	42	27	16	7	-	-
		DGNI	362	171	107	75	56	42	32	23	15
		WW	22.7	18.9	16.6	15.2	12.8	10.3	6.8	3.6	1.0
		WWNI	32.7	29.9	25.6	24.2	22.5	19.3	15.9	10.2	4.3
	D	TW	20.6	17.9	15.0	11.8	8.9	6.4	4.4	2.6	1.0
		TWNI	31.0	27.2	24.5	22.2	19.3	15.6	11.1	6.8	2.8
		DG	22.9	18.6	14.1	9.4	5.7	2.7	.5	-	-
		DGNI	34.9	30.0	27.2	23.7	20.0	15.1	10.5	6.3	2.4

PHOENIX ARIZONA		SSF =	.1	.2	.3	.4	.5	.6	.7	.8	.9
		WW	467	219	139	100	75	58	45	34	22
33.4 N	L	WWNI	620	293	188	136	104	82	65	51	36
1552 DD	C	TW	436	202	126	87	63	46	34	25	16
T(JAN)=51	R	TWNI	583	275	176	126	95	73	56	42	28
		DG	555	256	157	107	75	53	37	24	12
		DGNI	673	316	201	143	107	82	62	45	29
		WW	42.5	39.7	37.2	32.9	28.2	23.2	17.0	9.9	3.7
		WWNI	56.5	54.3	50.3	48.3	40.8	35.1	28.6	18.8	7.9
	D	TW	40.1	35.6	31.1	25.6	20.3	15.2	10.9	7.0	3.1
		TWNI	53.8	50.0	47.3	41.3	35.6	29.0	20.6	12.6	5.1
		DG	50.4	44.2	37.5	29.5	21.9	15.1	9.4	4.8	1.4
		DGNI	62.2	57.3	52.7	46.2	39.2	30.0	21.1	12.8	5.3

PRESCOTT ARIZONA		SSF =	.1	.2	.3	.4	.5	.6	.7	.8	.9
		WW	189	89	56	40	30	23	17	12	7
34.6 N	L	WWNI	286	135	88	63	48	39	31	25	18
4456 DD	C	TW	183	85	53	36	25	18	13	9	5
T(JAN)=37	R	TWNI	269	128	82	59	45	35	27	20	14
		DG	198	89	52	32	20	10	-	-	-
		DGNI	300	142	90	64	49	37	28	20	13
		WW	17.1	16.5	14.3	13.1	11.3	8.9	5.7	2.9	.7
		WWNI	26.2	25.4	23.7	21.5	20.3	18.0	14.6	9.3	3.9
	D	TW	16.9	14.9	12.7	10.0	7.6	5.5	3.5	2.0	.9
		TWNI	24.9	23.8	21.7	20.2	17.7	14.1	10.0	6.2	2.5
		DG	17.6	14.6	10.6	6.8	3.6	1.0	-	-	-
		DGNI	27.7	25.8	23.7	21.4	17.8	13.4	9.3	5.5	2.1

TUCSON ARIZONA		SSF =	.1	.2	.3	.4	.5	.6	.7	.8	.9
		WW	425	199	127	92	70	55	43	32	21
32.1 N	L	WWNI	571	268	172	125	97	77	62	49	36
1752 DD	C	TW	394	185	116	81	59	43	32	23	16
T(JAN)=51	R	TWNI	533	252	161	116	89	69	54	40	27
		DG	498	231	143	98	69	49	34	22	12
		DGNI	613	289	185	132	100	77	59	43	29
		WW	39.0	36.4	33.8	31.6	27.6	23.3	16.7	9.8	3.7
		WWNI	53.0	49.5	46.6	44.5	41.0	35.4	29.5	18.8	8.0
	D	TW	36.9	33.2	29.0	23.9	19.1	14.7	10.4	6.7	3.3
		TWNI	49.8	46.1	43.6	39.9	35.1	28.2	20.2	12.5	5.2
		DG	46.0	40.7	34.7	27.2	20.3	14.2	8.8	4.5	1.4
		DGNI	57.3	52.6	49.4	44.2	37.9	29.0	20.6	12.7	5.3

WINSLOW ARIZONA		SSF =	.1	.2	.3	.4	.5	.6	.7	.8	.9
		WW	165	78	49	34	25	19	14	9	4
35.0 N	L	WWNI	255	123	79	57	43	34	27	21	15
4733 DD	C	TW	162	75	46	31	22	15	10	7	4
T(JAN)=33	R	TWNI	242	116	74	53	40	31	24	18	12
		DG	168	74	42	24	12	-	-	-	-
		DGNI	268	127	81	57	43	32	24	17	11
		WW	15.3	14.0	12.2	10.8	8.9	6.8	4.1	1.9	.2
		WWNI	23.8	23.4	20.8	19.2	17.6	15.3	11.9	7.8	3.2
	D	TW	14.6	13.1	10.8	8.4	6.2	4.3	2.7	1.5	.6
		TWNI	22.6	21.4	19.5	17.8	15.1	12.2	8.6	5.3	2.1
		DG	14.7	11.6	7.7	4.2	1.2	-	-	-	-
		DGNI	25.1	23.1	20.9	18.4	15.2	11.3	7.7	4.5	1.6

YUMA ARIZONA		SSF =	.1	.2	.3	.4	.5	.6	.7	.8	.9
		WW	728	342	219	155	116	89	70	52	35
32.7 N	L	WWNI	932	442	283	204	154	119	94	74	53
1010 DD	C	TW	670	313	194	135	97	72	53	38	25
T(JAN)=55	R	TWNI	877	413	264	187	140	108	82	61	40
		DG	887	412	254	174	124	89	64	44	26
		DGNI	1021	481	305	215	160	122	92	67	44
		WW	68.9	61.9	58.9	49.2	42.7	35.4	26.9	16.1	6.6
		WWNI	88.7	80.0	77.5	67.6	57.4	49.4	41.5	27.2	11.6
	D	TW	61.7	55.0	47.8	39.2	31.4	24.0	17.3	10.3	5.4
		TWNI	81.9	75.7	68.5	59.7	51.2	41.9	29.6	18.1	7.5
		DG	81.2	71.5	61.0	49.4	37.6	27.3	18.2	10.4	4.0
		DGNI	94.8	88.2	77.6	67.9	57.7	44.5	31.3	19.2	8.2

FORT SMITH ARKANSAS		SSF =	.1	.2	.3	.4	.5	.6	.7	.8	.9
		WW	158	74	45	31	23	17	13	8	-
35.3 N	L	WWNI	246	119	75	53	40	32	26	20	14
3336 DD	C	TW	153	70	43	29	20	14	9	6	3
T(JAN)=39	R	TWNI	233	111	70	50	37	29	22	17	11
		DG	156	67	36	20	-	-	-	-	-
		DGNI	258	121	76	53	40	30	22	16	10
		WW	15.0	12.4	10.8	9.7	8.0	6.1	3.6	1.6	-
		WWNI	23.5	21.8	18.8	17.6	16.5	14.2	11.7	7.4	3.1
	D	TW	14.0	12.0	9.9	7.5	5.4	3.7	2.3	1.2	.4
		TWNI	22.3	19.9	17.9	16.4	14.2	11.4	8.1	5.0	2.0
		DG	13.4	10.1	6.2	2.8	-	-	-	-	-
		DGNI	24.6	21.3	19.3	16.9	14.0	10.4	7.1	4.1	1.5

LITTLE ROCK ARKANSAS		SSF =	.1	.2	.3	.4	.5	.6	.7	.8	.9
		WW	157	72	44	31	22	17	12	8	-
34.7 N	L	WWNI	246	117	74	52	40	31	25	20	14
3354 DD	C	TW	153	69	42	28	19	13	9	6	3
T(JAN)=39	R	TWNI	232	110	69	49	37	29	22	16	11
		DG	154	65	35	18	-	-	-	-	-
		DGNI	257	120	75	53	39	29	22	15	10
		WW	14.6	12.1	10.5	9.4	7.7	5.8	3.4	1.5	-
		WWNI	23.0	21.4	18.5	17.2	16.1	13.9	11.4	7.3	3.0
	D	TW	13.8	11.7	9.6	7.3	5.2	3.5	2.2	1.2	.3
		TWNI	21.9	19.6	17.6	16.0	13.9	11.2	7.9	4.9	2.0
		DG	13.0	9.6	5.8	2.4	-	-	-	-	-
		DGNI	24.2	21.0	18.9	16.5	13.7	10.2	7.0	4.0	1.5

BAKERSFIELD CALIFORNIA		SSF =	.1	.2	.3	.4	.5	.6	.7	.8	.9
		WW	323	143	89	63	46	35	26	18	10
35.4 N	L	WWNI	449	203	127	91	69	54	42	32	22
2185 DD	C	TW	297	135	81	55	39	28	20	14	8
T(JAN)=48	R	TWNI	418	191	119	85	63	48	36	27	17
		DG	359	158	93	60	40	26	15	-	-
		DGNI	475	216	135	94	69	52	39	27	17
		WW	28.6	24.4	22.6	19.7	15.7	11.9	8.5	4.5	1.3
		WWNI	41.9	34.7	33.5	31.1	26.5	21.1	16.2	11.1	4.6
	D	TW	26.9	22.7	18.8	15.1	11.6	8.5	5.7	3.3	1.2
		TWNI	38.0	33.2	30.7	26.9	22.3	18.1	12.7	7.6	3.0
		DG	31.5	25.4	19.8	14.3	9.5	5.4	2.3	-	-
		DGNI	42.9	37.5	34.0	28.8	23.6	18.2	12.4	7.2	2.8

DAGGETT CALIFORNIA		SSF =	.1	.2	.3	.4	.5	.6	.7	.8	.9
		WW	360	165	103	73	55	42	32	24	15
34.9 N	L	WWNI	491	228	144	103	79	62	49	39	27
2203 DD	C	TW	334	153	94	65	46	34	25	17	11
T(JAN)=47	R	TWNI	461	214	135	96	72	56	43	32	21
		DG	411	185	112	74	51	35	23	13	-
		DGNI	526	244	153	108	81	61	46	33	22
		WW	33.0	28.9	26.3	24.2	20.3	16.2	11.6	6.5	2.2
		WWNI	46.6	40.2	37.7	35.4	32.0	26.8	21.3	14.2	5.9
	D	TW	30.6	26.6	22.6	18.3	14.4	10.8	7.4	4.6	1.9
		TWNI	42.6	38.2	35.1	31.6	27.0	22.0	15.6	9.5	3.9
		DG	36.9	31.1	25.3	19.0	13.5	8.7	4.8	1.9	-
		DGNI	48.3	43.0	39.6	34.5	29.0	22.3	15.5	9.3	3.8

FRESNO CALIFORNIA		SSF =	.1	.2	.3	.4	.5	.6	.7	.8	.9
		WW	272	122	74	51	37	27	20	13	7
36.8 N	L	WWNI	383	179	110	77	58	45	34	26	18
2650 DD	C	TW	252	114	68	45	31	22	15	10	5
T(JAN)=45	R	TWNI	361	166	102	72	53	40	30	22	14
		DG	295	128	72	44	27	15	-	-	-
		DGNI	408	186	114	79	57	42	31	22	13
		WW	25.8	19.8	17.4	15.1	11.7	8.5	5.8	2.8	.6
		WWNI	37.0	29.9	27.4	24.9	21.6	17.1	12.7	8.8	3.6
	D	TW	22.6	18.8	15.1	11.6	8.7	6.2	3.9	2.2	.7
		TWNI	34.1	28.4	25.3	22.0	18.0	14.4	10.2	6.0	2.3
		DG	25.6	19.6	14.1	9.3	5.4	2.2	-	-	-
		DGNI	37.7	31.8	27.9	23.2	18.6	14.3	9.6	5.5	2.0

LONG BEACH CALIFORNIA		SSF =	.1	.2	.3	.4	.5	.6	.7	.8	.9
		WW	527	249	156	112	85	67	53	40	27
33.8 N	L	WWNI	689	331	208	150	115	91	74	59	42
1606 DD	C	TW	488	228	142	99	72	53	39	29	19
T(JAN)=54	R	TWNI	650	307	195	139	106	83	64	48	32
		DG	630	291	180	123	87	63	44	30	17
		DGNI	753	354	224	159	120	92	70	52	34
		WW	50.5	43.9	40.6	37.1	33.3	28.8	20.7	12.4	4.9
		WWNI	66.1	59.1	54.6	51.7	47.0	42.1	34.8	22.4	9.6
	D	TW	45.3	40.4	35.5	29.2	23.0	17.8	13.0	8.3	4.2
		TWNI	62.3	55.2	51.2	46.5	41.5	33.3	23.9	14.8	6.2
		DG	57.9	50.8	43.8	34.6	26.2	18.8	12.3	6.8	2.4
		DGNI	71.4	63.6	58.1	52.3	45.0	34.6	24.7	15.3	6.5

LOS ANGELES CALIFORNIA		SSF =	.1	.2	.3	.4	.5	.6	.7	.8	.9
		WW	563	259	161	115	87	68	54	41	27
33.9 N	L	WWNI	737	342	215	153	117	93	75	60	43
1819 DD	C	TW	513	238	147	101	73	54	40	29	20
T(JAN)=54	R	TWNI	687	320	200	143	108	84	65	49	33
		DG	665	304	187	127	90	65	46	31	18
		DGNI	793	369	231	163	123	94	72	53	35
		WW	52.5	45.3	41.1	37.7	33.7	29.6	21.2	12.8	5.1
		WWNI	71.1	60.1	55.3	52.1	47.4	42.5	35.7	22.9	9.8
	D	TW	47.7	41.4	36.1	29.5	23.7	18.3	13.1	8.7	4.3
		TWNI	64.3	56.3	51.7	47.0	42.6	33.9	24.3	15.1	6.3
		DG	61.0	52.0	44.7	35.3	27.0	19.4	12.7	7.1	2.7
		DGNI	74.0	64.7	58.9	53.0	45.9	35.3	25.2	15.7	6.7

MOUNT SHASTA CALIFORNIA		SSF =	.1	.2	.3	.4	.5	.6	.7	.8	.9
		WW	142	62	37	25	18	13	9	5	-
41.3 N	L	WWNI	230	106	66	46	35	27	21	16	11
5890 DD	C	TW	137	61	36	23	15	10	7	4	-
T(JAN)=34	R	TWNI	216	99	62	43	32	25	19	14	9
		DG	130	50	23	-	-	-	-	-	-
		DGNI	237	107	66	46	33	25	18	12	7
		WW	12.4	10.1	8.5	7.0	5.4	3.6	1.9	.5	-
		WWNI	21.0	18.3	16.6	15.0	13.2	11.4	8.5	5.7	2.3
	D	TW	12.1	9.7	7.6	5.6	3.9	2.4	1.3	.5	-
		TWNI	20.0	17.2	15.4	13.6	11.6	9.2	6.5	3.9	1.5
		DG	10.4	6.5	2.7	-	-	-	-	-	-
		DGNI	21.4	18.3	15.9	13.7	11.0	8.1	5.4	2.9	.9

NEEDLES CALIFORNIA — 34.8 N, 1428 DD, T(JAN)=52

		SSF = .1	.2	.3	.4	.5	.6	.7	.8	.9
	WW	508	239	152	108	81	62	47	35	23
L	WWNI	668	317	202	147	111	86	68	53	37
C	TW	474	220	136	94	67	49	36	26	17
R	TWNI	630	297	189	134	100	77	59	44	29
	DG	608	280	171	116	81	57	40	26	13
	DGNI	729	342	217	153	114	86	65	47	30
	WW	47.4	42.8	40.9	34.3	29.0	23.6	17.5	10.2	3.9
	WWNI	62.3	57.4	54.6	49.8	41.8	34.9	28.8	19.1	8.1
D	TW	43.3	38.5	33.1	27.3	21.3	16.1	11.4	6.9	3.3
	TWNI	59.0	53.9	49.8	43.2	36.5	29.9	21.1	12.8	5.2
	DG	55.1	48.1	40.2	31.7	23.3	16.2	10.1	5.2	1.5
	DGNI	67.6	62.2	55.5	48.0	40.6	31.2	21.7	13.1	5.4

OAKLAND CALIFORNIA — 37.7 N, 2909 DD, T(JAN)=49

		SSF = .1	.2	.3	.4	.5	.6	.7	.8	.9
	WW	348	170	107	75	56	42	32	23	14
L	WWNI	472	234	150	106	80	62	49	38	27
C	TW	327	155	96	66	47	34	24	17	11
R	TWNI	446	219	139	98	73	56	43	31	20
	DG	405	189	115	76	51	35	22	12	-
	DGNI	512	249	157	110	82	61	46	33	21
	WW	34.8	31.7	26.6	23.1	20.0	15.3	11.0	6.1	2.0
	WWNI	47.2	44.8	39.1	33.8	30.6	26.4	19.8	13.5	5.6
D	TW	31.7	27.6	23.2	18.7	14.3	10.4	7.0	4.2	2.0
	TWNI	44.5	40.8	35.4	31.5	26.6	21.0	15.0	9.1	3.6
	DG	38.8	32.5	25.8	19.2	13.3	8.3	4.4	1.6	-
	DGNI	50.6	45.6	39.7	34.7	28.1	21.6	15.0	8.9	3.5

RED BLUFF CALIFORNIA — 40.1 N, 2688 DD, T(JAN)=45

		SSF = .1	.2	.3	.4	.5	.6	.7	.8	.9
	WW	258	115	69	48	35	26	19	13	6
L	WWNI	365	170	104	74	56	43	33	26	18
C	TW	239	108	64	43	30	21	15	10	5
R	TWNI	345	158	98	69	51	39	29	21	14
	DG	276	119	68	41	25	14	-	-	-
	DGNI	389	177	109	76	55	41	30	21	13
	WW	24.4	18.5	16.7	14.8	11.4	8.3	5.7	2.7	.6
	WWNI	35.5	28.4	26.0	24.1	21.6	17.1	12.6	8.7	3.5
D	TW	21.4	18.0	14.4	11.1	8.4	6.0	3.9	2.1	.7
	TWNI	32.6	27.0	24.3	21.6	17.7	14.2	10.1	6.0	2.3
	DG	23.9	18.4	13.2	8.6	4.9	1.8	-	-	-
	DGNI	35.9	30.1	26.7	22.6	18.2	14.0	9.4	5.4	1.9

SACRAMENTO CALIFORNIA — 38.5 N, 2843 DD, T(JAN)=45

		SSF = .1	.2	.3	.4	.5	.6	.7	.8	.9
	WW	258	119	73	50	36	26	19	13	6
L	WWNI	368	174	108	76	57	44	34	26	18
C	TW	244	110	66	44	31	22	15	10	5
R	TWNI	347	162	101	71	52	39	30	21	14
	DG	283	124	71	44	27	15	-	-	-
	DGNI	393	182	112	78	57	42	31	22	13
	WW	23.9	20.2	17.4	15.0	11.6	8.3	5.7	2.7	.5
	WWNI	34.2	30.7	27.1	24.4	21.7	17.1	12.4	8.7	3.5
D	TW	22.0	18.5	15.0	11.8	8.6	6.0	3.9	2.0	.6
	TWNI	32.6	28.4	25.2	22.1	17.9	14.1	10.1	6.0	2.3
	DG	24.7	19.4	13.9	9.2	5.1	2.0	-	-	-
	DGNI	36.3	31.4	27.6	23.2	18.4	14.1	9.4	5.4	1.9

SAN DIEGO CALIFORNIA — 32.7 N, 1507 DD, T(JAN)=55

		SSF = .1	.2	.3	.4	.5	.6	.7	.8	.9
	WW	612	284	179	129	98	77	61	46	31
L	WWNI	796	373	235	170	131	104	84	67	48
C	TW	562	261	163	113	82	61	46	33	23
R	TWNI	744	348	220	158	120	94	73	54	37
	DG	734	338	210	144	103	75	53	37	22
	DGNI	864	403	255	181	137	105	80	59	40
	WW	56.7	49.9	47.0	43.3	39.0	33.5	24.4	14.7	6.0
	WWNI	74.2	66.1	62.3	59.7	53.5	48.3	39.3	25.5	10.9
D	TW	51.9	46.0	40.5	33.6	26.8	20.9	15.3	9.9	4.9
	TWNI	69.7	62.0	58.3	52.9	47.6	38.0	27.3	16.9	7.1
	DG	67.0	58.8	51.2	41.1	31.7	23.1	15.5	8.9	3.5
	DGNI	80.7	71.9	66.4	60.1	51.4	39.9	28.5	17.7	7.6

SAN FRANCISCO CALIFORNIA — 37.6 N, 3042 DD, T(JAN)=48

		SSF = .1	.2	.3	.4	.5	.6	.7	.8	.9
	WW	332	163	103	72	53	40	31	22	13
L	WWNI	453	225	145	103	77	60	47	37	26
C	TW	313	149	92	63	45	33	23	16	10
R	TWNI	428	210	134	94	71	54	41	30	20
	DG	385	180	109	72	49	32	20	11	-
	DGNI	491	239	151	106	79	59	44	32	20
	WW	33.2	30.6	25.5	22.0	19.0	14.7	10.5	5.8	1.9
	WWNI	45.3	43.2	37.9	32.5	29.4	25.5	19.3	13.1	5.4
D	TW	30.4	26.5	22.3	17.9	13.6	9.9	6.7	4.0	1.9
	TWNI	42.8	39.6	34.2	30.4	25.7	20.4	14.6	8.8	3.6
	DG	37.0	30.9	24.5	18.1	12.4	7.7	3.9	1.3	-
	DGNI	48.7	44.0	38.2	33.4	27.1	20.8	14.4	8.6	3.4

SANTA MARIA CALIFORNIA — 34.9 N, 3053 DD, T(JAN)=50

		SSF = .1	.2	.3	.4	.5	.6	.7	.8	.9
	WW	327	164	107	77	58	46	35	26	17
L	WWNI	448	224	148	108	83	66	53	42	30
C	TW	311	152	97	67	49	36	27	19	12
R	TWNI	423	211	138	100	76	59	46	34	23
	DG	383	185	116	79	55	38	26	16	7
	DGNI	485	241	156	112	85	65	50	36	24
	WW	32.7	32.3	29.1	25.7	22.4	19.1	13.2	7.6	2.8
	WWNI	44.8	44.8	42.6	37.4	33.8	29.6	24.9	15.9	6.8
D	TW	30.9	28.2	24.8	20.2	15.9	12.0	8.4	5.1	2.6
	TWNI	42.3	41.6	37.8	34.0	30.3	23.9	17.1	10.5	4.4
	DG	37.7	33.6	28.3	21.5	15.5	10.3	6.0	2.7	.6
	DGNI	48.5	46.9	42.3	37.7	32.0	24.3	17.1	10.4	4.3

COLORADO SPRINGS COLORADO — 38.8 N, 6473 DD, T(JAN)=29

		SSF = .1	.2	.3	.4	.5	.6	.7	.8	.9
	WW	130	60	38	26	19	15	11	7	-
L	WWNI	211	102	65	47	36	29	23	18	13
C	TW	128	59	36	24	17	11	8	5	2
R	TWNI	201	96	61	44	33	26	20	15	10
	DG	120	51	26	10	-	-	-	-	-
	DGNI	220	104	66	47	35	26	20	14	9
	WW	12.5	10.5	9.2	8.1	6.9	4.9	2.8	1.1	-
	WWNI	20.4	18.7	17.4	16.0	14.6	13.2	10.8	6.8	2.8
D	TW	11.8	10.2	8.3	6.3	4.5	3.0	1.8	.8	.3
	TWNI	19.4	17.6	16.2	14.7	13.1	10.4	7.4	4.5	1.8
	DG	10.4	7.5	3.8	.4	-	-	-	-	-
	DGNI	20.9	19.0	16.9	15.2	12.5	9.2	6.3	3.6	1.3

DENVER COLORADO — 39.7 N, 6016 DD, T(JAN)=30

		SSF = .1	.2	.3	.4	.5	.6	.7	.8	.9
	WW	136	63	39	27	20	15	11	7	-
L	WWNI	218	105	67	48	37	29	24	19	13
C	TW	132	61	38	25	17	12	8	5	3
R	TWNI	207	99	63	45	34	27	21	15	10
	DG	127	54	28	13	-	-	-	-	-
	DGNI	227	108	68	48	36	27	20	14	9
	WW	12.9	11.1	9.6	8.4	7.1	5.1	3.0	1.2	-
	WWNI	21.8	19.1	18.2	16.2	15.0	13.6	10.8	6.9	2.9
D	TW	12.5	10.6	8.7	6.5	4.7	3.2	1.9	.9	.3
	TWNI	20.2	18.2	16.5	15.1	13.3	10.6	7.5	4.6	1.9
	DG	11.2	8.0	4.3	1.1	-	-	-	-	-
	DGNI	21.8	19.6	17.4	15.6	12.8	9.5	6.5	3.7	1.3

EAGLE COLORADO — 39.6 N, 8426 DD, T(JAN)=18

		SSF = .1	.2	.3	.4	.5	.6	.7	.8	.9
	WW	95	43	26	17	12	8	5	-	-
L	WWNI	172	81	52	37	28	22	17	13	9
C	TW	96	43	25	16	10	7	4	-	-
R	TWNI	163	77	49	34	26	20	15	11	7
	DG	72	22	-	-	-	-	-	-	-
	DGNI	174	81	51	35	26	19	14	9	5
	WW	8.5	7.1	5.8	4.7	3.3	1.9	.6	-	-
	WWNI	16.0	15.2	13.4	12.2	10.8	9.3	7.1	4.7	1.9
D	TW	8.7	7.0	5.4	3.7	2.3	1.3	.5	-	-
	TWNI	15.3	13.8	12.5	11.1	9.5	7.6	5.3	3.2	1.2
	DG	5.4	1.5	-	-	-	-	-	-	-
	DGNI	16.2	14.3	12.6	10.8	8.6	6.2	4.0	2.1	.6

GRAND JUNCTION COLORADO — 39.1 N, 5605 DD, T(JAN)=27

		SSF = .1	.2	.3	.4	.5	.6	.7	.8	.9
	WW	135	61	37	25	18	13	9	5	-
L	WWNI	222	104	66	46	35	28	22	17	12
C	TW	133	60	36	23	16	11	7	4	1
R	TWNI	210	98	61	43	33	25	19	14	9
	DG	125	50	24	-	-	-	-	-	-
	DGNI	229	106	66	46	34	25	18	13	8
	WW	12.0	10.3	8.7	7.5	5.7	4.1	2.2	.7	-
	WWNI	20.1	18.7	16.8	15.2	13.9	11.8	9.1	6.0	2.4
D	TW	11.8	9.8	7.9	5.9	4.1	2.6	1.5	.6	.1
	TWNI	19.2	17.4	15.6	14.1	11.9	9.6	6.8	4.1	1.6
	DG	10.2	6.7	3.0	-	-	-	-	-	-
	DGNI	20.9	18.4	16.3	14.1	11.5	8.5	5.7	3.2	1.1

PUEBLO COLORADO — 38.3 N, 5394 DD, T(JAN)=30

		SSF = .1	.2	.3	.4	.5	.6	.7	.8	.9
	WW	145	67	41	29	21	16	12	8	-
L	WWNI	234	109	70	50	38	30	24	19	14
C	TW	143	65	39	26	18	13	9	5	3
R	TWNI	220	103	66	47	36	28	21	16	11
	DG	140	59	32	16	-	-	-	-	-
	DGNI	242	113	71	50	38	28	21	15	9
	WW	12.8	11.8	10.2	9.0	7.5	5.5	3.2	1.4	-
	WWNI	21.2	20.3	18.6	16.9	15.4	13.9	11.2	7.1	3.0
D	TW	12.9	11.0	9.1	7.0	5.1	3.4	2.1	1.1	.4
	TWNI	20.2	18.9	17.2	15.7	13.7	11.0	7.8	4.8	1.9
	DG	11.8	8.8	5.1	1.9	-	-	-	-	-
	DGNI	22.1	20.1	18.1	16.2	13.4	9.9	6.8	3.9	1.4

HARTFORD CONNECTICUT — 41.9 N, 6350 DD, T(JAN)=25

		SSF = .1	.2	.3	.4	.5	.6	.7	.8	.9
	WW	62	26	14	9	5	-	-	-	-
L	WWNI	134	62	39	28	21	16	13	10	7
C	TW	66	28	16	9	5	-	-	-	-
R	TWNI	127	59	37	26	19	15	11	8	5
	DG	-	-	-	-	-	-	-	-	-
	DGNI	131	60	37	25	18	13	9	6	3
	WW	5.2	3.8	2.8	1.7	.6	-	-	-	-
	WWNI	12.4	11.3	10.0	8.9	7.9	6.6	5.1	3.3	1.3
D	TW	5.6	4.2	2.8	1.6	.5	-	-	-	-
	TWNI	11.6	10.6	9.3	8.2	6.9	5.5	3.8	2.3	.8
	DG	-	-	-	-	-	-	-	-	-
	DGNI	11.8	10.2	8.8	7.3	5.7	3.9	2.4	1.1	.2

WILMINGTON DELAWARE — 39.7 N, 4940 DD, T(JAN)=32

		SSF = .1	.2	.3	.4	.5	.6	.7	.8	.9
	WW	95	44	26	18	12	9	5	-	-
L	WWNI	171	83	52	37	28	22	18	14	9
C	TW	97	44	26	17	11	7	4	-	-
R	TWNI	163	78	49	35	26	20	15	11	7
	DG	74	24	-	-	-	-	-	-	-
	DGNI	174	82	51	36	26	19	14	10	5
	WW	8.9	7.2	6.0	4.8	3.5	2.1	.7	-	-
	WWNI	16.1	15.5	13.4	12.3	11.1	9.3	7.4	4.8	1.9
D	TW	8.7	7.2	5.5	3.9	2.5	1.4	.6	-	-
	TWNI	15.4	14.1	12.7	11.3	9.6	7.7	5.4	3.3	1.3
	DG	5.7	1.9	-	-	-	-	-	-	-
	DGNI	16.5	14.5	12.8	10.9	8.8	6.4	4.1	2.2	.7

WASHINGTON DC — 38.9 N, 5010 DD, T(JAN)=32

		SSF = .1	.2	.3	.4	.5	.6	.7	.8	.9
	WW	92	42	25	17	11	8	4	-	-
L	WWNI	169	80	51	36	27	21	17	13	9
C	TW	94	42	25	16	10	6	3	-	-
R	TWNI	160	76	48	34	25	19	15	11	7
	DG	69	19	-	-	-	-	-	-	-
	DGNI	171	80	50	35	25	19	13	9	5
	WW	8.4	6.9	5.5	4.4	3.1	1.8	.5	-	-
	WWNI	15.6	14.9	13.1	11.7	10.7	8.9	7.1	4.6	1.8
D	TW	8.4	6.8	5.2	3.6	2.2	1.2	.4	-	-
	TWNI	14.9	13.7	12.2	10.9	9.3	7.4	5.2	3.1	1.2
	DG	5.0	.9	-	-	-	-	-	-	-
	DGNI	15.9	14.0	12.3	10.4	8.4	6.0	3.9	2.1	.6

APALACHICOLA FLORIDA — 29.7 N, 1361 DD, T(JAN)=54

		SSF = .1	.2	.3	.4	.5	.6	.7	.8	.9
	WW	405	182	114	81	61	48	37	27	18
L	WWNI	547	249	156	113	87	68	55	44	31
C	TW	371	170	104	72	51	38	28	20	13
R	TWNI	511	233	147	105	79	62	48	36	24
	DG	464	208	126	84	59	41	28	17	8
	DGNI	585	267	168	119	89	68	52	38	25
	WW	36.6	31.0	29.9	26.7	23.1	19.7	13.9	8.0	2.9
	WWNI	50.2	42.8	41.4	39.2	34.7	30.4	25.3	16.3	6.9
D	TW	33.7	29.1	25.2	20.5	16.3	12.4	8.8	5.6	2.5
	TWNI	46.7	40.7	38.4	34.5	31.0	24.7	17.6	10.9	4.5
	DG	41.1	34.7	29.2	22.5	16.4	11.1	6.6	3.1	.6
	DGNI	53.0	46.7	43.0	38.3	32.9	25.2	17.8	10.8	4.5

DAYTONA BEACH
FLORIDA

29.2 N
902 DD
T(JAN)=58

	SSF =	.1	.2	.3	.4	.5	.6	.7	.8	.9
	WW	603	281	180	130	100	79	62	48	32
L	WWNI	784	367	235	172	133	106	86	69	49
C	TW	556	260	164	114	83	62	47	34	23
R	TWNI	735	344	221	159	122	96	74	56	38
	DG	725	337	211	146	105	76	55	38	23
	DGNI	852	399	256	183	139	107	82	61	41
	WW	55.5	50.8	49.4	45.2	39.8	35.4	25.3	15.3	6.3
	WWNI	73.8	66.4	65.2	61.3	55.2	49.0	41.9	26.4	11.4
D	TW	50.7	46.6	41.7	34.3	27.9	21.7	15.9	10.5	5.4
	TWNI	67.6	63.0	59.8	54.4	49.3	39.1	28.1	17.5	7.3
	DG	65.9	59.8	52.8	42.3	32.8	24.1	16.2	9.4	3.7
	DGNI	78.1	73.2	68.4	61.6	53.3	41.1	29.4	18.3	7.9

JACKSONVILLE
FLORIDA

30.5 N
1327 DD
T(JAN)=55

	SSF =	.1	.2	.3	.4	.5	.6	.7	.8	.9
	WW	425	191	120	87	66	52	41	31	20
L	WWNI	573	260	164	119	92	74	60	47	34
C	TW	390	179	111	77	55	41	30	22	15
R	TWNI	533	244	154	111	85	66	52	39	26
	DG	489	221	136	92	65	46	32	21	11
	DGNI	611	279	177	126	96	73	56	41	27
	WW	37.8	32.9	32.3	29.8	26.0	22.6	15.8	9.3	3.5
	WWNI	52.8	44.9	44.0	43.2	38.5	33.9	28.8	18.1	7.7
D	TW	35.4	31.2	27.3	22.4	18.0	13.9	10.0	6.4	3.1
	TWNI	48.4	42.9	41.4	37.7	33.9	27.1	19.4	12.0	5.0
	DG	43.5	37.7	32.3	25.2	18.8	13.1	8.1	4.1	1.2
	DGNI	55.2	49.2	46.6	41.9	36.1	27.8	19.7	12.1	5.1

MIAMI
FLORIDA

25.8 N
206 DD
T(JAN)=67

	SSF =	.1	.2	.3	.4	.5	.6	.7	.8	.9
	WW	2285	1138	731	516	389	307	242	186	130
L	WWNI	2804	1401	906	641	483	382	307	242	174
C	TW	2102	1013	641	449	329	245	184	135	94
R	TWNI	2638	1304	832	589	446	344	265	198	133
	DG	2912	1403	886	620	452	336	249	181	121
	DGNI	3119	1534	975	692	520	398	303	224	151
	WW	229.	218.	189.	166.	153.	131.	101.	64.	28.
	WWNI	280.	274.	240.	205.	188.	178.	136.	92.	39.
D	TW	206.	185.	163.	137.	110.	84.	63.	42.	23.
	TWNI	264.	247.	214.	192.	169.	136.	97.	61.	25.
	DG	286.	255.	226.	188.	149.	114.	81.	52.	25.
	DGNI	312.	287.	251.	228.	190.	149.	107.	68.	30.

ORLANDO
FLORIDA

28.5 N
733 DD
T(JAN)=60

	SSF =	.1	.2	.3	.4	.5	.6	.7	.8	.9
	WW	723	346	224	163	124	98	78	60	41
L	WWNI	927	446	289	211	163	130	105	84	60
C	TW	663	319	202	142	104	78	58	43	30
R	TWNI	867	418	270	195	149	117	91	68	46
	DG	879	421	265	185	134	99	72	51	32
	DGNI	1010	487	314	226	171	132	101	75	50
	WW	69.8	64.7	61.8	56.4	49.8	43.6	31.9	19.6	8.2
	WWNI	90.9	83.1	81.9	75.0	67.1	60.0	50.3	32.1	13.8
D	TW	64.0	58.4	52.1	43.2	35.1	27.4	20.1	12.8	7.0
	TWNI	83.7	78.8	73.7	66.7	59.7	47.7	34.2	21.2	8.9
	DG	84.3	76.5	67.6	54.9	43.0	32.0	22.0	13.1	5.6
	DGNI	97.3	92.0	84.6	76.2	65.8	50.7	36.3	22.6	9.8

TALLAHASSEE
FLORIDA

30.4 N
1563 DD
T(JAN)=53

	SSF =	.1	.2	.3	.4	.5	.6	.7	.8	.9
	WW	359	162	101	72	55	43	33	24	16
L	WWNI	491	226	141	102	79	62	50	40	28
C	TW	330	152	93	64	46	34	25	18	12
R	TWNI	458	211	133	95	72	56	43	32	22
	DG	405	183	110	74	51	35	23	14	5
	DGNI	523	240	151	107	80	62	47	34	22
	WW	33.2	27.3	26.3	24.0	20.8	17.8	12.3	7.0	2.5
	WWNI	46.5	38.5	36.9	36.6	32.0	28.0	23.7	15.0	6.4
D	TW	30.2	26.2	22.5	18.2	14.5	11.0	7.8	4.9	2.1
	TWNI	42.7	36.7	35.1	31.5	28.2	22.6	16.2	10.0	4.1
	DG	36.2	30.5	25.3	19.2	13.8	9.1	5.2	2.2	.3
	DGNI	48.2	41.9	39.0	34.7	29.9	22.9	16.1	9.8	4.0

TAMPA
FLORIDA

28.0 N
718 DD
T(JAN)=60

	SSF =	.1	.2	.3	.4	.5	.6	.7	.8	.9
	WW	720	343	223	162	124	98	78	59	41
L	WWNI	926	441	288	211	162	129	105	83	60
C	TW	663	317	201	141	104	78	58	43	30
R	TWNI	866	415	269	195	149	117	90	68	46
	DG	878	418	264	184	134	98	72	50	32
	DGNI	1008	484	312	225	171	132	101	75	50
	WW	67.7	64.8	61.7	56.0	49.6	43.5	31.8	19.5	8.2
	WWNI	87.2	83.2	82.5	74.9	66.5	59.8	50.9	32.0	13.8
D	TW	63.4	57.8	51.8	43.2	35.1	27.3	20.1	12.8	6.5
	TWNI	82.1	78.7	73.7	66.5	59.6	47.6	34.1	21.2	8.9
	DG	83.5	75.8	67.2	54.9	42.9	31.9	21.9	13.1	5.6
	DGNI	95.7	91.5	84.2	76.1	65.6	50.6	36.2	22.5	9.8

WEST PALM BEACH
FLORIDA

26.7 N
299 DD
T(JAN)=65

	SSF =	.1	.2	.3	.4	.5	.6	.7	.8	.9
	WW	1523	761	494	350	264	208	164	126	88
L	WWNI	1886	943	619	441	331	261	211	167	120
C	TW	1407	682	433	304	223	167	125	91	63
R	TWNI	1775	884	568	403	305	237	183	136	92
	DG	1930	935	592	415	302	224	165	119	79
	DGNI	2093	1038	663	471	355	272	207	153	103
	WW	152.	151.	130.	112.	103.	90.	68.	43.	19.
	WWNI	189.	189.	168.	141.	128.	121.	96.	64.	28.
D	TW	139.	125.	111.	93.	75.	58.	42.	28.	15.
	TWNI	178.	171.	150.	132.	118.	94.	67.	42.	18.
	DG	192.	171.	152.	125.	99.	75.	53.	34.	16.
	DGNI	209.	197.	172.	155.	132.	102.	73.	46.	21.

ATLANTA
GEORGIA

33.6 N
3095 DD
T(JAN)=42

	SSF =	.1	.2	.3	.4	.5	.6	.7	.8	.9
	WW	172	79	48	33	25	19	14	9	4
L	WWNI	264	126	79	56	42	33	27	21	15
C	TW	166	75	46	31	21	15	10	7	4
R	TWNI	249	118	74	52	39	30	23	17	11
	DG	173	74	41	23	10	-	-	-	-
	DGNI	276	129	80	56	42	32	23	17	10
	WW	15.8	13.5	11.5	10.2	8.5	6.6	4.0	1.8	.2
	WWNI	24.6	22.9	20.0	18.3	16.9	14.6	12.1	7.7	3.2
D	TW	15.0	12.8	10.5	8.1	5.8	4.1	2.5	1.4	.5
	TWNI	23.3	21.0	18.8	17.0	14.8	11.9	8.4	5.2	2.1
	DG	14.8	11.2	7.3	3.7	.8	-	-	-	-
	DGNI	25.9	22.6	20.3	17.6	14.8	11.0	7.5	4.4	1.6

AUGUSTA GEORGIA			SSF = .1	.2	.3	.4	.5	.6	.7	.8	.9
		WW	213	101	62	43	32	25	19	13	7
33.4 N	L	WWNI	312	152	95	67	51	41	33	26	18
2547 DD	C	TW	204	94	58	39	27	20	14	9	6
T(JAN)=46	R	TWNI	295	141	89	63	48	37	28	21	14
		DG	228	101	59	37	22	12	-	-	-
		DGNI	332	157	98	69	52	39	29	21	13
		WW	20.5	17.2	15.0	13.8	11.6	9.4	6.0	3.1	.8
		WWNI	30.0	27.9	23.8	22.4	21.0	18.0	15.0	9.6	4.0
	D	TW	18.7	16.3	13.7	10.7	8.0	5.7	3.8	2.2	.9
		TWNI	28.5	25.2	22.7	20.7	18.0	14.5	10.4	6.4	2.6
		DG	20.3	16.3	12.0	7.8	4.3	1.6	-	-	-
		DGNI	32.0	27.7	25.1	21.9	18.5	13.9	9.7	5.7	2.2

MACON GEORGIA			SSF = .1	.2	.3	.4	.5	.6	.7	.8	.9
		WW	244	114	70	49	37	28	21	15	9
32.7 N	L	WWNI	349	169	105	74	57	45	36	28	20
2240 DD	C	TW	230	106	65	44	31	22	16	11	7
T(JAN)=48	R	TWNI	330	156	98	69	52	41	31	23	16
		DG	266	119	70	44	28	17	9	-	-
		DGNI	373	175	109	77	57	43	33	24	15
		WW	23.8	19.1	17.1	15.7	13.4	11.0	7.2	3.8	1.1
		WWNI	34.0	29.7	26.0	25.0	23.1	19.9	16.7	10.6	4.4
	D	TW	21.2	18.4	15.5	12.2	9.1	6.7	4.6	2.8	1.1
		TWNI	32.1	27.5	25.0	22.7	19.9	16.1	11.5	7.1	2.9
		DG	23.7	19.2	14.7	10.0	6.1	3.1	.9	-	-
		DGNI	35.9	30.5	27.8	24.3	20.7	15.6	10.9	6.5	2.6

SAVANNAH GEORGIA			SSF = .1	.2	.3	.4	.5	.6	.7	.8	.9
		WW	278	131	81	57	43	33	26	19	11
32.1 N	L	WWNI	389	189	118	84	65	51	41	32	23
1952 DD	C	TW	260	122	75	51	37	26	19	13	9
T(JAN)=50	R	TWNI	367	175	110	79	59	46	36	27	18
		DG	309	141	84	55	37	24	14	6	-
		DGNI	417	197	124	88	66	50	38	27	18
		WW	27.5	22.5	20.4	18.7	15.9	13.3	9.0	4.9	1.6
		WWNI	38.9	33.7	30.0	28.9	26.2	22.6	19.1	12.1	5.1
	D	TW	24.5	21.4	18.2	14.3	11.1	8.2	5.7	3.5	1.5
		TWNI	36.1	31.2	28.7	25.8	22.8	18.4	13.1	8.1	3.3
		DG	28.4	23.5	18.7	13.4	8.9	5.2	2.4	.4	-
		DGNI	40.5	35.0	32.0	28.0	23.9	18.1	12.7	7.7	3.1

BOISE IDAHO			SSF = .1	.2	.3	.4	.5	.6	.7	.8	.9
		WW	132	58	34	22	15	10	5	-	-
43.6 N	L	WWNI	215	101	63	44	32	25	19	14	9
5833 DD	C	TW	127	56	32	20	13	8	4	-	-
T(JAN)=29	R	TWNI	203	95	59	41	30	22	16	12	7
		DG	117	42	-	-	-	-	-	-	-
		DGNI	221	102	62	42	30	22	15	10	5
		WW	11.8	9.5	7.2	5.4	3.6	2.1	.7	-	-
		WWNI	21.1	17.5	15.8	13.4	11.3	9.2	6.5	4.4	1.7
	D	TW	11.4	8.9	6.5	4.6	3.0	1.6	.6	-	-
		TWNI	19.2	16.7	14.3	12.1	9.8	7.5	5.3	3.1	1.1
		DG	9.3	4.6	-	-	-	-	-	-	-
		DGNI	20.5	17.4	14.5	11.9	9.0	6.6	4.1	2.1	.6

LEWISTON IDAHO			SSF = .1	.2	.3	.4	.5	.6	.7	.8	.9
		WW	110	46	26	15	9	4	-	-	-
46.4 N	L	WWNI	192	88	54	37	27	20	15	11	7
5464 DD	C	TW	108	46	25	15	9	4	-	-	-
T(JAN)=31	R	TWNI	181	82	50	34	25	18	13	9	6
		DG	85	-	-	-	-	-	-	-	-
		DGNI	194	87	52	34	24	17	11	7	3
		WW	9.2	7.1	4.9	3.0	1.5	.3	-	-	-
		WWNI	18.5	14.8	13.3	10.9	8.7	6.8	4.8	3.3	1.3
	D	TW	9.3	6.7	4.6	2.9	1.5	.4	-	-	-
		TWNI	16.5	14.0	11.9	9.7	7.7	5.8	4.1	2.3	.8
		DG	5.8	-	-	-	-	-	-	-	-
		DGNI	17.4	14.4	11.6	9.0	6.7	4.7	2.8	1.3	.2

POCATELLO IDAHO			SSF = .1	.2	.3	.4	.5	.6	.7	.8	.9
		WW	105	47	28	18	12	8	3	-	-
42.9 N	L	WWNI	182	88	55	39	29	22	17	13	9
7063 DD	C	TW	104	47	27	17	11	6	3	-	-
T(JAN)=23	R	TWNI	173	82	51	36	26	20	15	11	7
		DG	84	27	-	-	-	-	-	-	-
		DGNI	186	87	54	37	26	19	14	9	5
		WW	9.9	7.8	5.9	4.4	2.8	1.5	.2	-	-
		WWNI	17.7	15.6	14.2	12.1	10.4	8.5	6.1	4.1	1.6
	D	TW	9.5	7.5	5.5	3.7	2.3	1.1	.3	-	-
		TWNI	16.8	14.7	12.8	11.0	9.1	7.0	4.9	2.9	1.1
		DG	6.5	2.1	-	-	-	-	-	-	-
		DGNI	17.6	15.3	12.8	10.5	8.1	5.8	3.7	1.8	.5

CHICAGO ILLINOIS			SSF = .1	.2	.3	.4	.5	.6	.7	.8	.9
		WW	73	31	17	11	6	-	-	-	-
41.8 N	L	WWNI	149	69	43	30	22	17	14	10	7
6127 DD	C	TW	78	33	18	11	6	-	-	-	-
T(JAN)=24	R	TWNI	141	65	40	28	21	16	12	9	5
		DG	34	-	-	-	-	-	-	-	-
		DGNI	148	67	41	28	20	14	10	7	3
		WW	6.3	4.6	3.4	2.2	1.1	-	-	-	-
		WWNI	13.1	12.7	10.6	9.3	8.4	6.9	5.3	3.5	1.3
	D	TW	6.5	4.9	3.4	2.0	.9	-	-	-	-
		TWNI	12.6	11.4	10.0	8.8	7.3	5.8	4.0	2.4	.9
		DG	1.3	-	-	-	-	-	-	-	-
		DGNI	13.1	11.3	9.6	7.9	6.2	4.3	2.6	1.3	.2

MOLINE ILLINOIS			SSF = .1	.2	.3	.4	.5	.6	.7	.8	.9
		WW	72	30	17	10	6	-	-	-	-
41.4 N	L	WWNI	147	68	42	30	22	17	13	10	7
6395 DD	C	TW	76	32	18	10	6	-	-	-	-
T(JAN)=21	R	TWNI	139	64	40	28	21	16	12	9	5
		DG	29	-	-	-	-	-	-	-	-
		DGNI	145	66	40	27	19	14	10	6	3
		WW	6.2	4.4	3.2	2.1	1.0	-	-	-	-
		WWNI	13.1	12.3	10.4	9.1	8.3	6.9	5.3	3.4	1.3
	D	TW	6.4	4.7	3.3	1.9	.8	-	-	-	-
		TWNI	12.6	11.2	9.7	8.7	7.2	5.7	4.0	2.4	.9
		DG	.8	-	-	-	-	-	-	-	-
		DGNI	13.0	11.0	9.4	7.8	6.1	4.2	2.6	1.2	.2

SPRINGFIELD ILLINOIS		SSF =	.1	.2	.3	.4	.5	.6	.7	.8	.9
		WW	90	39	22	14	10	6	-	-	-
39.8 N	L	WWNI	168	78	49	34	25	20	16	12	8
5558 DD	C	TW	92	39	23	14	9	5	2	-	-
T(JAN)=27	R	TWNI	159	74	46	32	24	18	14	10	6
		DG	61	-	-	-	-	-	-	-	-
		DGNI	169	77	47	32	23	17	12	8	4
		WW	7.9	5.9	4.7	3.6	2.4	1.1	-	-	-
		WWNI	15.0	14.1	12.0	10.6	9.8	8.2	6.4	4.2	1.7
	D	TW	7.8	6.1	4.5	2.9	1.7	.8	.1	-	-
		TWNI	14.4	12.8	11.2	10.0	8.5	6.8	4.8	2.9	1.1
		DG	4.0	-	-	-	-	-	-	-	-
		DGNI	15.2	13.0	11.2	9.4	7.6	5.4	3.4	1.8	.5

EVANSVILLE INDIANA		SSF =	.1	.2	.3	.4	.5	.6	.7	.8	.9
		WW	104	46	26	17	12	8	5	-	-
38.0 N	L	WWNI	182	86	54	37	28	22	17	13	9
4629 DD	C	TW	104	45	26	17	11	7	4	-	-
T(JAN)=33	R	TWNI	173	81	50	35	26	20	15	11	7
		DG	81	25	-	-	-	-	-	-	-
		DGNI	186	85	52	36	26	19	14	9	5
		WW	9.3	7.2	5.6	4.5	3.3	1.9	.6	-	-
		WWNI	16.9	15.6	13.3	11.6	10.9	9.1	7.2	4.7	1.9
	D	TW	9.1	7.2	5.5	3.7	2.3	1.3	.5	-	-
		TWNI	16.1	14.2	12.3	11.1	9.4	7.5	5.3	3.2	1.2
		DG	5.9	1.9	-	-	-	-	-	-	-
		DGNI	17.1	14.5	12.5	10.6	8.6	6.2	4.0	2.1	.6

FORT WAYNE INDIANA		SSF =	.1	.2	.3	.4	.5	.6	.7	.8	.9
		WW	62	25	13	8	-	-	-	-	-
41.0 N	L	WWNI	136	62	39	27	20	16	12	9	6
6209 DD	C	TW	67	28	15	8	4	-	-	-	-
T(JAN)=25	R	TWNI	128	59	37	25	19	14	11	8	5
		DG	-	-	-	-	-	-	-	-	-
		DGNI	132	60	36	24	17	12	9	5	2
		WW	5.1	3.6	2.4	1.3	-	-	-	-	-
		WWNI	12.1	11.3	9.6	8.3	7.4	6.2	4.7	3.1	1.2
	D	TW	5.5	4.0	2.6	1.3	.2	-	-	-	-
		TWNI	11.5	10.3	8.9	7.9	6.5	5.2	3.6	2.1	.8
		DG	-	-	-	-	-	-	-	-	-
		DGNI	11.7	9.9	8.4	6.9	5.3	3.6	2.2	1.0	.1

INDIANAPOLIS INDIANA		SSF =	.1	.2	.3	.4	.5	.6	.7	.8	.9
		WW	74	31	17	11	7	-	-	-	-
39.7 N	L	WWNI	148	69	43	30	23	17	14	11	7
5577 DD	C	TW	77	33	18	11	6	2	-	-	-
T(JAN)=28	R	TWNI	141	65	41	28	21	16	12	9	6
		DG	35	-	-	-	-	-	-	-	-
		DGNI	147	67	41	28	20	14	10	7	3
		WW	6.5	4.6	3.4	2.3	1.3	-	-	-	-
		WWNI	13.4	12.6	10.6	9.3	8.5	7.1	5.5	3.6	1.4
	D	TW	6.6	4.9	3.4	2.1	.9	.1	-	-	-
		TWNI	12.9	11.4	9.9	8.9	7.4	5.9	4.1	2.4	.9
		DG	1.4	-	-	-	-	-	-	-	-
		DGNI	13.4	11.3	9.6	8.0	6.3	4.4	2.7	1.3	.3

SOUTH BEND INDIANA		SSF =	.1	.2	.3	.4	.5	.6	.7	.8	.9
		WW	60	24	12	6	-	-	-	-	-
41.7 N	L	WWNI	137	61	38	26	19	15	11	9	6
6462 DD	C	TW	66	26	14	7	-	-	-	-	-
T(JAN)=24	R	TWNI	129	58	36	25	18	13	10	7	5
		DG	-	-	-	-	-	-	-	-	-
		DGNI	132	58	35	23	16	12	8	5	-
		WW	4.8	3.2	1.9	.7	-	-	-	-	-
		WWNI	11.8	10.8	9.1	7.8	6.9	5.6	4.2	2.8	1.0
	D	TW	5.3	3.6	2.2	1.0	-	-	-	-	-
		TWNI	11.2	9.9	8.5	7.4	6.1	4.8	3.3	1.9	.7
		DG	-	-	-	-	-	-	-	-	-
		DGNI	11.3	9.4	7.8	6.3	4.8	3.2	1.9	.8	-

BURLINGTON IOWA		SSF =	.1	.2	.3	.4	.5	.6	.7	.8	.9
		WW	83	36	20	13	8	5	-	-	-
40.8 N	L	WWNI	159	74	46	32	24	19	15	11	8
6149 DD	C	TW	85	36	21	13	7	4	-	-	-
T(JAN)=23	R	TWNI	151	70	43	30	22	17	13	9	6
		DG	50	-	-	-	-	-	-	-	-
		DGNI	159	72	44	30	22	16	11	7	4
		WW	7.2	5.4	4.1	3.0	1.9	.6	-	-	-
		WWNI	14.3	13.4	11.4	10.0	9.2	7.6	5.9	3.9	1.5
	D	TW	7.3	5.6	4.0	2.6	1.3	.5	-	-	-
		TWNI	13.7	12.2	10.6	9.5	7.9	6.4	4.4	2.6	1.0
		DG	3.0	-	-	-	-	-	-	-	-
		DGNI	14.4	12.2	10.5	8.8	6.9	4.9	3.1	1.5	.4

DES MOINES IOWA		SSF =	.1	.2	.3	.4	.5	.6	.7	.8	.9
		WW	78	33	18	11	7	3	-	-	-
41.5 N	L	WWNI	155	71	44	31	23	18	14	11	7
6710 DD	C	TW	81	34	19	11	6	3	-	-	-
T(JAN)=19	R	TWNI	146	67	41	29	21	16	12	9	6
		DG	41	-	-	-	-	-	-	-	-
		DGNI	153	69	42	29	20	15	11	7	3
		WW	6.6	4.8	3.6	2.5	1.5	.2	-	-	-
		WWNI	13.9	12.7	10.8	9.5	8.6	7.3	5.6	3.7	1.4
	D	TW	6.8	5.1	3.6	2.2	1.0	.2	-	-	-
		TWNI	13.2	11.6	10.1	9.0	7.5	6.1	4.2	2.5	1.0
		DG	2.0	-	-	-	-	-	-	-	-
		DGNI	13.7	11.5	9.8	8.2	6.5	4.5	2.8	1.4	.3

MASON CITY IOWA		SSF =	.1	.2	.3	.4	.5	.6	.7	.8	.9
		WW	68	27	14	8	4	-	-	-	-
43.1 N	L	WWNI	143	65	40	28	21	16	12	9	6
7901 DD	C	TW	72	29	16	9	4	-	-	-	-
T(JAN)=14	R	TWNI	136	62	38	26	19	15	11	8	5
		DG	-	-	-	-	-	-	-	-	-
		DGNI	141	63	37	25	18	13	9	6	2
		WW	5.5	3.7	2.6	1.5	.3	-	-	-	-
		WWNI	13.0	11.3	9.7	8.5	7.5	6.3	4.8	3.2	1.2
	D	TW	5.9	4.2	2.7	1.4	.4	-	-	-	-
		TWNI	12.2	10.6	9.1	8.0	6.7	5.3	3.7	2.2	.8
		DG	-	-	-	-	-	-	-	-	-
		DGNI	12.3	10.2	8.5	7.1	5.4	3.7	2.3	1.0	.1

SIOUX CITY IOWA		SSF =	.1	.2	.3	.4	.5	.6	.7	.8	.9
		WW	77	32	17	11	7	-	-	-	-
42.4 N	L	WWNI	153	70	44	30	22	17	14	10	7
6953 DD	C	TW	80	33	18	11	6	-	-	-	-
T(JAN)=18	R	TWNI	144	66	41	28	21	16	12	9	6
		DG	37	-	-	-	-	-	-	-	-
		DGNI	151	68	41	28	20	14	10	7	3
		WW	6.5	4.6	3.3	2.3	1.2	-	-	-	-
		WWNI	14.0	12.4	10.6	9.2	8.4	7.0	5.4	3.5	1.4
	D	TW	6.7	5.0	3.4	2.0	.9	-	-	-	-
		TWNI	13.2	11.4	9.9	8.8	7.3	5.9	4.1	2.4	.9
		DG	1.6	-	-	-	-	-	-	-	-
		DGNI	13.6	11.3	9.5	7.9	6.2	4.3	2.7	1.3	.3

DODGE CITY KANSAS		SSF =	.1	.2	.3	.4	.5	.6	.7	.8	.9
		WW	140	64	39	27	19	15	10	6	-
37.8 N	L	WWNI	228	107	67	48	36	29	23	18	13
5046 DD	C	TW	138	62	37	25	17	11	8	5	2
T(JAN)=31	R	TWNI	215	100	63	45	34	26	20	15	10
		DG	132	54	27	10	-	-	-	-	-
		DGNI	236	109	68	47	35	26	20	14	8
		WW	12.5	10.6	9.1	8.1	6.6	4.7	2.7	1.0	-
		WWNI	20.6	19.6	17.1	15.6	14.4	12.8	10.3	6.6	2.7
	D	TW	12.2	10.3	8.4	6.3	4.5	2.9	1.7	.8	.2
		TWNI	19.7	17.8	16.0	14.6	12.7	10.2	7.2	4.4	1.8
		DG	10.8	7.5	3.9	.3	-	-	-	-	-
		DGNI	21.6	18.9	16.8	14.9	12.3	9.1	6.2	3.5	1.2

GOODLAND KANSAS		SSF =	.1	.2	.3	.4	.5	.6	.7	.8	.9
		WW	120	55	33	23	17	12	8	5	-
39.4 N	L	WWNI	202	95	61	43	33	26	21	16	12
6119 DD	C	TW	118	54	32	21	14	10	6	4	1
T(JAN)=28	R	TWNI	190	90	57	40	31	24	18	13	9
		DG	105	42	18	-	-	-	-	-	-
		DGNI	206	97	60	42	31	24	17	12	7
		WW	10.8	9.4	7.8	6.7	5.4	3.7	1.9	.6	-
		WWNI	19.7	17.2	16.0	14.2	12.9	11.7	9.3	5.9	2.5
	D	TW	10.9	9.0	7.2	5.3	3.7	2.4	1.3	.5	.1
		TWNI	18.1	16.4	14.6	13.2	11.6	9.2	6.5	4.0	1.6
		DG	8.7	5.4	1.7	-	-	-	-	-	-
		DGNI	19.3	17.1	15.2	13.4	10.9	8.0	5.4	3.0	1.0

TOPEKA KANSAS		SSF =	.1	.2	.3	.4	.5	.6	.7	.8	.9
		WW	112	50	29	20	14	10	6	-	-
39.1 N	L	WWNI	194	91	57	40	30	23	19	14	10
5243 DD	C	TW	112	49	29	18	12	8	5	2	-
T(JAN)=28	R	TWNI	183	85	53	37	28	21	16	12	8
		DG	93	33	-	-	-	-	-	-	-
		DGNI	198	91	56	39	28	21	15	10	6
		WW	10.0	7.9	6.4	5.4	4.0	2.6	1.1	-	-
		WWNI	17.7	16.5	14.2	12.6	11.6	9.9	7.9	5.1	2.1
	D	TW	9.8	7.9	6.1	4.3	2.8	1.7	.8	.2	-
		TWNI	16.9	14.9	13.2	11.9	10.1	8.1	5.7	3.5	1.4
		DG	7.1	3.4	-	-	-	-	-	-	-
		DGNI	18.2	15.6	13.5	11.6	9.4	6.8	4.5	2.4	.8

WICHITA KANSAS		SSF =	.1	.2	.3	.4	.5	.6	.7	.8	.9
		WW	138	63	38	26	19	14	10	6	-
37.6 N	L	WWNI	224	106	66	47	35	28	22	17	12
4687 DD	C	TW	135	61	36	24	16	11	7	4	2
T(JAN)=31	R	TWNI	212	99	62	44	33	25	19	14	9
		DG	128	52	25	-	-	-	-	-	-
		DGNI	232	107	66	46	34	26	19	13	8
		WW	12.6	10.3	8.7	7.6	6.1	4.4	2.4	.8	-
		WWNI	20.7	19.1	16.8	15.1	14.1	12.2	9.8	6.3	2.6
	D	TW	12.0	10.1	8.1	5.9	4.1	2.7	1.5	.7	.1
		TWNI	19.8	17.5	15.6	14.2	12.2	9.8	7.0	4.3	1.7
		DG	10.5	7.1	3.3	-	-	-	-	-	-
		DGNI	21.4	18.7	16.4	14.3	11.8	8.7	5.9	3.3	1.1

LEXINGTON KENTUCKY		SSF =	.1	.2	.3	.4	.5	.6	.7	.8	.9
		WW	96	42	24	16	11	7	4	-	-
38.0 N	L	WWNI	174	82	51	36	27	21	17	13	9
4729 DD	C	TW	98	42	25	15	10	6	3	-	-
T(JAN)=33	R	TWNI	165	77	48	34	25	19	15	11	7
		DG	71	17	-	-	-	-	-	-	-
		DGNI	177	81	50	34	25	18	13	9	5
		WW	8.6	6.6	5.2	4.1	2.9	1.6	.4	-	-
		WWNI	16.0	14.9	12.7	11.2	10.4	8.7	7.0	4.5	1.8
	D	TW	8.5	6.7	5.0	3.4	2.0	1.1	.3	-	-
		TWNI	15.3	13.5	11.8	10.6	9.0	7.2	5.1	3.1	1.2
		DG	4.9	.4	-	-	-	-	-	-	-
		DGNI	16.2	13.8	11.9	10.1	8.2	5.8	3.8	2.0	.6

LOUISVILLE KENTUCKY		SSF =	.1	.2	.3	.4	.5	.6	.7	.8	.9
		WW	98	44	25	17	12	8	4	-	-
38.2 N	L	WWNI	176	84	52	37	27	21	17	13	9
4645 DD	C	TW	99	44	25	16	10	6	3	-	-
T(JAN)=33	R	TWNI	167	78	49	34	25	19	15	11	7
		DG	74	21	-	-	-	-	-	-	-
		DGNI	179	83	51	35	25	19	14	9	5
		WW	8.9	6.9	5.4	4.3	3.1	1.8	.5	-	-
		WWNI	16.3	15.3	13.1	11.4	10.6	8.9	7.1	4.6	1.8
	D	TW	8.7	6.9	5.2	3.5	2.2	1.2	.4	-	-
		TWNI	15.6	13.8	12.1	10.8	9.2	7.4	5.2	3.1	1.2
		DG	5.3	1.2	-	-	-	-	-	-	-
		DGNI	16.5	14.2	12.2	10.3	8.4	6.0	3.9	2.1	.6

BATON ROUGE LOUISIANA		SSF =	.1	.2	.3	.4	.5	.6	.7	.8	.9
		WW	312	142	87	62	46	36	28	20	12
30.5 N	L	WWNI	430	203	125	90	69	54	43	34	24
1670 DD	C	TW	287	133	81	55	39	28	21	15	9
T(JAN)=51	R	TWNI	405	188	118	84	63	49	38	28	19
		DG	346	156	93	61	41	27	17	8	-
		DGNI	460	213	133	94	70	53	40	29	19
		WW	30.3	23.6	22.1	20.0	17.2	14.1	9.7	5.4	1.7
		WWNI	43.0	34.4	31.8	31.4	27.5	23.9	19.6	12.7	5.3
	D	TW	26.6	22.8	19.3	15.2	12.0	9.0	6.2	3.9	1.6
		TWNI	39.0	32.7	30.5	27.2	24.1	19.4	13.8	8.5	3.5
		DG	31.2	25.5	20.4	14.8	10.2	6.3	3.1	.9	-
		DGNI	43.5	37.0	33.8	29.7	25.2	19.3	13.5	8.1	3.3

LAKE CHARLES LOUISIANA		SSF =	.1	.2	.3	.4	.5	.6	.7	.8	.9
		WW	330	144	89	63	48	37	28	21	13
30.1 N	L	WWNI	457	205	127	92	70	55	44	35	25
1498 DD	C	TW	301	136	83	56	40	29	21	15	10
T(JAN)=52	R	TWNI	425	192	120	85	64	50	39	29	19
		DG	364	160	95	62	42	28	18	9	-
		DGNI	482	218	136	96	71	54	41	30	19
		WW	29.3	23.9	23.1	20.6	17.8	14.4	10.0	5.6	1.8
		WWNI	43.2	34.2	33.1	31.7	28.2	24.7	19.7	12.9	5.4
	D	TW	27.1	23.0	19.4	15.8	12.4	9.3	6.5	4.0	1.7
		TWNI	38.8	32.9	30.9	27.9	24.8	19.9	14.1	8.7	3.5
		DG	31.8	25.9	20.9	15.5	10.7	6.7	3.4	1.0	-
		DGNI	43.6	37.3	34.3	30.6	25.8	19.8	13.9	8.3	3.3

NEW ORLEANS LOUISIANA		SSF =	.1	.2	.3	.4	.5	.6	.7	.8	.9
		WW	374	169	104	74	56	44	34	25	16
30.0 N	L	WWNI	506	235	146	105	80	64	51	40	29
1465 DD	C	TW	342	158	97	66	47	35	25	18	12
T(JAN)=53	R	TWNI	474	219	137	98	74	57	44	33	22
		DG	423	191	115	76	52	36	24	14	6
		DGNI	542	249	156	110	83	63	48	35	23
		WW	35.5	28.2	26.8	24.6	21.3	17.7	12.5	7.1	2.5
		WWNI	50.6	39.6	37.4	36.6	32.7	28.6	23.2	15.1	6.4
	D	TW	31.5	27.2	23.2	18.6	14.8	11.2	8.0	5.0	2.2
		TWNI	45.0	37.9	35.6	32.1	28.6	22.9	16.4	10.1	4.1
		DG	38.1	31.8	26.1	19.7	14.2	9.5	5.4	2.3	.3
		DGNI	50.6	43.2	39.8	35.4	30.2	23.2	16.4	10.0	4.1

SHREVEPORT LOUISIANA		SSF =	.1	.2	.3	.4	.5	.6	.7	.8	.9
		WW	250	115	70	50	37	29	22	15	9
32.5 N	L	WWNI	356	170	105	75	57	45	36	29	20
2167 DD	C	TW	234	108	66	45	32	23	16	11	7
T(JAN)=47	R	TWNI	336	158	99	70	53	41	32	23	16
		DG	271	120	71	45	29	18	9	-	-
		DGNI	380	177	110	78	58	44	33	24	15
		WW	23.9	19.2	17.5	15.9	13.6	10.9	7.3	3.9	1.1
		WWNI	34.8	29.7	26.5	25.6	23.3	20.1	16.5	10.6	4.5
	D	TW	21.5	18.6	15.6	12.2	9.3	6.9	4.7	2.8	1.2
		TWNI	32.2	27.6	25.4	22.9	20.1	16.2	11.6	7.1	2.9
		DG	24.0	19.5	14.9	10.2	6.4	3.3	.9	-	-
		DGNI	35.9	30.8	27.9	24.6	20.8	15.8	11.0	6.6	2.6

CARIBOU MAINE		SSF =	.1	.2	.3	.4	.5	.6	.7	.8	.9
		WW	48	18	8	-	-	-	-	-	-
46.9 N	L	WWNI	119	55	34	24	18	13	10	7	5
9632 DD	C	TW	54	21	10	3	-	-	-	-	-
T(JAN)=11	R	TWNI	113	52	32	22	16	12	9	6	4
		DG	-	-	-	-	-	-	-	-	-
		DGNI	114	51	31	20	14	10	6	4	-
		WW	3.8	2.3	.9	-	-	-	-	-	-
		WWNI	11.1	9.5	8.4	7.4	6.2	4.9	3.5	2.2	.8
	D	TW	4.5	2.8	1.4	.1	-	-	-	-	-
		TWNI	10.5	8.9	7.9	6.7	5.4	4.1	2.9	1.6	.6
		DG	-	-	-	-	-	-	-	-	-
		DGNI	10.3	8.5	7.0	5.4	3.9	2.6	1.4	.4	-

PORTLAND MAINE		SSF =	.1	.2	.3	.4	.5	.6	.7	.8	.9
		WW	60	24	13	7	-	-	-	-	-
43.6 N	L	WWNI	133	61	38	27	20	15	12	9	6
7498 DD	C	TW	65	27	14	8	3	-	-	-	-
T(JAN)=21	R	TWNI	126	58	36	25	19	14	11	8	5
		DG	-	-	-	-	-	-	-	-	-
		DGNI	130	58	35	24	17	12	8	5	2
		WW	4.8	3.4	2.3	1.2	-	-	-	-	-
		WWNI	12.2	10.5	9.7	8.4	7.3	6.1	4.7	3.0	1.2
	D	TW	5.4	3.8	2.4	1.2	.2	-	-	-	-
		TWNI	11.6	10.0	8.8	7.7	6.5	5.1	3.6	2.1	.8
		DG	-	-	-	-	-	-	-	-	-
		DGNI	11.5	9.8	8.2	6.8	5.2	3.6	2.1	.9	.1

BALTIMORE MARYLAND		SSF =	.1	.2	.3	.4	.5	.6	.7	.8	.9
		WW	101	46	28	19	13	9	6	-	-
39.2 N	L	WWNI	178	86	55	39	29	23	18	14	10
4729 DD	C	TW	102	46	27	18	12	7	4	2	-
T(JAN)=33	R	TWNI	169	81	51	36	27	21	16	12	8
		DG	81	29	-	-	-	-	-	-	-
		DGNI	182	86	54	37	27	20	15	10	6
		WW	9.5	7.7	6.4	5.2	3.9	2.4	1.0	-	-
		WWNI	16.8	16.1	14.0	12.7	11.6	9.7	7.8	5.0	2.0
	D	TW	9.1	7.6	5.9	4.2	2.7	1.6	.7	.1	-
		TWNI	16.1	14.7	13.1	11.8	10.0	8.0	5.6	3.4	1.3
		DG	6.4	2.8	-	-	-	-	-	-	-
		DGNI	17.1	15.2	13.4	11.4	9.3	6.7	4.4	2.4	.7

BOSTON MASSACHUSETTS		SSF =	.1	.2	.3	.4	.5	.6	.7	.8	.9
		WW	81	35	20	13	9	5	-	-	-
42.4 N	L	WWNI	159	73	46	32	24	19	15	11	8
5621 DD	C	TW	85	36	21	13	8	4	-	-	-
T(JAN)=29	R	TWNI	150	69	43	30	23	17	13	10	6
		DG	49	-	-	-	-	-	-	-	-
		DGNI	158	72	44	30	22	16	11	8	4
		WW	6.8	5.3	4.2	3.1	2.0	.8	-	-	-
		WWNI	14.1	13.2	11.5	10.3	9.3	7.7	6.1	4.0	1.6
	D	TW	7.2	5.5	4.1	2.6	1.4	.5	-	-	-
		TWNI	13.5	12.1	10.8	9.6	8.1	6.5	4.5	2.7	1.0
		DG	2.9	-	-	-	-	-	-	-	-
		DGNI	14.0	12.1	10.6	8.9	7.1	5.0	3.2	1.6	.4

ALPENA MICHIGAN		SSF =	.1	.2	.3	.4	.5	.6	.7	.8	.9
		WW	53	19	8	-	-	-	-	-	-
45.1 N	L	WWNI	126	57	35	24	17	13	10	7	5
8518 DD	C	TW	58	22	10	-	-	-	-	-	-
T(JAN)=18	R	TWNI	119	54	33	22	16	12	9	6	4
		DG	-	-	-	-	-	-	-	-	-
		DGNI	121	53	31	20	14	10	6	3	-
		WW	4.0	2.2	.7	-	-	-	-	-	-
		WWNI	11.6	9.5	8.5	7.1	5.8	4.6	3.3	2.2	.8
	D	TW	4.7	2.8	1.4	-	-	-	-	-	-
		TWNI	11.0	9.0	7.7	6.4	5.2	4.0	2.7	1.6	.5
		DG	-	-	-	-	-	-	-	-	-
		DGNI	10.8	8.5	6.7	5.2	3.7	2.4	1.3	.4	-

DETROIT MICHIGAN			SSF = .1	.2	.3	.4	.5	.6	.7	.8	.9
		WW	62	26	14	8	-	-	-	-	-
42.4 N	L	WWNI	137	63	39	27	20	16	12	9	6
6228 DD	C	TW	68	28	15	8	3	-	-	-	-
T(JAN)=25	R	TWNI	130	60	37	26	19	14	11	8	5
		DG	-	-	-	-	-	-	-	-	-
		DGNI	134	60	36	25	17	12	9	5	2
		WW	5.2	3.6	2.5	1.2	-	-	-	-	-
		WWNI	12.0	11.4	9.6	8.4	7.4	6.1	4.6	3.0	1.1
	D	TW	5.6	4.0	2.6	1.3	.2	-	-	-	-
		TWNI	11.5	10.3	9.0	7.9	6.5	5.1	3.6	2.1	.8
		DG	-	-	-	-	-	-	-	-	-
		DGNI	11.8	10.0	8.4	6.8	5.2	3.6	2.1	.9	.1

FLINT MICHIGAN			SSF = .1	.2	.3	.4	.5	.6	.7	.8	.9
		WW	54	21	10	-	-	-	-	-	-
43.0 N	L	WWNI	129	58	36	25	18	14	11	8	5
7041 DD	C	TW	60	24	12	5	-	-	-	-	-
T(JAN)=22	R	TWNI	122	55	34	23	17	13	9	7	4
		DG	-	-	-	-	-	-	-	-	-
		DGNI	124	54	32	21	15	10	7	4	-
		WW	4.1	2.6	1.3	-	-	-	-	-	-
		WWNI	11.6	9.8	8.7	7.3	6.3	5.2	3.8	2.5	.9
	D	TW	4.8	3.1	1.7	.5	-	-	-	-	-
		TWNI	10.8	9.4	8.0	6.9	5.6	4.4	3.0	1.8	.6
		DG	-	-	-	-	-	-	-	-	-
		DGNI	10.6	8.8	7.1	5.7	4.2	2.8	1.6	.6	-

GRAND RAPIDS MICHIGAN			SSF = .1	.2	.3	.4	.5	.6	.7	.8	.9
		WW	57	22	10	-	-	-	-	-	-
42.9 N	L	WWNI	133	59	37	25	19	14	11	8	5
6801 DD	C	TW	63	25	12	6	-	-	-	-	-
T(JAN)=23	R	TWNI	125	56	34	24	17	13	10	7	4
		DG	-	-	-	-	-	-	-	-	-
		DGNI	128	56	33	22	15	11	7	4	-
		WW	4.4	2.8	1.5	-	-	-	-	-	-
		WWNI	11.5	10.3	8.8	7.5	6.5	5.2	3.8	2.5	.9
	D	TW	5.0	3.2	1.9	.6	-	-	-	-	-
		TWNI	10.9	9.6	8.2	7.0	5.7	4.4	3.1	1.8	.6
		DG	-	-	-	-	-	-	-	-	-
		DGNI	10.9	9.0	7.4	5.9	4.3	2.9	1.6	.6	-

SAULT STE. MARIE MICHIGAN			SSF = .1	.2	.3	.4	.5	.6	.7	.8	.9
		WW	45	15	-	-	-	-	-	-	-
46.5 N	L	WWNI	116	53	33	22	16	12	9	6	4
9193 DD	C	TW	51	19	8	-	-	-	-	-	-
T(JAN)=14	R	TWNI	110	50	31	21	15	11	8	6	3
		DG	-	-	-	-	-	-	-	-	-
		DGNI	110	49	29	19	12	8	5	2	-
		WW	3.4	1.5	-	-	-	-	-	-	-
		WWNI	10.8	9.0	7.9	6.6	5.3	4.0	2.8	1.8	.7
	D	TW	4.1	2.2	.8	-	-	-	-	-	-
		TWNI	10.2	8.5	7.2	5.9	4.7	3.5	2.4	1.3	.4
		DG	-	-	-	-	-	-	-	-	-
		DGNI	10.0	7.9	6.1	4.5	3.1	1.9	.9	.2	-

TRAVERSE CITY MICHIGAN			SSF = .1	.2	.3	.4	.5	.6	.7	.8	.9
		WW	51	18	-	-	-	-	-	-	-
44.7 N	L	WWNI	127	56	34	23	17	13	9	7	4
7698 DD	C	TW	58	21	9	-	-	-	-	-	-
T(JAN)=21	R	TWNI	120	53	32	22	16	11	8	6	4
		DG	-	-	-	-	-	-	-	-	-
		DGNI	121	52	30	20	13	9	6	3	-
		WW	3.6	1.9	-	-	-	-	-	-	-
		WWNI	11.3	9.1	8.1	6.7	5.5	4.3	3.1	2.0	.7
	D	TW	4.4	2.6	1.1	-	-	-	-	-	-
		TWNI	10.6	8.9	7.4	6.2	4.9	3.7	2.6	1.4	.5
		DG	-	-	-	-	-	-	-	-	-
		DGNI	10.2	8.1	6.4	4.9	3.4	2.2	1.1	.3	-

DULUTH MINNESOTA			SSF = .1	.2	.3	.4	.5	.6	.7	.8	.9
		WW	42	15	-	-	-	-	-	-	-
46.8 N	L	WWNI	113	52	32	22	16	12	9	7	4
9756 DD	C	TW	49	18	8	-	-	-	-	-	-
T(JAN)= 9	R	TWNI	107	49	30	21	15	11	8	6	3
		DG	-	-	-	-	-	-	-	-	-
		DGNI	107	48	28	18	12	8	5	3	-
		WW	3.2	1.5	-	-	-	-	-	-	-
		WWNI	10.2	9.0	7.9	6.6	5.4	4.2	3.0	2.0	.7
	D	TW	3.9	2.2	.8	-	-	-	-	-	-
		TWNI	9.8	8.4	7.2	6.0	4.8	3.6	2.5	1.4	.5
		DG	-	-	-	-	-	-	-	-	-
		DGNI	9.6	7.8	6.1	4.6	3.3	2.0	1.0	.2	-

INTER. FALLS MINNESOTA			SSF = .1	.2	.3	.4	.5	.6	.7	.8	.9
		WW	38	10	-	-	-	-	-	-	-
48.6 N	L	WWNI	109	49	30	21	15	11	8	6	4
10547 DD	C	TW	45	16	-	-	-	-	-	-	-
T(JAN)= 2	R	TWNI	104	47	28	19	14	10	7	5	3
		DG	-	-	-	-	-	-	-	-	-
		DGNI	103	45	26	16	11	7	4	-	-
		WW	2.7	.4	-	-	-	-	-	-	-
		WWNI	9.8	8.4	7.2	5.7	4.5	3.5	2.5	1.6	.5
	D	TW	3.5	1.6	-	-	-	-	-	-	-
		TWNI	9.3	7.9	6.6	5.3	4.1	3.1	2.1	1.2	.4
		DG	-	-	-	-	-	-	-	-	-
		DGNI	9.0	7.1	5.3	3.8	2.5	1.4	.6	-	-

MINNEAPOLIS MINNESOTA			SSF = .1	.2	.3	.4	.5	.6	.7	.8	.9
		WW	54	21	10	-	-	-	-	-	-
44.9 N	L	WWNI	126	58	36	25	18	14	11	8	5
8159 DD	C	TW	59	24	12	5	-	-	-	-	-
T(JAN)=12	R	TWNI	120	55	34	23	17	13	9	7	4
		DG	-	-	-	-	-	-	-	-	-
		DGNI	122	54	32	22	15	10	7	4	-
		WW	4.3	2.5	1.3	-	-	-	-	-	-
		WWNI	11.6	10.1	8.6	7.3	6.4	5.3	3.8	2.5	.9
	D	TW	4.8	3.2	1.7	.5	-	-	-	-	-
		TWNI	10.9	9.4	8.0	6.9	5.7	4.4	3.1	1.8	.6
		DG	-	-	-	-	-	-	-	-	-
		DGNI	10.8	8.8	7.2	5.7	4.3	2.8	1.6	.6	-

ROCHESTER MINNESOTA		SSF =	.1	.2	.3	.4	.5	.6	.7	.8	.9
		WW	53	20	9	-	-	-	-	-	-
43.9 N	L	WWNI	126	57	35	24	18	14	11	8	5
8227 DD	C	TW	59	23	12	5	-	-	-	-	-
T(JAN)=13	R	TWNI	120	54	33	23	17	13	9	7	4
		DG	-	-	-	-	-	-	-	-	-
		DGNI	121	54	32	21	15	10	7	4	-
		WW	4.1	2.5	1.3	-	-	-	-	-	-
		WWNI	11.3	9.8	8.5	7.3	6.4	5.4	4.0	2.6	1.0
	D	TW	4.7	3.1	1.7	.5	-	-	-	-	-
		TWNI	10.7	9.3	7.9	6.9	5.7	4.5	3.1	1.8	.7
		DG	-	-	-	-	-	-	-	-	-
		DGNI	10.6	8.7	7.1	5.8	4.3	2.9	1.6	.6	-

JACKSON MISSISSIPPI		SSF =	.1	.2	.3	.4	.5	.6	.7	.8	.9
		WW	230	108	66	46	34	26	20	14	8
32.3 N	L	WWNI	331	161	100	71	54	43	34	27	19
2300 DD	C	TW	218	101	62	42	29	21	15	10	6
T(JAN)=47	R	TWNI	313	149	94	66	50	39	30	22	15
		DG	248	111	65	41	26	15	5	-	-
		DGNI	354	167	104	73	55	41	31	22	14
		WW	22.4	18.2	16.1	14.7	12.5	10.0	6.6	3.4	.9
		WWNI	32.3	29.0	25.0	23.8	22.0	18.9	15.6	10.0	4.2
	D	TW	20.1	17.5	14.6	11.5	8.6	6.2	4.2	2.5	1.0
		TWNI	30.5	26.5	24.0	21.7	19.0	15.3	10.9	6.7	2.7
		DG	22.2	17.9	13.4	8.9	5.2	2.4	.2	-	-
		DGNI	34.1	29.3	26.5	23.1	19.5	14.8	10.3	6.1	2.4

MERIDIAN MISSISSIPPI		SSF =	.1	.2	.3	.4	.5	.6	.7	.8	.9
		WW	216	103	63	44	33	25	19	13	7
32.3 N	L	WWNI	315	154	97	69	53	41	33	26	19
2388 DD	C	TW	206	96	59	40	28	20	14	10	6
T(JAN)=47	R	TWNI	298	144	90	64	48	38	29	21	14
		DG	233	104	61	38	24	13	-	-	-
		DGNI	335	160	100	71	53	40	30	21	14
		WW	21.0	17.8	15.4	14.2	11.9	9.5	6.2	3.2	.8
		WWNI	30.6	28.8	24.4	22.8	21.4	18.3	15.1	9.7	4.1
	D	TW	19.2	16.8	14.1	11.0	8.2	5.9	4.0	2.3	.9
		TWNI	28.9	25.9	23.3	21.1	18.3	14.8	10.5	6.5	2.6
		DG	20.9	17.0	12.6	8.2	4.6	1.9	-	-	-
		DGNI	32.6	28.4	25.7	22.4	18.8	14.2	9.8	5.8	2.3

COLUMBIA MISSOURI		SSF =	.1	.2	.3	.4	.5	.6	.7	.8	.9
		WW	102	45	26	17	12	8	5	-	-
38.8 N	L	WWNI	182	85	53	37	28	22	17	13	9
5083 DD	C	TW	103	45	26	16	10	6	3	-	-
T(JAN)=29	R	TWNI	173	80	50	35	26	20	15	11	7
		DG	79	23	-	-	-	-	-	-	-
		DGNI	185	84	51	36	26	19	14	9	5
		WW	9.1	7.0	5.5	4.5	3.2	1.9	.6	-	-
		WWNI	16.5	15.2	13.1	11.6	10.7	9.1	7.1	4.6	1.9
	D	TW	8.9	7.1	5.3	3.6	2.2	1.2	.4	-	-
		TWNI	15.8	13.9	12.2	10.9	9.3	7.5	5.2	3.2	1.2
		DG	5.6	1.6	-	-	-	-	-	-	-
		DGNI	16.8	14.3	12.4	10.5	8.5	6.1	4.0	2.1	.6

KANSAS CITY MISSOURI		SSF =	.1	.2	.3	.4	.5	.6	.7	.8	.9
		WW	102	45	26	17	12	8	5	-	-
39.3 N	L	WWNI	182	85	53	37	28	22	17	14	9
5357 DD	C	TW	103	45	26	17	11	7	4	-	-
T(JAN)=27	R	TWNI	172	80	50	35	26	20	15	11	7
		DG	79	25	-	-	-	-	-	-	-
		DGNI	184	84	52	36	26	19	14	10	5
		WW	9.1	7.1	5.7	4.6	3.4	2.0	.7	-	-
		WWNI	16.5	15.5	13.3	11.8	10.9	9.2	7.4	4.8	1.9
	D	TW	8.9	7.2	5.5	3.8	2.3	1.3	.5	-	-
		TWNI	15.8	14.0	12.4	11.1	9.5	7.6	5.3	3.2	1.3
		DG	5.7	1.9	-	-	-	-	-	-	-
		DGNI	16.9	14.5	12.5	10.7	8.7	6.2	4.1	2.2	.6

SAINT LOUIS MISSOURI		SSF =	.1	.2	.3	.4	.5	.6	.7	.8	.9
		WW	113	50	29	19	14	10	6	-	-
38.7 N	L	WWNI	194	91	57	40	30	23	18	14	10
4750 DD	C	TW	113	49	29	18	12	8	4	2	-
T(JAN)=31	R	TWNI	184	86	53	37	28	21	16	12	8
		DG	94	32	-	-	-	-	-	-	-
		DGNI	199	91	56	38	28	21	15	10	6
		WW	10.2	8.0	6.3	5.2	3.9	2.5	1.0	-	-
		WWNI	17.8	16.4	14.1	12.4	11.5	9.7	7.8	5.0	2.0
	D	TW	9.8	7.9	6.1	4.2	2.7	1.6	.7	.2	-
		TWNI	17.0	15.0	13.1	11.7	10.0	8.0	5.7	3.4	1.3
		DG	7.0	3.3	-	-	-	-	-	-	-
		DGNI	18.2	15.5	13.4	11.4	9.3	6.7	4.4	2.4	.7

SPRINGFIELD MISSOURI		SSF =	.1	.2	.3	.4	.5	.6	.7	.8	.9
		WW	123	55	33	22	16	11	8	4	-
37.2 N	L	WWNI	207	97	61	43	32	25	20	16	11
4570 DD	C	TW	122	54	32	21	14	9	6	3	-
T(JAN)=33	R	TWNI	196	91	57	40	30	23	18	13	8
		DG	108	41	16	-	-	-	-	-	-
		DGNI	213	97	60	42	31	23	17	12	7
		WW	11.0	8.9	7.3	6.3	4.9	3.3	1.7	.4	-
		WWNI	18.7	17.6	15.2	13.6	12.6	10.9	8.8	5.7	2.3
	D	TW	10.6	8.8	6.9	5.0	3.4	2.1	1.1	.4	-
		TWNI	17.9	15.9	14.1	12.8	11.0	8.9	6.3	3.8	1.5
		DG	8.4	5.0	1.0	-	-	-	-	-	-
		DGNI	19.2	16.8	14.7	12.7	10.4	7.6	5.1	2.8	.9

BILLINGS MONTANA		SSF =	.1	.2	.3	.4	.5	.6	.7	.8	.9
		WW	91	41	23	15	9	6	-	-	-
45.8 N	L	WWNI	165	80	50	35	26	20	15	12	8
7265 DD	C	TW	92	40	23	14	9	5	-	-	-
T(JAN)=22	R	TWNI	158	75	47	32	24	18	14	10	6
		DG	64	-	-	-	-	-	-	-	-
		DGNI	168	78	48	33	23	17	12	8	4
		WW	8.5	6.5	4.7	3.4	2.0	.8	-	-	-
		WWNI	16.0	14.2	12.8	10.8	9.3	7.8	5.8	3.9	1.5
	D	TW	8.2	6.4	4.6	2.9	1.7	.6	-	-	-
		TWNI	15.2	13.4	11.5	9.9	8.3	6.4	4.5	2.7	1.0
		DG	4.4	-	-	-	-	-	-	-	-
		DGNI	15.8	13.6	11.3	9.4	7.2	5.1	3.2	1.6	.4

CUT BANK MONTANA		SSF =	.1	.2	.3	.4	.5	.6	.7	.8	.9
		WW	77	33	18	10	5	-	-	-	-
48.6 N	L	WWNI	153	71	44	31	23	17	13	9	6
9033 DD	C	TW	80	34	18	11	5	-	-	-	-
T(JAN)=16	R	TWNI	144	67	42	29	21	15	11	8	5
		DG	40	-	-	-	-	-	-	-	-
		DGNI	151	69	42	28	20	14	9	6	-
		WW	6.5	5.0	3.2	1.8	.4	-	-	-	-
		WWNI	14.1	12.7	11.1	9.4	7.7	6.1	4.3	2.9	1.1
	D	TW	7.0	5.0	3.3	1.8	.6	-	-	-	-
		TWNI	13.3	11.7	10.1	8.5	6.9	5.1	3.6	2.0	.7
		DG	1.7	-	-	-	-	-	-	-	-
		DGNI	13.8	11.7	9.5	7.5	5.6	3.8	2.2	.9	-

DILLON MONTANA		SSF =	.1	.2	.3	.4	.5	.6	.7	.8	.9
		WW	90	40	24	15	10	6	-	-	-
45.2 N	L	WWNI	168	78	50	35	26	20	16	12	8
8354 DD	C	TW	92	40	23	15	9	5	-	-	-
T(JAN)=20	R	TWNI	158	74	47	33	24	18	14	10	6
		DG	64	-	-	-	-	-	-	-	-
		DGNI	168	78	48	33	24	17	12	8	4
		WW	7.8	6.6	5.1	3.7	2.3	1.0	-	-	-
		WWNI	15.2	14.4	12.9	11.4	9.7	8.1	5.9	4.0	1.5
	D	TW	8.1	6.3	4.7	3.1	1.8	.8	-	-	-
		TWNI	14.6	13.3	11.9	10.3	8.6	6.6	4.7	2.7	1.0
		DG	4.5	-	-	-	-	-	-	-	-
		DGNI	15.3	13.6	11.7	9.7	7.5	5.4	3.4	1.7	.4

GLASGOW MONTANA		SSF =	.1	.2	.3	.4	.5	.6	.7	.8	.9
		WW	60	23	10	-	-	-	-	-	-
48.2 N	L	WWNI	130	61	38	26	19	14	10	8	5
8969 DD	C	TW	64	25	13	5	-	-	-	-	-
T(JAN)= 9	R	TWNI	124	58	35	24	17	13	9	7	4
		DG	-	-	-	-	-	-	-	-	-
		DGNI	128	58	34	22	15	10	7	4	-
		WW	5.0	2.9	1.2	-	-	-	-	-	-
		WWNI	12.4	10.6	9.0	7.2	6.1	4.9	3.5	2.3	.8
	D	TW	5.3	3.4	1.8	.4	-	-	-	-	-
		TWNI	11.7	10.0	8.2	6.8	5.5	4.1	2.9	1.6	.6
		DG	-	-	-	-	-	-	-	-	-
		DGNI	11.6	9.4	7.3	5.6	4.0	2.6	1.4	.5	-

GREAT FALLS MONTANA		SSF =	.1	.2	.3	.4	.5	.6	.7	.8	.9
		WW	86	38	21	13	7	-	-	-	-
47.5 N	L	WWNI	161	77	48	33	25	19	14	10	7
7652 DD	C	TW	88	38	21	13	7	3	-	-	-
T(JAN)=20	R	TWNI	153	72	45	31	23	17	12	9	5
		DG	56	-	-	-	-	-	-	-	-
		DGNI	163	75	46	31	22	15	11	7	3
		WW	7.8	5.9	4.0	2.6	1.1	-	-	-	-
		WWNI	14.9	13.7	12.1	10.1	8.5	6.9	4.8	3.3	1.2
	D	TW	7.6	5.8	4.0	2.4	1.2	.2	-	-	-
		TWNI	14.3	12.7	10.9	9.2	7.5	5.6	4.0	2.3	.8
		DG	3.4	-	-	-	-	-	-	-	-
		DGNI	15.0	12.9	10.5	8.4	6.3	4.4	2.6	1.2	.2

HELENA MONTANA		SSF =	.1	.2	.3	.4	.5	.6	.7	.8	.9
		WW	77	34	19	11	6	-	-	-	-
46.6 N	L	WWNI	151	71	45	31	23	17	13	10	6
8190 DD	C	TW	80	34	19	11	6	-	-	-	-
T(JAN)=18	R	TWNI	144	67	42	29	21	16	12	8	5
		DG	41	-	-	-	-	-	-	-	-
		DGNI	151	69	42	29	20	14	10	6	2
		WW	6.9	5.2	3.5	2.1	.7	-	-	-	-
		WWNI	13.7	13.0	11.4	9.7	8.0	6.4	4.6	3.0	1.1
	D	TW	6.9	5.1	3.5	2.0	.8	-	-	-	-
		TWNI	13.1	11.9	10.4	8.7	7.1	5.3	3.8	2.1	.8
		DG	2.0	-	-	-	-	-	-	-	-
		DGNI	13.8	11.9	9.8	7.8	5.8	4.0	2.4	1.0	.1

LEWISTOWN MONTANA		SSF =	.1	.2	.3	.4	.5	.6	.7	.8	.9
		WW	76	33	19	11	6	-	-	-	-
47.0 N	L	WWNI	153	71	45	31	23	18	13	10	7
8586 DD	C	TW	80	34	19	11	6	-	-	-	-
T(JAN)=19	R	TWNI	144	67	42	29	21	16	12	9	5
		DG	41	-	-	-	-	-	-	-	-
		DGNI	151	69	42	29	20	14	10	6	3
		WW	6.5	5.2	3.5	2.2	.9	-	-	-	-
		WWNI	13.8	12.8	11.3	9.7	8.1	6.6	4.8	3.2	1.2
	D	TW	6.9	5.1	3.5	2.1	.9	-	-	-	-
		TWNI	13.2	11.8	10.3	8.8	7.3	5.5	3.9	2.2	.8
		DG	1.9	-	-	-	-	-	-	-	-
		DGNI	13.6	11.8	9.9	8.0	6.0	4.2	2.5	1.1	.2

MILES CITY MONTANA		SSF =	.1	.2	.3	.4	.5	.6	.7	.8	.9
		WW	78	33	18	10	5	-	-	-	-
46.4 N	L	WWNI	151	72	44	31	22	17	13	10	6
7889 DD	C	TW	80	34	18	10	5	-	-	-	-
T(JAN)=15	R	TWNI	144	67	41	28	21	15	11	8	5
		DG	39	-	-	-	-	-	-	-	-
		DGNI	151	69	41	28	19	14	9	6	2
		WW	6.8	4.7	3.1	1.8	.6	-	-	-	-
		WWNI	14.7	12.5	10.8	9.0	7.7	6.3	4.6	3.1	1.2
	D	TW	6.9	5.0	3.2	1.8	.7	-	-	-	-
		TWNI	13.7	11.7	9.9	8.3	6.9	5.3	3.7	2.2	.8
		DG	1.6	-	-	-	-	-	-	-	-
		DGNI	14.0	11.5	9.3	7.5	5.6	3.9	2.3	1.0	.1

MISSOULA MONTANA		SSF =	.1	.2	.3	.4	.5	.6	.7	.8	.9
		WW	66	27	13	-	-	-	-	-	-
46.9 N	L	WWNI	140	65	40	28	20	15	11	8	5
7931 DD	C	TW	71	28	15	7	-	-	-	-	-
T(JAN)=21	R	TWNI	133	61	38	26	18	13	10	7	4
		DG	-	-	-	-	-	-	-	-	-
		DGNI	138	62	37	24	16	11	7	4	-
		WW	5.7	3.7	1.8	-	-	-	-	-	-
		WWNI	12.4	11.4	9.7	8.2	6.4	4.9	3.4	2.2	.8
	D	TW	5.8	3.9	2.2	.8	-	-	-	-	-
		TWNI	12.0	10.5	9.0	7.3	5.7	4.2	2.9	1.6	.5
		DG	-	-	-	-	-	-	-	-	-
		DGNI	12.4	10.2	8.1	6.1	4.3	2.8	1.5	.5	-

GRAND ISLAND NEBRASKA		SSF =	.1	.2	.3	.4	.5	.6	.7	.8	.9
		WW	98	43	25	16	11	7	4	-	-
41.0 N	L	WWNI	178	82	52	36	27	21	17	13	9
6425 DD	C	TW	99	43	25	16	10	6	3	-	-
T(JAN)=22	R	TWNI	168	78	48	34	25	19	15	11	7
		DG	73	18	-	-	-	-	-	-	-
		DGNI	179	82	50	34	25	18	13	9	5
		WW	8.4	6.7	5.2	4.2	3.0	1.6	.4	-	-
		WWNI	16.4	14.7	12.8	11.3	10.2	8.8	7.0	4.5	1.8
	D	TW	8.6	6.8	5.0	3.5	2.1	1.1	.3	-	-
		TWNI	15.4	13.6	11.9	10.6	9.1	7.3	5.1	3.1	1.2
		DG	5.1	.7	-	-	-	-	-	-	-
		DGNI	16.2	13.8	11.9	10.2	8.2	5.9	3.8	2.0	.6

NORTH OMAHA NEBRASKA		SSF =	.1	.2	.3	.4	.5	.6	.7	.8	.9
		WW	85	36	20	13	9	5	-	-	-
41.4 N	L	WWNI	163	75	47	33	24	19	15	12	8
6601 DD	C	TW	87	37	21	13	8	4	-	-	-
T(JAN)=20	R	TWNI	154	71	44	30	23	17	13	10	6
		DG	52	-	-	-	-	-	-	-	-
		DGNI	162	73	44	31	22	16	12	8	4
		WW	7.2	5.4	4.1	3.2	2.1	.8	-	-	-
		WWNI	14.8	13.3	11.4	10.1	9.2	7.9	6.2	4.0	1.6
	D	TW	7.4	5.7	4.1	2.6	1.4	.6	-	-	-
		TWNI	13.9	12.2	10.7	9.6	8.1	6.5	4.6	2.7	1.1
		DG	3.1	-	-	-	-	-	-	-	-
		DGNI	14.5	12.2	10.5	8.9	7.1	5.0	3.2	1.6	.4

NORTH PLATTE NEBRASKA		SSF =	.1	.2	.3	.4	.5	.6	.7	.8	.9
		WW	103	46	27	18	13	9	5	-	-
41.1 N	L	WWNI	181	85	54	38	29	22	18	14	10
6743 DD	C	TW	103	46	27	17	11	7	4	1	-
T(JAN)=23	R	TWNI	172	80	50	35	27	20	16	11	7
		DG	81	27	-	-	-	-	-	-	-
		DGNI	184	85	53	37	27	20	14	10	6
		WW	9.1	7.5	6.0	5.0	3.6	2.2	.9	-	-
		WWNI	17.3	15.1	13.9	12.3	10.9	9.7	7.6	4.9	2.0
	D	TW	9.2	7.4	5.7	3.9	2.6	1.5	.6	.1	-
		TWNI	16.1	14.4	12.7	11.3	9.8	7.8	5.5	3.3	1.3
		DG	6.1	2.3	-	-	-	-	-	-	-
		DGNI	17.0	14.8	12.8	11.1	9.0	6.5	4.2	2.3	.7

SCOTTSBLUFF NEBRASKA		SSF =	.1	.2	.3	.4	.5	.6	.7	.8	.9
		WW	103	47	28	19	13	9	6	-	-
41.9 N	L	WWNI	180	86	55	39	29	23	18	14	10
6774 DD	C	TW	103	46	28	18	12	7	4	2	-
T(JAN)=25	R	TWNI	171	81	51	36	27	21	16	12	8
		DG	82	29	-	-	-	-	-	-	-
		DGNI	183	86	54	37	27	20	15	10	6
		WW	9.5	7.9	6.3	5.3	3.9	2.4	1.0	-	-
		WWNI	17.8	15.4	14.4	12.6	11.3	10.1	7.8	5.1	2.0
	D	TW	9.4	7.7	5.9	4.1	2.7	1.6	.7	.1	-
		TWNI	16.5	14.7	13.0	11.7	10.1	8.1	5.7	3.4	1.3
		DG	6.5	2.9	-	-	-	-	-	-	-
		DGNI	17.4	15.3	13.3	11.5	9.3	6.7	4.4	2.4	.7

ELKO NEVADA		SSF =	.1	.2	.3	.4	.5	.6	.7	.8	.9
		WW	113	52	31	21	15	10	7	-	-
40.8 N	L	WWNI	195	92	58	42	31	24	19	15	10
7483 DD	C	TW	114	51	30	20	13	8	5	2	-
T(JAN)=23	R	TWNI	184	87	55	39	29	22	17	12	8
		DG	97	36	-	-	-	-	-	-	-
		DGNI	199	93	58	40	29	22	16	11	6
		WW	10.2	8.7	7.2	5.8	4.2	2.7	1.2	-	-
		WWNI	17.9	17.0	15.2	13.6	12.0	10.3	7.6	5.1	2.0
	D	TW	10.1	8.3	6.5	4.7	3.1	1.8	.8	.2	-
		TWNI	17.1	15.6	14.1	12.4	10.5	8.3	5.8	3.5	1.3
		DG	7.8	4.2	-	-	-	-	-	-	-
		DGNI	18.4	16.4	14.3	12.3	9.7	7.1	4.6	2.5	.7

ELY NEVADA		SSF =	.1	.2	.3	.4	.5	.6	.7	.8	.9
		WW	111	52	32	22	16	12	8	4	-
39.3 N	L	WWNI	191	91	59	42	32	26	20	16	11
7814 DD	C	TW	111	51	31	20	14	9	6	3	1
T(JAN)=24	R	TWNI	180	86	55	39	30	23	18	13	9
		DG	96	38	16	-	-	-	-	-	-
		DGNI	195	92	58	41	31	23	17	12	7
		WW	10.1	9.1	7.8	6.7	5.2	3.5	1.8	.4	-
		WWNI	17.8	17.1	15.8	14.4	13.0	11.5	8.9	5.8	2.3
	D	TW	10.2	8.7	7.0	5.1	3.6	2.2	1.2	.5	.1
		TWNI	16.9	15.9	14.6	13.2	11.4	9.1	6.4	3.9	1.5
		DG	8.0	4.9	1.2	-	-	-	-	-	-
		DGNI	18.3	16.8	15.1	13.2	10.7	7.9	5.2	2.9	1.0

LAS VEGAS NEVADA		SSF =	.1	.2	.3	.4	.5	.6	.7	.8	.9
		WW	311	144	90	64	48	37	28	20	12
36.1 N	L	WWNI	430	204	129	92	70	56	44	35	24
2601 DD	C	TW	291	134	83	56	40	29	21	15	9
T(JAN)=44	R	TWNI	405	191	120	86	64	50	38	28	19
		DG	350	158	95	62	42	28	17	9	-
		DGNI	461	216	136	96	71	54	41	29	19
		WW	29.2	25.4	22.7	20.9	17.5	14.0	9.8	5.4	1.7
		WWNI	41.4	36.4	33.2	31.2	28.6	24.0	19.1	12.7	5.3
	D	TW	26.6	23.4	19.8	15.8	12.3	9.2	6.3	3.9	1.7
		TWNI	38.2	34.1	31.2	28.2	24.0	19.6	13.9	8.5	3.4
		DG	31.5	26.5	21.2	15.4	10.6	6.5	3.2	.9	-
		DGNI	43.0	38.2	35.0	30.5	25.7	19.6	13.6	8.2	3.2

LOVELOCK NEVADA		SSF =	.1	.2	.3	.4	.5	.6	.7	.8	.9
		WW	160	73	45	31	22	16	12	7	-
40.1 N	L	WWNI	248	118	74	53	40	31	25	19	13
5990 DD	C	TW	153	70	42	28	19	13	9	5	3
T(JAN)=29	R	TWNI	235	110	70	49	37	28	22	16	10
		DG	156	66	35	18	-	-	-	-	-
		DGNI	259	121	75	53	39	29	21	15	9
		WW	14.8	12.5	10.7	9.1	7.3	5.2	3.0	1.2	-
		WWNI	24.1	20.8	19.7	17.3	15.4	13.4	10.1	6.7	2.7
	D	TW	14.0	11.8	9.5	7.1	5.2	3.5	2.0	1.0	.3
		TWNI	22.3	19.9	17.8	15.8	13.5	10.8	7.6	4.6	1.8
		DG	13.5	9.6	5.7	2.2	-	-	-	-	-
		DGNI	24.2	21.5	18.8	16.3	13.2	9.9	6.6	3.7	1.3

		SSF =	.1	.2	.3	.4	.5	.6	.7	.8	.9
RENO NEVADA		WW	166	77	47	33	24	18	13	8	-
39.5 N	L	WWNI	258	122	77	55	42	33	26	20	14
6022 DD	C	TW	161	74	45	30	21	14	10	6	3
T(JAN)=32	R	TWNI	243	115	73	51	39	30	23	17	11
		DG	166	71	39	22	-	-	-	-	-
		DGNI	269	126	79	56	41	31	23	16	10
		WW	15.3	13.2	11.7	10.0	8.2	5.9	3.6	1.5	-
		WWNI	24.0	22.3	20.1	18.5	16.5	14.4	10.9	7.2	2.9
	D	TW	14.6	12.5	10.2	7.8	5.7	3.9	2.4	1.1	.4
		TWNI	22.8	20.6	18.8	16.8	14.4	11.5	8.1	4.9	1.9
		DG	14.4	10.8	6.9	3.3	-	-	-	-	-
		DGNI	25.2	22.4	19.9	17.5	14.1	10.6	7.2	4.1	1.4

		SSF =	.1	.2	.3	.4	.5	.6	.7	.8	.9
TONOPAH NEVADA		WW	163	75	47	33	24	18	13	9	4
38.1 N	L	WWNI	251	120	76	55	42	33	26	21	15
5900 DD	C	TW	156	72	44	30	21	14	10	6	3
T(JAN)=30	R	TWNI	238	113	72	51	39	30	23	17	11
		DG	161	70	39	22	9	-	-	-	-
		DGNI	263	124	78	55	41	31	23	16	10
		WW	15.3	13.1	11.5	10.1	8.5	6.3	3.8	1.7	.2
		WWNI	24.6	21.5	20.5	18.4	16.8	14.8	11.5	7.5	3.1
	D	TW	14.5	12.5	10.3	7.8	5.8	4.0	2.5	1.3	.4
		TWNI	22.8	20.5	18.6	16.9	14.7	11.8	8.3	5.1	2.0
		DG	14.2	10.8	6.9	3.4	.5	-	-	-	-
		DGNI	24.8	22.4	19.9	17.6	14.5	10.8	7.4	4.3	1.5

		SSF =	.1	.2	.3	.4	.5	.6	.7	.8	.9
WINNEMUCCA NEVADA		WW	131	60	37	25	18	13	9	5	-
40.9 N	L	WWNI	214	103	65	47	35	28	22	17	12
6629 DD	C	TW	129	59	35	23	16	11	7	4	1
T(JAN)=28	R	TWNI	203	96	61	43	32	25	19	14	9
		DG	121	49	24	-	-	-	-	-	-
		DGNI	222	104	65	46	34	25	18	13	7
		WW	12.2	10.3	8.8	7.3	5.7	3.9	2.1	.6	-
		WWNI	20.2	18.8	16.9	15.4	13.7	11.8	8.8	5.9	2.4
	D	TW	11.7	9.9	7.9	5.8	4.0	2.6	1.5	.6	.1
		TWNI	19.2	17.4	15.8	14.0	11.9	9.4	6.7	4.0	1.6
		DG	10.2	6.7	2.9	-	-	-	-	-	-
		DGNI	21.0	18.6	16.4	14.2	11.3	8.4	5.6	3.1	1.0

		SSF =	.1	.2	.3	.4	.5	.6	.7	.8	.9
CONCORD NEW HAMPSHIRE		WW	56	23	12	6	-	-	-	-	-
43.2 N	L	WWNI	129	59	37	26	19	15	12	9	6
7360 DD	C	TW	62	25	13	7	-	-	-	-	-
T(JAN)=21	R	TWNI	123	56	35	24	18	14	10	7	5
		DG	-	-	-	-	-	-	-	-	-
		DGNI	125	56	34	23	16	12	8	5	-
		WW	4.4	3.1	2.0	.9	-	-	-	-	-
		WWNI	11.5	10.2	9.4	8.1	7.1	5.9	4.5	2.9	1.1
	D	TW	5.1	3.6	2.2	1.0	-	-	-	-	-
		TWNI	11.1	9.8	8.6	7.5	6.3	5.0	3.4	2.0	.7
		DG	-	-	-	-	-	-	-	-	-
		DGNI	11.0	9.4	7.9	6.5	5.0	3.4	2.0	.9	-

		SSF =	.1	.2	.3	.4	.5	.6	.7	.8	.9
NEWARK NEW JERSEY		WW	91	42	25	16	11	8	4	-	-
40.7 N	L	WWNI	167	80	51	36	27	21	17	13	9
5034 DD	C	TW	93	42	25	16	10	6	3	-	-
T(JAN)=31	R	TWNI	158	76	48	34	25	19	15	11	7
		DG	67	18	-	-	-	-	-	-	-
		DGNI	169	79	49	34	25	19	13	9	5
		WW	8.5	6.7	5.5	4.3	3.1	1.7	.5	-	-
		WWNI	15.7	15.1	12.9	11.8	10.6	8.9	7.1	4.6	1.8
	D	TW	8.3	6.8	5.2	3.5	2.1	1.2	.4	-	-
		TWNI	15.0	13.6	12.2	10.9	9.3	7.4	5.2	3.1	1.2
		DG	5.0	.8	-	-	-	-	-	-	-
		DGNI	16.0	13.9	12.3	10.4	8.4	6.0	3.9	2.1	.6

		SSF =	.1	.2	.3	.4	.5	.6	.7	.8	.9
ALBUQUERQUE NEW MEXICO		WW	178	84	53	37	28	21	16	11	6
35.0 N	L	WWNI	270	130	83	60	46	37	30	23	17
4292 DD	C	TW	171	80	50	34	24	17	12	8	5
T(JAN)=35	R	TWNI	255	122	78	56	43	33	26	19	13
		DG	183	82	47	29	16	-	-	-	-
		DGNI	283	135	86	61	46	35	26	19	12
		WW	16.8	14.9	13.3	12.1	10.3	8.0	5.0	2.5	.5
		WWNI	26.3	24.5	22.1	20.5	19.1	16.9	13.6	8.7	3.6
	D	TW	16.0	14.2	11.9	9.2	7.0	4.9	3.2	1.8	.8
		TWNI	24.5	22.5	20.6	19.1	16.6	13.3	9.5	5.8	2.4
		DG	16.5	13.3	9.3	5.6	2.6	-	-	-	-
		DGNI	27.1	24.6	22.3	20.0	16.6	12.5	8.6	5.1	1.9

		SSF =	.1	.2	.3	.4	.5	.6	.7	.8	.9
CLAYTON NEW MEXICO		WW	160	73	46	32	24	18	14	9	4
36.4 N	L	WWNI	251	117	75	54	41	33	27	21	15
5212 DD	C	TW	154	71	43	29	20	14	10	6	4
T(JAN)=33	R	TWNI	235	111	70	50	38	30	23	17	11
		DG	157	68	38	21	9	-	-	-	-
		DGNI	260	122	77	54	41	31	23	17	10
		WW	14.3	13.1	11.3	10.2	8.8	6.6	4.0	1.9	.3
		WWNI	23.6	21.3	20.1	18.1	16.9	15.3	12.5	7.9	3.3
	D	TW	14.2	12.2	10.1	7.8	5.8	4.1	2.5	1.4	.5
		TWNI	22.0	20.3	18.3	16.9	15.1	11.9	8.5	5.2	2.1
		DG	13.6	10.6	6.8	3.4	.6	-	-	-	-
		DGNI	24.1	21.8	19.7	17.8	14.7	11.0	7.6	4.4	1.7

		SSF =	.1	.2	.3	.4	.5	.6	.7	.8	.9
FARMINGTON NEW MEXICO		WW	145	66	41	29	21	16	11	7	-
36.7 N	L	WWNI	234	108	70	50	38	30	24	19	13
5713 DD	C	TW	142	64	39	26	18	13	8	5	3
T(JAN)=29	R	TWNI	220	103	65	47	35	27	21	16	10
		DG	138	58	31	15	-	-	-	-	-
		DGNI	241	112	70	50	37	28	21	15	9
		WW	12.7	11.7	10.1	9.0	7.3	5.4	3.1	1.3	-
		WWNI	21.3	19.8	18.7	16.8	15.5	13.7	10.6	6.9	2.9
	D	TW	12.8	10.9	9.0	6.9	5.0	3.4	2.0	1.0	.3
		TWNI	20.1	18.7	17.0	15.6	13.5	10.8	7.7	4.7	1.9
		DG	11.6	8.6	4.9	1.7	-	-	-	-	-
		DGNI	21.9	20.0	18.1	16.1	13.2	9.8	6.6	3.8	1.3

LOS ALAMOS NEW MEXICO		SSF = .1	.2	.3	.4	.5	.6	.7	.8	.9
		WW 128	58	36	26	19	14	10	7	-
35.9 N	L	WWNI 210	100	64	46	35	28	23	18	13
6359 DD	C	TW 125	58	35	24	16	11	8	5	2
T(JAN)=29	R	TWNI 199	94	60	43	33	26	20	15	10
		DG 116	49	25	-	-	-	-	-	-
		DGNI 217	102	64	46	34	26	19	14	9
		WW 11.9	10.0	9.2	8.2	7.0	4.9	2.8	1.1	-
		WWNI 19.9	18.1	16.8	16.1	14.8	13.4	10.9	6.8	2.9
	D	TW 11.5	10.0	8.1	6.2	4.4	3.0	1.8	.9	.3
		TWNI 18.9	17.0	16.0	14.7	13.1	10.4	7.4	4.6	1.9
		DG 9.9	7.1	3.5	-	-	-	-	-	-
		DGNI 20.5	18.2	16.7	15.1	12.4	9.2	6.3	3.6	1.3

ROSWELL NEW MEXICO		SSF = .1	.2	.3	.4	.5	.6	.7	.8	.9
		WW 200	95	59	42	32	24	18	13	7
33.4 N	L	WWNI 299	142	92	66	50	40	32	25	18
3697 DD	C	TW 194	89	55	38	27	19	14	9	6
T(JAN)=38	R	TWNI 283	134	86	61	47	36	28	21	14
		DG 214	95	56	35	22	12	-	-	-
		DGNI 316	149	95	67	51	38	29	21	13
		WW 18.2	17.3	14.9	13.7	11.7	9.3	5.9	3.0	.7
		WWNI 27.3	27.0	24.6	22.1	20.9	18.3	14.8	9.5	4.0
	D	TW 17.6	15.7	13.4	10.5	8.0	5.7	3.7	2.2	1.0
		TWNI 26.0	24.9	22.5	20.8	18.0	14.5	10.3	6.3	2.6
		DG 18.8	15.7	11.6	7.5	4.2	1.5	-	-	-
		DGNI 29.1	27.1	24.6	22.0	18.4	13.8	9.6	5.7	2.2

TRUTH OR CONSEQ. NEW MEXICO		SSF = .1	.2	.3	.4	.5	.6	.7	.8	.9
		WW 229	109	69	49	37	29	22	16	9
33.2 N	L	WWNI 332	160	103	74	57	45	37	29	21
3392 DD	C	TW 220	102	64	44	31	23	16	11	7
T(JAN)=40	R	TWNI 314	150	96	69	53	41	32	24	16
		DG 251	114	69	44	29	18	9	-	-
		DGNI 354	168	107	76	58	44	33	24	15
		WW 21.5	19.7	17.6	16.3	14.1	11.3	7.5	4.0	1.2
		WWNI 31.2	30.1	27.7	25.2	24.0	21.0	16.8	10.9	4.6
	D	TW 20.4	18.3	15.6	12.4	9.6	6.9	4.8	2.8	1.3
		TWNI 29.7	27.7	25.4	23.7	20.4	16.4	11.7	7.2	3.0
		DG 22.8	19.4	15.0	10.4	6.5	3.4	1.0	-	-
		DGNI 33.5	30.8	28.1	25.2	21.2	16.0	11.2	6.7	2.6

TUCUMCARI NEW MEXICO		SSF = .1	.2	.3	.4	.5	.6	.7	.8	.9
		WW 194	90	56	40	30	23	17	12	7
35.2 N	L	WWNI 291	138	88	63	48	39	31	25	18
4047 DD	C	TW 187	86	53	36	25	18	13	9	5
T(JAN)=37	R	TWNI 274	130	82	59	45	35	27	20	13
		DG 203	90	53	32	20	10	-	-	-
		DGNI 306	144	91	65	49	37	28	20	13
		WW 17.7	16.0	14.3	12.9	11.2	8.8	5.6	2.9	.7
		WWNI 27.2	25.4	23.3	21.5	20.0	17.9	14.6	9.3	3.9
	D	TW 17.0	15.0	12.7	9.9	7.6	5.5	3.6	2.0	.9
		TWNI 25.4	23.6	21.7	20.0	17.6	14.0	10.0	6.2	2.5
		DG 17.8	14.7	10.6	6.7	3.5	1.0	-	-	-
		DGNI 28.2	25.8	23.5	21.2	17.7	13.3	9.2	5.5	2.1

ZUNI NEW MEXICO		SSF = .1	.2	.3	.4	.5	.6	.7	.8	.9
		WW 144	66	41	29	22	16	12	8	-
35.1 N	L	WWNI 231	109	69	50	38	31	25	19	14
5815 DD	C	TW 140	64	39	26	18	13	9	6	3
T(JAN)=30	R	TWNI 218	103	65	47	36	28	21	16	11
		DG 137	59	32	16	-	-	-	-	-
		DGNI 238	112	71	50	38	28	21	15	9
		WW 13.1	11.6	10.2	9.2	7.7	5.7	3.3	1.4	-
		WWNI 22.5	19.5	18.9	17.1	15.8	14.2	11.4	7.2	3.0
	D	TW 12.9	11.1	9.1	6.9	5.1	3.5	2.2	1.1	.4
		TWNI 20.5	18.6	17.2	15.8	13.9	11.1	7.9	4.8	2.0
		DG 11.9	8.8	5.1	2.0	-	-	-	-	-
		DGNI 22.2	20.2	18.2	16.4	13.5	10.1	6.9	4.0	1.4

ALBANY NEW YORK		SSF = .1	.2	.3	.4	.5	.6	.7	.8	.9
		WW 56	23	12	6	-	-	-	-	-
42.7 N	L	WWNI 129	59	37	26	19	15	11	9	6
6888 DD	C	TW 62	25	13	7	-	-	-	-	-
T(JAN)=21	R	TWNI 123	56	35	24	18	14	10	7	5
		DG -	-	-	-	-	-	-	-	-
		DGNI 125	56	34	23	16	12	8	5	-
		WW 4.5	3.1	2.0	.8	-	-	-	-	-
		WWNI 11.8	10.4	9.2	8.0	7.0	5.8	4.4	2.9	1.1
	D	TW 5.1	3.6	2.2	1.0	-	-	-	-	-
		TWNI 11.0	9.8	8.5	7.5	6.2	4.9	3.4	2.0	.7
		DG -	-	-	-	-	-	-	-	-
		DGNI 11.0	9.3	7.8	6.4	4.9	3.3	1.9	.8	-

BINGHAMTON NEW YORK		SSF = .1	.2	.3	.4	.5	.6	.7	.8	.9
		WW 42	15	-	-	-	-	-	-	-
42.2 N	L	WWNI 115	52	32	22	16	12	10	7	5
7285 DD	C	TW 50	19	8	-	-	-	-	-	-
T(JAN)=22	R	TWNI 109	49	30	21	15	11	8	6	4
		DG -	-	-	-	-	-	-	-	-
		DGNI 109	48	28	19	13	9	6	3	-
		WW 3.1	1.5	-	-	-	-	-	-	-
		WWNI 10.3	8.7	7.7	6.6	5.7	4.7	3.6	2.3	.8
	D	TW 3.8	2.3	.9	-	-	-	-	-	-
		TWNI 9.7	8.4	7.2	6.1	5.1	4.0	2.8	1.6	.6
		DG -	-	-	-	-	-	-	-	-
		DGNI 9.3	7.6	6.2	4.9	3.6	2.3	1.2	.4	-

BUFFALO NEW YORK		SSF = .1	.2	.3	.4	.5	.6	.7	.8	.9
		WW 46	17	6	-	-	-	-	-	-
42.9 N	L	WWNI 121	54	33	23	17	13	10	7	5
6927 DD	C	TW 54	20	9	-	-	-	-	-	-
T(JAN)=24	R	TWNI 115	51	31	21	16	12	9	6	4
		DG -	-	-	-	-	-	-	-	-
		DGNI 115	50	29	20	13	9	6	3	-
		WW 3.4	1.8	.4	-	-	-	-	-	-
		WWNI 10.7	9.2	8.0	6.8	5.8	4.7	3.5	2.3	.8
	D	TW 4.2	2.6	1.2	-	-	-	-	-	-
		TWNI 10.0	8.7	7.4	6.4	5.2	4.0	2.8	1.6	.6
		DG -	-	-	-	-	-	-	-	-
		DGNI 9.7	8.0	6.5	5.1	3.8	2.4	1.3	.4	-

MASSENA NEW YORK			SSF = .1	.2	.3	.4	.5	.6	.7	.8	.9
		WW	42	14	-	-	-	-	-	-	-
44.9 N	L	WWNI	115	51	32	22	16	12	9	7	5
8237 DD	C	TW	50	18	8	-	-	-	-	-	-
T(JAN)=14	R	TWNI	109	49	30	21	15	11	8	6	4
		DG	-	-	-	-	-	-	-	-	-
		DGNI	109	47	28	18	13	9	6	3	-
		WW	3.0	1.4	-	-	-	-	-	-	-
		WWNI	10.1	8.7	7.8	6.5	5.6	4.4	3.3	2.1	.8
	D	TW	3.8	2.2	.8	-	-	-	-	-	-
		TWNI	9.7	8.3	7.1	6.1	5.0	3.8	2.6	1.5	.5
		DG	-	-	-	-	-	-	-	-	-
		DGNI	9.3	7.6	6.1	4.8	3.4	2.2	1.1	.3	-

NYC (CNTRL. PARK) NEW YORK			SSF = .1	.2	.3	.4	.5	.6	.7	.8	.9
		WW	85	37	22	14	10	6	-	-	-
40.8 N	L	WWNI	159	76	48	34	25	20	16	12	8
4848 DD	C	TW	87	38	22	14	8	5	2	-	-
T(JAN)=32	R	TWNI	151	72	45	32	24	18	14	10	6
		DG	56	-	-	-	-	-	-	-	-
		DGNI	161	74	46	32	23	17	12	8	4
		WW	8.0	5.8	4.7	3.6	2.4	1.1	-	-	-
		WWNI	15.0	13.8	11.8	10.9	9.8	8.2	6.5	4.2	1.7
	D	TW	7.6	6.1	4.5	3.0	1.7	.8	.1	-	-
		TWNI	14.3	12.6	11.3	10.1	8.5	6.8	4.8	2.9	1.1
		DG	3.8	-	-	-	-	-	-	-	-
		DGNI	15.2	12.9	11.2	9.4	7.6	5.4	3.5	1.8	.5

ROCHESTER NEW YORK			SSF = .1	.2	.3	.4	.5	.6	.7	.8	.9
		WW	50	18	8	-	-	-	-	-	-
43.1 N	L	WWNI	124	55	34	24	17	13	10	8	5
6719 DD	C	TW	56	22	10	-	-	-	-	-	-
T(JAN)=24	R	TWNI	117	53	32	22	16	12	9	6	4
		DG	-	-	-	-	-	-	-	-	-
		DGNI	118	52	31	20	14	10	6	4	-
		WW	3.8	2.1	.8	-	-	-	-	-	-
		WWNI	10.9	9.6	8.3	7.0	6.0	4.8	3.7	2.4	.9
	D	TW	4.4	2.8	1.4	-	-	-	-	-	-
		TWNI	10.2	9.0	7.6	6.6	5.3	4.2	2.9	1.7	.6
		DG	-	-	-	-	-	-	-	-	-
		DGNI	10.1	8.3	6.8	5.3	3.9	2.6	1.4	.5	-

SYRACUSE NEW YORK			SSF = .1	.2	.3	.4	.5	.6	.7	.8	.9
		WW	50	18	8	-	-	-	-	-	-
43.1 N	L	WWNI	124	55	34	24	17	13	10	8	5
6678 DD	C	TW	56	22	11	4	-	-	-	-	-
T(JAN)=24	R	TWNI	117	53	32	22	16	12	9	6	4
		DG	-	-	-	-	-	-	-	-	-
		DGNI	118	52	31	20	14	10	7	4	-
		WW	3.8	2.1	.9	-	-	-	-	-	-
		WWNI	10.9	9.7	8.3	7.0	6.1	5.0	3.8	2.4	.9
	D	TW	4.4	2.8	1.4	.1	-	-	-	-	-
		TWNI	10.3	9.0	7.7	6.6	5.4	4.3	2.9	1.7	.6
		DG	-	-	-	-	-	-	-	-	-
		DGNI	10.1	8.4	6.8	5.4	4.0	2.6	1.4	.5	-

ASHEVILLE NORTH CAROLINA			SSF = .1	.2	.3	.4	.5	.6	.7	.8	.9
		WW	138	64	39	27	20	15	11	7	-
35.4 N	L	WWNI	225	106	68	49	37	29	23	18	13
4237 DD	C	TW	136	62	38	25	17	12	8	5	2
T(JAN)=38	R	TWNI	212	100	64	45	34	26	20	15	10
		DG	131	55	28	12	-	-	-	-	-
		DGNI	232	109	68	48	36	27	20	14	9
		WW	12.5	11.2	9.5	8.2	6.7	4.9	2.8	1.1	-
		WWNI	21.0	19.5	18.0	16.2	14.6	12.7	10.4	6.6	2.7
	D	TW	12.4	10.5	8.6	6.5	4.6	3.1	1.8	.8	.2
		TWNI	19.8	18.3	16.5	14.9	12.9	10.4	7.3	4.5	1.8
		DG	11.1	7.9	4.2	.8	-	-	-	-	-
		DGNI	21.6	19.4	17.3	15.2	12.6	9.3	6.3	3.6	1.3

CAPE HATTERAS NORTH CAROLINA			SSF = .1	.2	.3	.4	.5	.6	.7	.8	.9
		WW	214	97	60	42	31	23	18	12	7
35.3 N	L	WWNI	312	148	93	66	50	39	31	25	18
2731 DD	C	TW	200	92	56	38	26	19	13	9	5
T(JAN)=45	R	TWNI	294	138	87	61	46	36	27	20	13
		DG	223	98	56	34	21	10	-	-	-
		DGNI	329	154	96	67	50	38	28	20	13
		WW	20.4	16.5	14.9	12.5	10.9	8.7	5.6	2.8	.6
		WWNI	31.2	25.7	24.1	21.9	19.1	17.2	14.3	9.1	3.8
	D	TW	18.6	15.7	13.1	10.0	7.6	5.5	3.6	1.9	.8
		TWNI	28.5	24.5	22.2	19.6	17.4	14.0	9.9	6.1	2.5
		DG	20.0	15.4	11.2	7.0	3.7	1.1	-	-	-
		DGNI	31.3	27.2	23.8	21.0	17.7	13.3	9.2	5.4	2.1

CHARLOTTE NORTH CAROLINA			SSF = .1	.2	.3	.4	.5	.6	.7	.8	.9
		WW	177	82	50	35	25	19	14	9	4
35.2 N	L	WWNI	270	129	81	57	43	34	27	21	15
3218 DD	C	TW	170	78	47	31	22	15	10	7	4
T(JAN)=42	R	TWNI	255	121	76	53	40	31	24	18	12
		DG	180	77	43	24	12	-	-	-	-
		DGNI	283	133	82	58	43	32	24	17	11
		WW	16.2	14.0	11.8	10.5	8.7	6.8	4.1	1.9	.3
		WWNI	25.3	23.7	20.4	18.5	17.2	14.8	12.3	7.9	3.3
	D	TW	15.5	13.2	10.8	8.3	6.1	4.2	2.6	1.4	.5
		TWNI	23.9	21.6	19.1	17.3	15.1	12.1	8.6	5.3	2.1
		DG	15.5	11.7	7.8	4.1	1.2	-	-	-	-
		DGNI	26.5	23.2	20.7	18.0	15.1	11.2	7.7	4.5	1.7

GREENSBORO NORTH CAROLINA			SSF = .1	.2	.3	.4	.5	.6	.7	.8	.9
		WW	151	70	43	29	22	16	12	7	-
36.1 N	L	WWNI	239	113	72	51	39	31	24	19	14
3825 DD	C	TW	147	67	41	27	19	13	9	5	3
T(JAN)=39	R	TWNI	226	107	67	48	36	28	21	16	10
		DG	147	62	33	16	-	-	-	-	-
		DGNI	249	117	73	51	38	28	21	15	9
		WW	13.8	12.2	10.1	8.9	7.2	5.4	3.1	1.3	-
		WWNI	22.6	20.9	18.8	16.6	15.4	13.2	10.9	7.0	2.9
	D	TW	13.5	11.4	9.3	7.0	5.1	3.4	2.0	1.0	.3
		TWNI	21.2	19.4	17.2	15.6	13.5	10.8	7.7	4.7	1.9
		DG	12.5	9.1	5.3	1.9	-	-	-	-	-
		DGNI	23.1	20.7	18.3	15.9	13.3	9.8	6.6	3.8	1.4

RALEIGH-DURHAM
NORTH CAROLINA

		SSF =	.1	.2	.3	.4	.5	.6	.7	.8	.9
		WW	155	73	45	31	23	17	12	8	-
35.9 N	L	WWNI	244	117	75	53	40	31	25	20	14
3514 DD	C	TW	151	69	42	28	19	13	9	6	3
T(JAN)=40	R	TWNI	230	110	69	49	37	29	22	16	11
		DG	153	65	36	19	-	-	-	-	-
		DGNI	254	120	75	53	39	29	22	15	9
		WW	14.4	12.8	10.6	9.3	7.6	5.8	3.4	1.4	-
		WWNI	23.4	21.8	19.3	17.2	15.8	13.6	11.2	7.2	3.0
	D	TW	13.9	11.9	9.7	7.3	5.4	3.6	2.2	1.1	.3
		TWNI	21.8	20.1	17.8	16.1	13.9	11.2	7.9	4.8	1.9
		DG	13.3	9.8	5.9	2.5	-	-	-	-	-
		DGNI	23.9	21.4	19.0	16.5	13.7	10.2	6.9	4.0	1.4

BISMARCK
NORTH DAKOTA

		SSF =	.1	.2	.3	.4	.5	.6	.7	.8	.9
		WW	60	24	11	5	-	-	-	-	-
46.8 N	L	WWNI	132	61	38	26	19	14	11	8	5
9044 DD	C	TW	64	26	13	7	-	-	-	-	-
T(JAN)= 8	R	TWNI	125	58	36	24	18	13	10	7	4
		DG	-	-	-	-	-	-	-	-	-
		DGNI	129	58	34	23	16	11	7	4	-
		WW	4.9	3.1	1.6	.3	-	-	-	-	-
		WWNI	12.3	10.6	9.2	7.7	6.6	5.4	3.9	2.6	1.0
	D	TW	5.3	3.6	2.1	.8	-	-	-	-	-
		TWNI	11.6	10.0	8.4	7.1	5.8	4.5	3.1	1.8	.7
		DG	-	-	-	-	-	-	-	-	-
		DGNI	11.6	9.5	7.7	6.1	4.5	3.0	1.7	.7	-

FARGO
NORTH DAKOTA

		SSF =	.1	.2	.3	.4	.5	.6	.7	.8	.9
		WW	48	17	-	-	-	-	-	-	-
46.9 N	L	WWNI	119	55	34	23	17	12	9	7	5
9271 DD	C	TW	54	20	9	-	-	-	-	-	-
T(JAN)= 6	R	TWNI	114	52	31	21	15	11	8	6	4
		DG	-	-	-	-	-	-	-	-	-
		DGNI	114	51	29	19	13	9	6	3	-
		WW	3.7	1.6	-	-	-	-	-	-	-
		WWNI	10.9	9.4	7.9	6.5	5.5	4.5	3.2	2.1	.8
	D	TW	4.3	2.5	1.0	-	-	-	-	-	-
		TWNI	10.3	8.8	7.3	6.1	4.9	3.8	2.6	1.5	.5
		DG	-	-	-	-	-	-	-	-	-
		DGNI	10.1	8.1	6.3	4.8	3.4	2.2	1.1	.3	-

MINOT
NORTH DAKOTA

		SSF =	.1	.2	.3	.4	.5	.6	.7	.8	.9
		WW	54	19	7	-	-	-	-	-	-
48.3 N	L	WWNI	126	58	35	24	17	13	10	7	5
9407 DD	C	TW	59	22	10	-	-	-	-	-	-
T(JAN)= 8	R	TWNI	120	55	33	22	16	12	9	6	4
		DG	-	-	-	-	-	-	-	-	-
		DGNI	122	54	31	20	14	9	6	3	-
		WW	4.2	2.1	.4	-	-	-	-	-	-
		WWNI	11.5	9.9	8.3	6.6	5.5	4.5	3.2	2.1	.8
	D	TW	4.7	2.8	1.2	-	-	-	-	-	-
		TWNI	10.9	9.3	7.6	6.2	5.0	3.8	2.6	1.5	.5
		DG	-	-	-	-	-	-	-	-	-
		DGNI	10.8	8.6	6.6	5.0	3.5	2.3	1.2	.3	-

AKRON-CANTON
OHIO

		SSF =	.1	.2	.3	.4	.5	.6	.7	.8	.9
		WW	61	25	13	7	-	-	-	-	-
40.9 N	L	WWNI	137	62	39	27	20	15	12	9	6
6224 DD	C	TW	67	27	14	8	-	-	-	-	-
T(JAN)=26	R	TWNI	130	59	36	25	18	14	10	7	5
		DG	-	-	-	-	-	-	-	-	-
		DGNI	133	59	35	24	17	12	8	5	2
		WW	4.9	3.3	2.2	1.0	-	-	-	-	-
		WWNI	12.0	10.9	9.3	8.1	7.2	5.9	4.5	2.9	1.1
	D	TW	5.4	3.8	2.4	1.1	-	-	-	-	-
		TWNI	11.4	10.1	8.7	7.6	6.3	5.0	3.5	2.0	.8
		DG	-	-	-	-	-	-	-	-	-
		DGNI	11.5	9.7	8.1	6.6	5.1	3.4	2.0	.9	.1

CINCINNATI
OHIO

		SSF =	.1	.2	.3	.4	.5	.6	.7	.8	.9
		WW	81	36	20	13	8	5	-	-	-
39.1 N	L	WWNI	156	74	46	32	24	19	15	11	8
5070 DD	C	TW	84	36	21	13	8	4	-	-	-
T(JAN)=31	R	TWNI	148	70	43	30	22	17	13	9	6
		DG	49	-	-	-	-	-	-	-	-
		DGNI	157	72	44	30	22	16	11	8	4
		WW	7.3	5.4	4.1	3.0	1.9	.7	-	-	-
		WWNI	14.3	13.6	11.5	10.0	9.2	7.7	6.1	3.9	1.6
	D	TW	7.3	5.6	4.1	2.6	1.4	.5	-	-	-
		TWNI	13.7	12.2	10.7	9.5	8.0	6.4	4.5	2.7	1.0
		DG	3.0	-	-	-	-	-	-	-	-
		DGNI	14.4	12.3	10.5	8.8	7.0	4.9	3.1	1.6	.4

CLEVELAND
OHIO

		SSF =	.1	.2	.3	.4	.5	.6	.7	.8	.9
		WW	59	23	11	5	-	-	-	-	-
41.4 N	L	WWNI	135	60	37	26	19	14	11	8	6
6154 DD	C	TW	65	26	13	7	-	-	-	-	-
T(JAN)=27	R	TWNI	128	57	35	24	18	13	10	7	4
		DG	-	-	-	-	-	-	-	-	-
		DGNI	130	57	34	23	16	11	8	5	-
		WW	4.6	3.0	1.7	.5	-	-	-	-	-
		WWNI	11.7	10.6	8.9	7.7	6.7	5.4	4.2	2.7	1.0
	D	TW	5.1	3.4	2.1	.8	-	-	-	-	-
		TWNI	11.1	9.7	8.3	7.2	5.9	4.7	3.2	1.9	.7
		DG	-	-	-	-	-	-	-	-	-
		DGNI	11.2	9.2	7.6	6.1	4.6	3.1	1.8	.7	-

COLUMBUS
OHIO

		SSF =	.1	.2	.3	.4	.5	.6	.7	.8	.9
		WW	67	28	15	9	5	-	-	-	-
40.0 N	L	WWNI	141	65	41	29	21	16	13	10	7
5702 DD	C	TW	71	30	16	9	5	-	-	-	-
T(JAN)=28	R	TWNI	133	62	38	27	20	15	11	8	5
		DG	-	-	-	-	-	-	-	-	-
		DGNI	138	63	38	26	19	13	9	6	3
		WW	5.7	4.1	2.9	1.8	.7	-	-	-	-
		WWNI	12.7	11.8	10.1	8.8	7.9	6.5	5.1	3.3	1.3
	D	TW	6.0	4.4	3.0	1.7	.6	-	-	-	-
		TWNI	12.1	10.8	9.4	8.3	6.9	5.5	3.8	2.3	.9
		DG	-	-	-	-	-	-	-	-	-
		DGNI	12.5	10.6	8.9	7.3	5.7	4.0	2.4	1.1	.2

DAYTON OHIO		SSF =	.1	.2	.3	.4	.5	.6	.7	.8	.9
		WW	72	30	17	10	6	-	-	-	-
39.9 N	L	WWNI	146	68	43	30	22	17	13	10	7
5641 DD	C	TW	76	32	18	10	6	-	-	-	-
T(JAN)=28	R	TWNI	139	64	40	28	21	16	12	9	6
		DG	30	-	-	-	-	-	-	-	-
		DGNI	145	66	40	27	20	14	10	7	3
		WW	6.2	4.4	3.3	2.2	1.1	-	-	-	-
		WWNI	13.1	12.4	10.4	9.2	8.4	6.9	5.4	3.5	1.4
	D	TW	6.4	4.8	3.3	2.0	.8	-	-	-	-
		TWNI	12.6	11.2	9.8	8.7	7.3	5.8	4.1	2.4	.9
		DG	.9	-	-	-	-	-	-	-	-
		DGNI	13.1	11.1	9.4	7.8	6.2	4.3	2.7	1.3	.3

TOLEDO OHIO		SSF =	.1	.2	.3	.4	.5	.6	.7	.8	.9
		WW	62	25	13	7	-	-	-	-	-
41.6 N	L	WWNI	138	62	39	27	20	15	12	9	6
6381 DD	C	TW	68	27	15	8	3	-	-	-	-
T(JAN)=25	R	TWNI	130	59	36	25	19	14	10	8	5
		DG	-	-	-	-	-	-	-	-	-
		DGNI	134	60	36	24	17	12	8	5	2
		WW	5.0	3.4	2.3	1.0	-	-	-	-	-
		WWNI	12.1	11.1	9.4	8.2	7.2	5.9	4.5	2.9	1.1
	D	TW	5.5	3.8	2.4	1.2	.0	-	-	-	-
		TWNI	11.5	10.2	8.8	7.7	6.4	5.0	3.5	2.0	.8
		DG	-	-	-	-	-	-	-	-	-
		DGNI	11.6	9.8	8.2	6.7	5.1	3.5	2.0	.9	.1

YOUNGSTOWN OHIO		SSF =	.1	.2	.3	.4	.5	.6	.7	.8	.9
		WW	52	20	9	-	-	-	-	-	-
41.3 N	L	WWNI	126	57	35	24	18	14	10	8	5
6426 DD	C	TW	58	23	11	5	-	-	-	-	-
T(JAN)=26	R	TWNI	120	54	33	23	17	12	9	7	4
		DG	-	-	-	-	-	-	-	-	-
		DGNI	121	53	32	21	15	10	7	4	-
		WW	4.0	2.4	1.1	-	-	-	-	-	-
		WWNI	11.2	9.9	8.4	7.2	6.3	5.1	3.9	2.6	1.0
	D	TW	4.6	3.0	1.6	.4	-	-	-	-	-
		TWNI	10.5	9.2	7.9	6.8	5.6	4.4	3.1	1.8	.6
		DG	-	-	-	-	-	-	-	-	-
		DGNI	10.4	8.6	7.0	5.6	4.2	2.8	1.6	.6	-

OKLAHOMA CITY OKLAHOMA		SSF =	.1	.2	.3	.4	.5	.6	.7	.8	.9
		WW	161	74	45	31	23	17	13	8	-
35.4 N	L	WWNI	251	119	75	53	40	32	26	20	14
3695 DD	C	TW	156	71	43	28	20	14	9	6	3
T(JAN)=37	R	TWNI	237	111	70	50	37	29	22	16	11
		DG	158	67	36	19	-	-	-	-	-
		DGNI	262	122	76	53	40	30	22	16	10
		WW	14.7	12.5	10.7	9.6	8.0	6.0	3.6	1.6	-
		WWNI	23.5	21.3	19.0	17.4	16.3	14.1	11.6	7.4	3.1
	D	TW	14.0	11.9	9.8	7.5	5.4	3.7	2.3	1.2	.4
		TWNI	22.2	19.9	17.8	16.3	14.2	11.4	8.1	5.0	2.0
		DG	13.4	10.0	6.1	2.8	-	-	-	-	-
		DGNI	24.4	21.3	19.1	16.8	14.0	10.4	7.1	4.1	1.5

TULSA OKLAHOMA		SSF =	.1	.2	.3	.4	.5	.6	.7	.8	.9
		WW	147	67	41	28	21	15	11	7	-
36.2 N	L	WWNI	235	111	70	49	38	30	24	19	13
3680 DD	C	TW	144	65	39	26	18	12	8	5	3
T(JAN)=37	R	TWNI	222	104	65	46	35	27	21	15	10
		DG	141	58	30	14	-	-	-	-	-
		DGNI	245	113	70	49	37	28	20	14	9
		WW	13.5	11.2	9.7	8.6	7.0	5.2	3.0	1.2	-
		WWNI	21.8	19.8	17.7	16.4	15.2	13.1	10.7	6.8	2.8
	D	TW	12.9	10.8	8.8	6.7	4.8	3.2	1.9	1.0	.2
		TWNI	20.8	18.5	16.7	15.2	13.2	10.6	7.5	4.6	1.8
		DG	11.7	8.4	4.7	1.3	-	-	-	-	-
		DGNI	22.7	19.7	17.8	15.5	12.9	9.5	6.5	3.7	1.3

ASTORIA OREGON		SSF =	.1	.2	.3	.4	.5	.6	.7	.8	.9
		WW	122	59	36	24	17	11	7	-	-
46.1 N	L	WWNI	201	100	64	46	34	26	20	15	10
5295 DD	C	TW	121	56	34	22	14	9	6	2	-
T(JAN)=41	R	TWNI	191	94	60	42	31	23	17	13	8
		DG	112	46	19	-	-	-	-	-	-
		DGNI	208	101	64	44	32	23	17	11	6
		WW	12.2	10.2	8.3	6.4	4.5	2.6	1.1	-	-
		WWNI	20.1	19.7	16.6	14.7	12.4	10.1	7.2	4.6	1.8
	D	TW	11.5	9.6	7.3	5.3	3.5	2.0	.9	.2	-
		TWNI	19.1	17.5	15.5	13.2	10.8	8.2	5.8	3.4	1.2
		DG	9.9	5.8	1.7	-	-	-	-	-	-
		DGNI	20.8	18.4	15.8	13.0	10.0	7.3	4.6	2.4	.7

BURNS OREGON		SSF =	.1	.2	.3	.4	.5	.6	.7	.8	.9
		WW	100	44	26	17	11	7	-	-	-
43.6 N	L	WWNI	179	84	52	37	28	21	16	12	8
7212 DD	C	TW	101	44	25	16	10	6	2	-	-
T(JAN)=25	R	TWNI	170	79	49	34	25	19	14	10	6
		DG	76	19	-	-	-	-	-	-	-
		DGNI	182	83	51	35	25	18	13	8	4
		WW	8.9	7.0	5.5	4.0	2.4	1.2	-	-	-
		WWNI	16.3	15.0	13.3	11.7	9.9	8.2	5.9	4.0	1.5
	D	TW	8.8	6.9	5.1	3.4	2.0	1.0	.1	-	-
		TWNI	15.6	13.8	12.3	10.5	8.8	6.7	4.8	2.8	1.0
		DG	5.5	.8	-	-	-	-	-	-	-
		DGNI	16.6	14.3	12.2	10.0	7.7	5.6	3.5	1.7	.4

MEDFORD OREGON		SSF =	.1	.2	.3	.4	.5	.6	.7	.8	.9
		WW	134	57	34	22	15	9	5	-	-
42.4 N	L	WWNI	222	100	62	44	32	25	19	14	9
4930 DD	C	TW	130	56	32	20	13	8	4	-	-
T(JAN)=37	R	TWNI	209	94	58	40	29	22	16	12	7
		DG	118	42	-	-	-	-	-	-	-
		DGNI	227	102	62	42	30	21	15	10	5
		WW	11.3	9.4	7.4	5.4	3.5	1.9	.6	-	-
		WWNI	20.2	16.9	15.9	13.6	11.4	9.0	6.3	4.2	1.6
	D	TW	11.2	8.7	6.4	4.5	2.9	1.6	.6	-	-
		TWNI	18.7	16.2	14.2	12.1	9.8	7.4	5.2	3.0	1.1
		DG	9.0	4.5	-	-	-	-	-	-	-
		DGNI	19.9	17.2	14.4	11.8	8.9	6.5	4.1	2.0	.5

NORTH BEND OREGON		SSF =	.1	.2	.3	.4	.5	.6	.7	.8	.9
		WW	172	86	55	39	29	21	15	10	5
43.4 N	L	WWNI	260	130	86	63	48	37	29	23	16
4688 DD	C	TW	168	81	51	34	24	17	12	8	4
T(JAN)=45	R	TWNI	247	123	80	58	44	33	25	19	12
		DG	181	84	49	30	17	-	-	-	-
		DGNI	275	137	88	63	47	35	26	18	11
		WW	17.2	16.8	14.4	11.9	9.9	7.0	4.5	2.0	.3
		WWNI	26.0	26.0	24.4	21.4	18.3	15.9	11.8	7.9	3.2
	D	TW	16.6	14.7	12.1	9.5	7.0	4.9	2.9	1.5	.5
		TWNI	24.7	24.2	21.9	19.1	16.3	12.6	9.0	5.4	2.1
		DG	17.4	14.0	9.7	5.7	2.5	-	-	-	-
		DGNI	27.5	26.3	23.2	20.2	16.1	12.1	8.1	4.6	1.7

PENDLETON OREGON		SSF =	.1	.2	.3	.4	.5	.6	.7.	.8	.9
		WW	117	49	28	17	10	5	-	-	-
45.7 N	L	WWNI	204	90	56	39	28	21	15	11	7
5240 DD	C	TW	115	48	26	16	9	5	-	-	-
T(JAN)=32	R	TWNI	190	85	52	35	25	19	14	10	6
		DG	94	22	-	-	-	-	-	-	-
		DGNI	205	91	54	36	25	17	12	8	3
		WW	9.5	7.5	5.3	3.3	1.7	.5	-	-	-
		WWNI	18.4	15.3	13.7	11.2	8.9	7.1	5.0	3.4	1.3
	D	TW	9.7	7.1	4.9	3.2	1.7	.6	-	-	-
		TWNI	16.9	14.4	12.2	10.0	8.0	6.0	4.2	2.4	.9
		DG	6.5	.9	-	-	-	-	-	-	-
		DGNI	17.8	14.8	12.0	9.5	7.0	4.9	2.9	1.4	.3

PORTLAND OREGON		SSF =	.1	.2	.3	.4	.5	.6	.7	.8	.9
		WW	135	56	32	20	13	8	-	-	-
45.6 N	L	WWNI	223	100	61	42	31	23	17	13	8
4792 DD	C	TW	130	55	31	19	12	7	3	-	-
T(JAN)=38	R	TWNI	209	94	57	39	28	21	15	11	7
		DG	117	38	-	-	-	-	-	-	-
		DGNI	228	100	60	40	28	20	14	9	4
		WW	11.4	8.5	6.6	4.6	2.8	1.4	-	-	-
		WWNI	20.2	16.6	14.8	12.7	10.5	8.2	5.8	3.8	1.5
	D	TW	11.2	8.1	6.0	4.0	2.4	1.2	.3	-	-
		TWNI	19.0	15.6	13.5	11.3	9.1	6.8	4.8	2.8	1.0
		DG	8.6	3.5	-	-	-	-	-	-	-
		DGNI	20.3	16.4	13.5	10.8	8.2	5.9	3.6	1.8	.4

REDMOND OREGON		SSF =	.1	.2	.3	.4	.5	.6	.7	.8	.9
		WW	125	56	33	22	15	10	6	-	-
44.3 N	L	WWNI	209	98	61	43	32	25	19	15	10
6643 DD	C	TW	122	54	32	20	13	9	5	2	-
T(JAN)=30	R	TWNI	196	92	57	40	30	22	17	12	8
		DG	110	41	12	-	-	-	-	-	-
		DGNI	213	98	61	42	30	22	16	11	6
		WW	11.1	9.0	7.5	5.9	4.2	2.5	1.1	-	-
		WWNI	19.8	17.4	15.4	14.0	12.0	9.9	7.3	4.9	1.9
	D	TW	11.0	8.7	6.6	4.8	3.1	1.8	.9	.1	-
		TWNI	18.4	16.1	14.5	12.5	10.4	8.1	5.7	3.4	1.3
		DG	8.8	4.7	.2	-	-	-	-	-	-
		DGNI	19.8	16.9	14.7	12.3	9.6	7.1	4.5	2.4	.7

SALEM OREGON		SSF =	.1	.2	.3	.4	.5	.6	.7	.8	.9
		WW	137	57	34	22	15	9	5	-	-
44.9 N	L	WWNI	226	101	62	43	32	24	18	14	9
4852 DD	C	TW	133	56	32	20	13	8	4	-	-
T(JAN)=39	R	TWNI	212	95	58	40	29	22	16	11	7
		DG	121	41	-	-	-	-	-	-	-
		DGNI	231	102	62	42	30	21	15	10	5
		WW	11.5	8.8	7.2	5.3	3.5	1.9	.6	-	-
		WWNI	20.3	16.9	15.1	13.4	11.3	8.9	6.3	4.2	1.6
	D	TW	11.3	8.5	6.4	4.5	2.8	1.5	.5	-	-
		TWNI	19.1	15.8	14.0	12.0	9.7	7.4	5.2	3.0	1.1
		DG	9.1	4.3	-	-	-	-	-	-	-
		DGNI	20.4	16.8	14.2	11.6	8.9	6.4	4.0	2.0	.5

ALLENTOWN PENNSYLVANIA		SSF =	.1	.2	.3	.4	.5	.6	.7	.8	.9
		WW	76	33	19	12	8	5	-	-	-
40.6 N	L	WWNI	152	71	45	32	24	18	15	11	8
5827 DD	C	TW	80	35	20	12	7	4	-	-	-
T(JAN)=28	R	TWNI	143	67	42	30	22	17	13	9	6
		DG	43	-	-	-	-	-	-	-	-
		DGNI	150	69	43	29	21	16	11	7	4
		WW	6.6	5.2	4.0	2.9	1.8	.6	-	-	-
		WWNI	13.9	12.9	11.4	10.1	9.1	7.6	5.9	3.9	1.5
	D	TW	6.9	5.4	3.9	2.5	1.3	.4	-	-	-
		TWNI	13.1	12.0	10.6	9.4	8.0	6.3	4.4	2.6	1.0
		DG	2.4	-	-	-	-	-	-	-	-
		DGNI	13.6	11.9	10.3	8.7	6.9	4.8	3.0	1.5	.4

ERIE PENNSYLVANIA		SSF =	.1	.2	.3	.4	.5	.6	.7	.8	.9
		WW	53	19	8	-	-	-	-	-	-
42.1 N	L	WWNI	129	57	35	24	17	13	10	7	5
6851 DD	C	TW	59	22	11	4	-	-	-	-	-
T(JAN)=25	R	TWNI	122	54	33	22	16	12	9	6	4
		DG	-	-	-	-	-	-	-	-	-
		DGNI	124	53	31	21	14	10	6	4	-
		WW	3.8	2.2	.8	-	-	-	-	-	-
		WWNI	11.5	9.4	8.3	7.1	6.0	4.8	3.6	2.3	.9
	D	TW	4.6	2.8	1.4	.1	-	-	-	-	-
		TWNI	10.7	9.1	7.7	6.6	5.3	4.1	2.9	1.6	.6
		DG	-	-	-	-	-	-	-	-	-
		DGNI	10.4	8.4	6.8	5.3	3.9	2.6	1.4	.5	-

HARRISBURG PENNSYLVANIA		SSF =	.1	.2	.3	.4	.5	.6	.7	.8	.9
		WW	83	37	22	14	10	6	-	-	-
40.2 N	L	WWNI	158	75	47	34	25	20	16	12	8
5224 DD	C	TW	86	38	22	14	8	5	2	-	-
T(JAN)=30	R	TWNI	150	71	44	31	23	18	14	10	6
		DG	55	-	-	-	-	-	-	-	-
		DGNI	158	74	46	32	23	17	12	8	4
		WW	7.5	5.9	4.6	3.5	2.4	1.1	-	-	-
		WWNI	14.7	13.8	12.1	10.7	9.8	8.2	6.5	4.2	1.7
	D	TW	7.6	6.0	4.4	2.9	1.7	.8	.1	-	-
		TWNI	14.1	12.7	11.2	10.1	8.5	6.8	4.8	2.8	1.1
		DG	3.6	-	-	-	-	-	-	-	-
		DGNI	14.8	12.8	11.2	9.4	7.5	5.3	3.4	1.8	.5

PHILADELPHIA PENNSYLVANIA
39.9 N
4865 DD
T(JAN)=32

	SSF =	.1	.2	.3	.4	.5	.6	.7	.8	.9
	WW	95	43	26	17	12	8	5	-	-
L	WWNI	171	82	52	37	28	22	17	13	9
C	TW	96	43	25	16	10	6	4	-	-
R	TWNI	162	77	49	34	26	20	15	11	7
	DG	72	22	-	-	-	-	-	-	-
	DGNI	174	81	51	35	26	19	14	9	5
	WW	8.8	6.9	5.7	4.5	3.3	1.9	.6	-	-
	WWNI	16.0	15.3	13.2	12.0	10.8	9.1	7.3	4.7	1.9
D	TW	8.6	7.0	5.3	3.7	2.3	1.3	.5	-	-
	TWNI	15.3	13.8	12.4	11.1	9.4	7.6	5.3	3.2	1.2
	DG	5.4	1.4	-	-	-	-	-	-	-
	DGNI	16.3	14.2	12.5	10.6	8.6	6.2	4.0	2.1	.6

PITTSBURGH PENNSYLVANIA
40.5 N
5930 DD
T(JAN)=28

	SSF =	.1	.2	.3	.4	.5	.6	.7	.8	.9
	WW	62	25	13	7	-	-	-	-	-
L	WWNI	137	62	39	27	20	15	12	9	6
C	TW	68	28	15	8	3	-	-	-	-
R	TWNI	130	59	37	25	19	14	11	8	5
	DG	-	-	-	-	-	-	-	-	-
	DGNI	134	60	36	24	17	12	8	5	2
	WW	5.1	3.5	2.3	1.1	-	-	-	-	-
	WWNI	12.2	11.1	9.6	8.2	7.2	5.9	4.6	3.0	1.1
D	TW	5.5	3.9	2.5	1.2	.1	-	-	-	-
	TWNI	11.5	10.3	8.9	7.7	6.4	5.0	3.5	2.1	.8
	DG	-	-	-	-	-	-	-	-	-
	DGNI	11.7	9.9	8.2	6.7	5.1	3.5	2.1	.9	.1

SCRANTON PENNSYLVANIA
41.3 N
6277 DD
T(JAN)=26

	SSF =	.1	.2	.3	.4	.5	.6	.7	.8	.9
	WW	63	26	14	8	4	-	-	-	-
L	WWNI	137	63	40	28	20	16	12	9	6
C	TW	68	28	15	9	4	-	-	-	-
R	TWNI	130	60	37	26	19	14	11	8	5
	DG	-	-	-	-	-	-	-	-	-
	DGNI	134	61	37	25	18	13	9	6	2
	WW	5.2	3.7	2.5	1.4	.3	-	-	-	-
	WWNI	12.5	11.2	9.8	8.5	7.5	6.3	4.9	3.1	1.2
D	TW	5.6	4.1	2.7	1.4	.4	-	-	-	-
	TWNI	11.7	10.5	9.1	8.0	6.7	5.3	3.7	2.2	.8
	DG	-	-	-	-	-	-	-	-	-
	DGNI	11.9	10.1	8.5	7.0	5.4	3.7	2.2	1.0	.1

PROVIDENCE RHODE ISLAND
41.7 N
5972 DD
T(JAN)=28

	SSF =	.1	.2	.3	.4	.5	.6	.7	.8	.9
	WW	77	34	20	13	8	5	-	-	-
L	WWNI	153	71	45	32	24	19	15	11	8
C	TW	81	35	20	12	7	4	-	-	-
R	TWNI	145	67	42	30	22	17	13	9	6
	DG	44	-	-	-	-	-	-	-	-
	DGNI	152	69	43	30	22	16	11	8	4
	WW	6.5	5.3	4.2	3.1	1.9	.7	-	-	-
	WWNI	14.0	12.7	11.6	10.3	9.2	7.8	6.1	3.9	1.6
D	TW	7.0	5.5	3.9	2.5	1.4	.5	-	-	-
	TWNI	13.2	12.1	10.7	9.6	8.1	6.5	4.5	2.7	1.0
	DG	2.5	-	-	-	-	-	-	-	-
	DGNI	13.6	11.9	10.5	8.8	7.0	5.0	3.1	1.6	.4

CHARLESTON SOUTH CAROLINA
32.9 N
2146 DD
T(JAN)=49

	SSF =	.1	.2	.3	.4	.5	.6	.7	.8	.9
	WW	252	118	72	51	38	29	22	16	9
L	WWNI	358	173	108	76	58	46	37	29	21
C	TW	238	110	67	46	32	23	17	11	7
R	TWNI	339	161	101	71	54	42	32	24	16
	DG	276	124	73	47	30	19	10	-	-
	DGNI	384	180	113	79	59	45	34	24	16
	WW	24.6	19.9	17.9	16.1	13.7	11.3	7.5	4.0	1.2
	WWNI	35.0	30.8	27.0	25.9	23.5	20.3	17.0	10.9	4.6
D	TW	21.9	19.0	16.0	12.6	9.5	7.0	4.8	2.9	1.3
	TWNI	33.0	28.4	25.9	23.2	20.5	16.5	11.8	7.2	3.0
	DG	24.8	20.1	15.5	10.7	6.7	3.5	1.1	-	-
	DGNI	36.8	31.6	28.4	24.9	21.2	16.0	11.2	6.7	2.7

COLUMBIA SOUTH CAROLINA
33.9 N
2598 DD
T(JAN)=45

	SSF =	.1	.2	.3	.4	.5	.6	.7	.8	.9
	WW	215	102	62	43	32	25	19	13	7
L	WWNI	315	153	96	68	52	41	33	26	18
C	TW	206	95	58	39	28	20	14	9	6
R	TWNI	298	142	89	63	48	37	28	21	14
	DG	231	102	60	37	23	12	-	-	-
	DGNI	335	158	99	69	52	39	29	21	13
	WW	20.6	17.2	15.1	13.7	11.6	9.4	6.1	3.1	.8
	WWNI	30.2	27.9	23.8	22.4	20.9	18.0	15.1	9.6	4.0
D	TW	18.9	16.4	13.7	10.7	7.9	5.7	3.9	2.3	.9
	TWNI	28.6	25.3	22.8	20.6	18.1	14.6	10.4	6.4	2.6
	DG	20.5	16.4	12.1	7.8	4.3	1.7	-	-	-
	DGNI	32.2	27.8	25.1	21.9	18.5	13.9	9.7	5.7	2.2

GREENVILLE SOUTH CAROLINA
34.9 N
3163 DD
T(JAN)=42

	SSF =	.1	.2	.3	.4	.5	.6	.7	.8	.9
	WW	178	83	51	35	26	20	14	10	4
L	WWNI	272	129	82	58	44	35	28	22	15
C	TW	172	78	48	32	22	16	11	7	4
R	TWNI	257	121	76	54	41	31	24	18	12
	DG	182	78	44	25	13	-	-	-	-
	DGNI	285	134	83	58	43	33	24	17	11
	WW	16.4	14.2	12.0	10.8	8.9	7.0	4.3	2.0	.3
	WWNI	25.4	23.8	20.7	18.8	17.5	15.1	12.5	8.0	3.3
D	TW	15.6	13.4	11.0	8.5	6.3	4.3	2.7	1.5	.5
	TWNI	24.0	21.8	19.4	17.6	15.3	12.3	8.7	5.4	2.2
	DG	15.7	12.0	8.1	4.4	1.4	-	-	-	-
	DGNI	26.7	23.4	21.0	18.3	15.3	11.4	7.8	4.6	1.7

HURON SOUTH DAKOTA
44.4 N
8054 DD
T(JAN)=13

	SSF =	.1	.2	.3	.4	.5	.6	.7	.8	.9
	WW	63	25	12	6	-	-	-	-	-
L	WWNI	136	62	39	27	19	15	11	9	6
C	TW	67	27	14	7	-	-	-	-	-
R	TWNI	129	59	36	25	18	14	10	7	5
	DG	-	-	-	-	-	-	-	-	-
	DGNI	133	59	35	24	16	12	8	5	-
	WW	5.0	3.2	1.9	.7	-	-	-	-	-
	WWNI	12.7	10.9	9.3	7.8	6.9	5.6	4.2	2.8	1.1
D	TW	5.5	3.7	2.2	1.0	-	-	-	-	-
	TWNI	11.8	10.1	8.5	7.4	6.1	4.8	3.3	1.9	.7
	DG	-	-	-	-	-	-	-	-	-
	DGNI	11.8	9.7	7.9	6.3	4.8	3.2	1.9	.8	-

PIERRE
SOUTH DAKOTA
44.4 N
7677 DD
T(JAN)=16

		SSF = .1	.2	.3	.4	.5	.6	.7	.8	.9
	WW	77	32	17	10	6	-	-	-	-
L	WWNI	153	70	44	30	22	17	13	10	7
C	TW	80	33	18	10	6	-	-	-	-
R	TWNI	144	67	41	28	21	16	12	8	5
	DG	36	-	-	-	-	-	-	-	-
	DGNI	151	68	41	28	20	14	10	6	3
	WW	6.4	4.5	3.1	2.0	.9	-	-	-	-
	WWNI	14.4	12.3	10.5	9.0	8.0	6.6	5.0	3.3	1.3
D	TW	6.8	4.8	3.2	1.9	.7	-	-	-	-
	TWNI	13.3	11.5	9.8	8.5	7.0	5.6	3.9	2.3	.9
	DG	1.3	-	-	-	-	-	-	-	-
	DGNI	13.6	11.2	9.3	7.6	5.9	4.1	2.5	1.2	.2

RAPID CITY
SOUTH DAKOTA
44.0 N
7324 DD
T(JAN)=22

		SSF = .1	.2	.3	.4	.5	.6	.7	.8	.9
	WW	91	40	23	15	10	6	-	-	-
L	WWNI	166	79	50	35	26	20	16	12	8
C	TW	92	40	23	14	9	5	2	-	-
R	TWNI	158	75	47	33	24	18	14	10	6
	DG	64	-	-	-	-	-	-	-	-
	DGNI	168	78	48	33	24	17	12	8	4
	WW	8.3	6.4	4.8	3.7	2.3	1.1	-	-	-
	WWNI	15.9	14.1	12.7	10.9	9.5	8.2	6.2	4.1	1.6
D	TW	8.2	6.4	4.7	3.0	1.8	.8	.1	-	-
	TWNI	15.1	13.4	11.5	10.1	8.5	6.7	4.7	2.8	1.1
	DG	4.4	-	-	-	-	-	-	-	-
	DGNI	15.7	13.5	11.4	9.6	7.5	5.4	3.4	1.7	.4

SIOUX FALLS
SOUTH DAKOTA
43.6 N
7838 DD
T(JAN)=14

		SSF = .1	.2	.3	.4	.5	.6	.7	.8	.9
	WW	68	28	15	8	4	-	-	-	-
L	WWNI	143	66	41	28	21	16	12	9	6
C	TW	72	30	16	9	4	-	-	-	-
R	TWNI	135	62	38	26	19	15	11	8	5
	DG	-	-	-	-	-	-	-	-	-
	DGNI	140	63	38	25	18	13	9	6	2
	WW	5.6	3.8	2.6	1.4	.3	-	-	-	-
	WWNI	13.2	11.5	9.9	8.5	7.5	6.2	4.7	3.1	1.2
D	TW	6.0	4.3	2.7	1.5	.4	-	-	-	-
	TWNI	12.3	10.7	9.1	8.0	6.6	5.3	3.7	2.1	.8
	DG	-	-	-	-	-	-	-	-	-
	DGNI	12.5	10.4	8.6	7.0	5.4	3.7	2.2	1.0	.1

CHATTANOOGA
TENNESSEE
35.0 N
3505 DD
T(JAN)=40

		SSF = .1	.2	.3	.4	.5	.6	.7	.8	.9
	WW	136	63	38	26	19	14	10	6	-
L	WWNI	220	106	67	47	35	28	22	17	12
C	TW	133	61	36	24	16	11	7	4	2
R	TWNI	208	99	62	44	33	25	19	14	9
	DG	127	52	25	-	-	-	-	-	-
	DGNI	228	108	67	47	34	26	19	13	8
	WW	12.7	10.6	8.7	7.6	6.0	4.3	2.4	.8	-
	WWNI	21.4	19.7	16.8	15.0	14.1	11.9	9.8	6.3	2.6
D	TW	12.2	10.2	8.1	6.0	4.2	2.7	1.5	.7	.1
	TWNI	19.9	17.9	15.6	14.2	12.2	9.8	6.9	4.2	1.7
	DG	10.6	7.1	3.3	-	-	-	-	-	-
	DGNI	21.6	18.8	16.5	14.3	11.8	8.7	5.8	3.3	1.1

KNOXVILLE
TENNESSEE
35.8 N
3478 DD
T(JAN)=41

		SSF = .1	.2	.3	.4	.5	.6	.7	.8	.9
	WW	143	66	39	27	20	14	10	6	-
L	WWNI	230	110	69	48	37	29	23	18	13
C	TW	140	63	38	25	17	11	8	5	2
R	TWNI	217	103	64	45	34	26	20	15	10
	DG	136	56	28	10	-	-	-	-	-
	DGNI	238	111	69	48	35	26	19	14	8
	WW	13.2	11.0	9.0	7.9	6.4	4.6	2.6	.9	-
	WWNI	21.9	20.1	17.2	15.4	14.5	12.2	10.0	6.5	2.7
D	TW	12.7	10.6	8.4	6.3	4.4	2.9	1.7	.8	.2
	TWNI	20.5	18.4	16.0	14.5	12.5	10.1	7.1	4.3	1.7
	DG	11.3	7.7	3.9	.3	-	-	-	-	-
	DGNI	22.4	19.4	17.0	14.7	12.2	9.0	6.0	3.4	1.2

MEMPHIS
TENNESSEE
35.0 N
3227 DD
T(JAN)=40

		SSF = .1	.2	.3	.4	.5	.6	.7	.8	.9
	WW	161	74	44	31	22	17	12	8	-
L	WWNI	251	119	75	52	40	31	25	20	14
C	TW	156	70	42	28	19	13	9	6	3
R	TWNI	237	112	70	49	37	28	22	16	11
	DG	159	66	35	18	-	-	-	-	-
	DGNI	263	122	75	53	39	29	22	15	9
	WW	14.8	12.1	10.4	9.2	7.6	5.7	3.3	1.4	-
	WWNI	23.3	21.7	18.4	17.0	15.9	13.6	11.1	7.2	3.0
D	TW	14.0	11.8	9.6	7.3	5.2	3.5	2.2	1.1	.3
	TWNI	22.2	19.6	17.5	15.9	13.7	11.1	7.9	4.8	1.9
	DG	13.3	9.7	5.8	2.4	-	-	-	-	-
	DGNI	24.6	21.0	18.8	16.3	13.6	10.1	6.9	3.9	1.4

NASHVILLE
TENNESSEE
36.1 N
3696 DD
T(JAN)=38

		SSF = .1	.2	.3	.4	.5	.6	.7	.8	.9
	WW	124	56	32	22	16	11	8	4	-
L	WWNI	207	98	61	42	32	25	20	16	11
C	TW	122	54	32	21	14	9	6	3	-
R	TWNI	195	92	57	40	30	23	17	13	8
	DG	108	41	15	-	-	-	-	-	-
	DGNI	212	98	60	42	31	23	17	11	7
	WW	11.2	8.8	7.2	6.1	4.8	3.2	1.6	.3	-
	WWNI	19.4	17.6	14.8	13.4	12.6	10.7	8.6	5.6	2.3
D	TW	10.8	8.8	6.8	4.9	3.3	2.0	1.1	.4	-
	TWNI	18.3	16.0	14.0	12.7	10.9	8.8	6.2	3.8	1.5
	DG	8.5	4.9	.8	-	-	-	-	-	-
	DGNI	19.8	16.7	14.6	12.5	10.3	7.5	5.0	2.8	.9

ABILENE
TEXAS
32.4 N
2610 DD
T(JAN)=44

		SSF = .1	.2	.3	.4	.5	.6	.7	.8	.9
	WW	236	111	69	49	37	28	22	15	9
L	WWNI	339	164	104	74	57	45	36	29	20
C	TW	224	104	64	44	31	22	16	11	7
R	TWNI	321	153	97	69	52	41	31	23	16
	DG-	257	116	69	44	29	18	9	-	-
	DGNI	362	171	108	77	57	44	33	24	15
	WW	22.6	19.5	17.3	16.0	13.7	11.0	7.3	3.9	1.1
	WWNI	32.6	30.0	26.7	25.1	23.6	20.3	16.6	10.7	4.5
D	TW	20.8	18.3	15.5	12.3	9.3	6.9	4.6	2.8	1.2
	TWNI	30.9	27.7	25.3	23.1	20.2	16.2	11.6	7.2	2.9
	DG	23.2	19.3	14.8	10.2	6.3	3.2	.9	-	-
	DGNI	34.5	30.6	28.1	24.7	20.9	15.8	11.0	6.6	2.6

AMARILLO TEXAS		SSF =	.1	.2	.3	.4	.5	.6	.7	.8	.9
		WW	178	83	52	36	27	21	16	11	5
35.2 N	L	WWNI	271	130	83	59	45	36	29	23	16
4183 DD	C	TW	173	79	49	33	23	16	11	8	5
T(JAN)=36	R	TWNI	256	122	77	55	42	33	25	19	12
		DG	183	80	46	27	15	-	-	-	-
		DGNI	285	134	85	60	45	34	25	18	12
		WW	16.5	14.5	12.8	11.6	9.9	7.7	4.8	2.4	.5
		WWNI	25.2	24.2	21.5	19.8	18.4	16.5	13.5	8.6	3.6
	D	TW	15.6	13.8	11.5	9.0	6.7	4.7	3.1	1.7	.7
		TWNI	24.0	22.1	20.1	18.5	16.3	13.0	9.3	5.7	2.3
		DG	15.9	12.8	8.9	5.2	2.2	-	-	-	-
		DGNI	26.6	24.0	21.7	19.4	16.2	12.1	8.4	5.0	1.9

AUSTIN TEXAS		SSF =	.1	.2	.3	.4	.5	.6	.7	.8	.9
		WW	311	143	89	63	48	37	28	21	13
30.3 N	L	WWNI	432	203	127	91	70	55	45	35	25
1737 DD	C	TW	288	134	82	56	40	29	21	15	10
T(JAN)=50	R	TWNI	404	189	119	85	64	50	39	29	19
		DG	348	157	95	62	42	28	18	9	-
		DGNI	460	214	135	95	71	54	41	30	19
		WW	29.0	24.4	23.0	20.8	18.0	14.7	10.2	5.7	1.9
		WWNI	41.6	35.3	33.0	31.8	28.5	25.0	20.1	13.1	5.5
	D	TW	26.7	23.2	19.6	15.9	12.4	9.3	6.5	4.1	1.7
		TWNI	38.3	33.3	31.2	28.1	25.0	20.0	14.3	8.8	3.6
		DG	31.4	26.2	21.2	15.6	10.8	6.7	3.5	1.1	-
		DGNI	43.4	37.7	34.7	30.8	26.0	20.0	14.0	8.4	3.4

BROWNSVILLE TEXAS		SSF =	.1	.2	.3	.4	.5	.6	.7	.8	.9
		WW	656	308	192	137	105	82	65	49	33
25.9 N	L	WWNI	846	402	252	180	138	111	89	70	50
650 DD	C	TW	609	280	175	121	88	65	48	35	24
T(JAN)=60	R	TWNI	797	373	235	168	128	99	77	57	38
		DG	799	365	226	155	111	80	58	40	24
		DGNI	929	432	272	194	146	112	85	63	42
		WW	60.7	53.8	49.0	46.4	41.9	34.5	25.8	15.6	6.3
		WWNI	78.3	71.7	64.3	62.0	57.7	50.8	40.0	26.6	11.3
	D	TW	54.4	49.4	43.2	35.8	28.2	22.0	16.2	10.7	5.3
		TWNI	74.1	66.0	61.2	56.5	49.2	39.9	28.6	17.7	7.3
		DG	71.5	63.6	54.7	44.2	34.0	24.8	16.7	9.7	3.9
		DGNI	86.3	76.2	70.7	63.6	53.8	42.0	30.0	18.6	7.9

CORPUS CHRISTI TEXAS		SSF =	.1	.2	.3	.4	.5	.6	.7	.8	.9
		WW'	469	225	146	105	80	62	49	37	24
27.8 N	L	WWNI	621	299	195	141	109	86	69	55	39
930 DD	C	TW	442	208	131	92	67	50	37	26	18
T(JAN)=56	R	TWNI	586	282	182	131	100	78	60	45	30
		DG	564	264	165	113	81	58	41	27	15
		DGNI	677	325	208	149	113	87	66	48	32
		WW	43.4	43.4	39.0	35.3	31.5	25.6	18.8	11.1	4.3
		WWNI	57.6	57.5	53.8	49.2	44.2	40.3	30.9	20.5	8.7
	D	TW	40.8	37.6	33.2	27.5	21.9	16.8	11.6	7.6	3.7
		TWNI	54.5	53.4	49.2	44.5	38.8	31.2	22.2	13.7	5.6
		DG	52.3	47.2	40.5	32.1	24.1	17.1	10.9	5.7	1.9
		DGNI	63.0	60.5	55.8	50.3	41.9	32.4	22.9	14.1	5.9

DALLAS TEXAS		SSF =	.1	.2	.3	.4	.5	.6	.7	.8	.9
		WW	237	109	68	48	36	28	21	15	9
32.8 N	L	WWNI	343	162	102	73	56	45	36	28	20
2290 DD	C	TW	223	103	64	43	31	22	16	11	7
T(JAN)=45	R	TWNI	321	151	95	68	52	40	31	23	15
		DG	255	114	68	43	28	17	8	-	-
		DGNI	362	169	106	75	57	43	32	23	15
		WW	22.2	18.6	17.4	15.8	13.5	10.8	7.2	3.8	1.1
		WWNI	33.4	28.7	26.3	25.6	23.2	20.2	16.3	10.5	4.4
	D	TW	20.7	17.9	15.2	12.0	9.3	6.7	4.6	2.8	1.1
		TWNI	30.6	26.8	25.1	22.8	20.1	16.1	11.5	7.1	2.9
		DG	22.9	18.7	14.4	9.9	6.1	3.1	.8	-	-
		DGNI	34.2	29.9	27.5	24.4	20.6	15.6	10.9	6.5	2.5

DEL RIO TEXAS		SSF =	.1	.2	.3	.4	.5	.6	.7	.8	.9
		WW	375	167	105	75	57	44	34	25	16
29.4 N	L	WWNI	512	232	146	106	81	64	51	40	29
1523 DD	C	TW	343	158	97	66	48	35	26	18	12
T(JAN)=51	R	TWNI	476	218	138	98	74	58	44	33	22
		DG	424	191	115	77	53	36	24	14	5
		DGNI	543	248	157	111	83	63	48	35	23
		WW	33.4	28.8	28.3	24.5	21.3	17.6	12.4	7.0	2.5
		WWNI	47.5	39.9	39.2	37.8	32.2	28.1	22.9	14.9	6.3
	D	TW	31.3	27.0	23.0	19.0	15.0	11.3	8.0	5.0	2.2
		TWNI	43.4	38.2	36.6	32.3	28.3	22.9	16.3	10.0	4.1
		DG	37.7	31.8	26.4	20.1	14.4	9.5	5.4	2.3	.3
		DGNI	49.3	43.7	40.2	35.8	30.4	23.3	16.3	9.8	4.0

EL PASO TEXAS		SSF =	.1	.2	.3	.4	.5	.6	.7	.8	.9
		WW	274	131	82	58	44	34	26	19	12
31.8 N	L	WWNI	384	189	119	85	65	52	42	33	24
2678 DD	C	TW	258	122	76	52	37	27	19	14	9
T(JAN)=44	R	TWNI	363	175	111	79	60	47	36	27	18
		DG	307	141	85	56	37	25	15	7	-
		DGNI	413	197	125	89	66	51	38	28	18
		WW	27.4	22.9	20.5	19.4	16.7	13.6	9.3	5.1	1.6
		WWNI	38.4	34.5	30.2	28.9	27.6	23.7	19.2	12.4	5.2
	D	TW	24.5	21.8	18.5	14.7	11.3	8.4	5.9	3.6	1.6
		TWNI	36.2	31.6	28.9	26.7	23.1	18.7	13.4	8.3	3.4
		DG	28.5	24.1	19.1	13.8	9.2	5.5	2.6	.6	-
		DGNI	40.6	35.4	32.5	28.8	24.4	18.5	13.0	7.9	3.2

FORT WORTH TEXAS		SSF =	.1	.2	.3	.4	.5	.6	.7	.8	.9
		WW	225	104	65	46	34	26	20	14	8
32.8 N	L	WWNI	328	155	98	71	54	43	34	27	19
2382 DD	C	TW	212	99	61	41	29	21	15	10	6
T(JAN)=45	R	TWNI	307	146	92	66	50	39	30	22	15
		DG	240	108	64	40	25	15	6	-	-
		DGNI	345	162	102	73	54	41	31	22	14
		WW	21.0	18.1	16.6	14.9	12.7	10.0	6.7	3.5	.9
		WWNI	32.1	28.1	25.6	24.6	22.2	19.2	15.5	10.1	4.2
	D	TW	19.8	17.2	14.5	11.4	8.8	6.3	4.3	2.5	1.1
		TWNI	29.4	26.1	24.3	21.9	19.3	15.4	11.0	6.7	2.7
		DG	21.6	17.6	13.4	9.0	5.4	2.4	.2	-	-
		DGNI	32.7	28.9	26.4	23.4	19.6	14.9	10.3	6.1	2.4

HOUSTON TEXAS		SSF =	.1	.2	.3	.4	.5	.6	.7	.8	.9
		WW	336	148	93	66	50	39	30	22	14
30.0 N	L	WWNI	466	209	131	95	73	58	46	36	26
1434 DD	C	TW	309	140	86	59	42	31	22	16	10
T(JAN)=52	R	TWNI	433	196	124	88	67	52	40	30	20
		DG	375	166	99	66	45	30	19	11	-
		DGNI	492	223	140	99	74	57	43	31	20
		WW	29.1	25.4	24.1	21.7	18.9	15.3	10.7	6.0	2.0
		WWNI	42.2	35.9	35.3	32.7	29.3	25.9	20.6	13.5	5.7
	D	TW	27.7	23.8	20.5	16.6	13.1	9.8	6.9	4.1	1.8
		TWNI	38.8	34.3	32.2	29.2	25.9	20.7	14.7	9.1	3.7
		DG	32.7	27.3	22.4	16.7	11.7	7.4	3.9	1.3	-
		DGNI	43.9	39.0	35.9	32.1	27.1	20.8	14.5	8.8	3.5

LAREDO TEXAS		SSF =	.1	.2	.3	.4	.5	.6	.7	.8	.9
		WW	528	252	163	115	87	67	52	39	26
27.5 N	L	WWNI	693	331	216	156	117	92	74	58	41
876 DD	C	TW	495	231	145	101	73	54	40	28	19
T(JAN)=56	R	TWNI	652	312	200	143	108	83	64	48	32
		DG	639	297	184	126	89	64	45	30	17
		DGNI	756	360	230	163	122	93	70	51	34
		WW	48.3	48.1	43.3	36.9	33.0	27.1	20.0	11.9	4.7
		WWNI	63.4	63.1	59.6	51.5	45.1	40.5	32.4	21.5	9.2
	D	TW	46.0	41.0	36.2	29.8	23.7	18.0	12.5	7.7	4.0
		TWNI	60.2	58.9	52.7	46.8	41.0	32.8	23.3	14.3	5.9
		DG	59.0	51.9	44.7	35.4	26.6	18.8	12.1	6.4	2.2
		DGNI	69.8	67.0	59.5	53.1	44.5	34.3	24.1	14.8	6.2

LUBBOCK TEXAS		SSF =	.1	.2	.3	.4	.5	.6	.7	.8	.9
		WW	206	97	61	43	32	25	19	13	8
33.6 N	L	WWNI	306	146	94	67	51	41	33	26	19
3545 DD	C	TW	198	92	57	39	28	20	14	10	6
T(JAN)=39	R	TWNI	288	137	87	63	48	37	29	21	14
		DG	220	99	58	37	23	13	-	-	-
		DGNI	323	153	97	69	52	39	30	21	14
		WW	19.1	17.4	15.3	14.1	12.2	9.7	6.3	3.3	.9
		WWNI	28.6	27.5	25.0	22.6	21.5	18.8	15.6	9.9	4.2
	D	TW	18.3	16.2	13.7	10.8	8.3	5.9	4.0	2.3	1.1
		TWNI	27.1	25.2	22.9	21.2	18.6	14.9	10.6	6.6	2.7
		DG	19.7	16.4	12.2	8.0	4.6	1.9	-	-	-
		DGNI	30.3	27.7	25.1	22.5	18.9	14.3	9.9	5.9	2.3

LUFKIN TEXAS		SSF =	.1	.2	.3	.4	.5	.6	.7	.8	.9
		WW	273	125	78	55	42	32	25	18	11
31.2 N	L	WWNI	385	182	113	81	63	50	40	32	22
1940 DD	C	TW	254	118	73	49	35	25	18	13	8
T(JAN)=49	R	TWNI	361	170	107	76	58	45	35	26	17
		DG	299	135	80	52	34	22	13	-	-
		DGNI	409	191	120	85	63	48	36	26	17
		WW	25.8	21.2	19.7	18.2	15.6	12.6	8.6	4.7	1.4
		WWNI	38.0	31.8	29.1	28.4	25.8	22.4	18.1	11.8	4.9
	D	TW	23.6	20.5	17.3	13.8	10.6	7.9	5.5	3.4	1.5
		TWNI	34.6	29.8	27.8	25.3	22.3	17.9	12.8	7.9	3.2
		DG	27.0	22.3	17.6	12.5	8.3	4.8	2.1	-	-
		DGNI	38.8	33.5	30.8	27.3	23.1	17.6	12.3	7.4	2.9

MIDLAND-ODESSA TEXAS		SSF =	.1	.2	.3	.4	.5	.6	.7	.8	.9
		WW	265	127	79	56	43	33	26	19	11
31.9 N	L	WWNI	373	182	115	83	64	51	41	32	23
2621 DD	C	TW	249	118	74	50	36	26	19	13	9
T(JAN)=44	R	TWNI	352	170	108	77	59	46	35	27	18
		DG	294	136	82	54	36	24	14	6	-
		DGNI	400	191	121	86	65	50	37	27	18
		WW	26.4	22.4	20.1	19.0	16.3	13.4	9.1	5.0	1.6
		WWNI	37.3	33.7	29.9	28.5	26.9	23.3	19.1	12.3	5.2
	D	TW	23.7	21.1	18.0	14.4	11.1	8.2	5.7	3.5	1.7
		TWNI	34.8	30.9	28.4	26.1	23.0	18.4	13.2	8.2	3.4
		DG	27.4	23.2	18.4	13.3	8.9	5.2	2.4	.5	-
		DGNI	39.0	34.6	31.9	28.3	24.0	18.2	12.8	7.7	3.1

PORT ARTHUR TEXAS		SSF =	.1	.2	.3	.4	.5	.6	.7	.8	.9
		WW	339	153	95	67	51	40	31	22	14
29.9 N	L	WWNI	466	215	134	96	74	59	47	37	27
1518 DD	C	TW	311	143	88	60	43	31	23	16	11
T(JAN)=52	R	TWNI	435	201	126	90	68	53	41	30	20
		DG	380	170	102	68	46	31	20	11	-
		DGNI	496	228	143	101	76	58	44	32	21
		WW	31.5	25.6	24.5	22.4	19.4	15.8	11.1	6.3	2.1
		WWNI	44.7	36.5	34.7	33.6	30.3	26.6	21.3	13.9	5.8
	D	TW	28.5	24.7	20.8	17.0	13.3	10.1	7.1	4.4	2.1
		TWNI	40.9	34.9	32.8	29.7	26.4	21.2	15.1	9.3	3.8
		DG	33.9	28.3	23.1	17.2	12.1	7.8	4.2	1.6	-
		DGNI	46.0	39.6	36.5	32.6	27.6	21.3	15.0	9.0	3.7

SAN ANGELO TEXAS		SSF =	.1	.2	.3	.4	.5	.6	.7	.8	.9
		WW	269	129	80	57	43	33	26	19	11
31.4 N	L	WWNI	377	185	117	84	65	51	41	33	23
2240 DD	C	TW	253	120	75	51	36	26	19	13	9
T(JAN)=46	R	TWNI	357	172	109	78	59	46	36	27	18
		DG	300	138	84	55	37	24	15	6	-
		DGNI	405	194	123	87	66	50	38	27	18
		WW	26.9	22.6	20.4	19.0	16.3	13.4	9.1	5.0	1.6
		WWNI	37.7	34.2	30.2	28.8	26.8	23.2	19.2	12.3	5.2
	D	TW	24.1	21.4	18.2	14.6	11.2	8.3	5.8	3.6	1.6
		TWNI	35.5	31.3	28.8	26.2	23.1	18.5	13.3	8.2	3.4
		DG	27.9	23.6	18.8	13.5	9.0	5.3	2.5	.5	-
		DGNI	39.7	35.0	32.3	28.4	24.1	18.3	12.9	7.8	3.1

SAN ANTONIO TEXAS		SSF =	.1	.2	.3	.4	.5	.6	.7	.8	.9
		WW	345	156	97	70	52	41	31	23	14
29.5 N	L	WWNI	474	219	137	99	76	60	48	38	27
1570 DD	C	TW	318	146	90	62	44	32	24	17	11
T(JAN)=51	R	TWNI	443	204	129	92	70	54	42	31	21
		DG	389	175	105	70	48	33	21	12	-
		DGNI	505	232	146	103	78	59	45	33	21
		WW	31.9	26.6	25.6	22.8	19.9	16.3	11.4	6.5	2.2
		WWNI	44.9	37.5	36.0	34.5	30.6	27.0	21.7	14.2	6.0
	D	TW	29.2	25.2	21.5	17.4	13.8	10.4	7.3	4.6	2.0
		TWNI	41.4	35.8	33.7	30.4	27.0	21.6	15.4	9.5	3.9
		DG	34.8	29.1	23.8	18.0	12.7	8.2	4.5	1.8	-
		DGNI	46.6	40.8	37.6	33.4	28.4	21.8	15.3	9.2	3.8

SHERMAN TEXAS — 33.7 N, 2864 DD, T(JAN)=42

		SSF = .1	.2	.3	.4	.5	.6	.7	.8	.9
	WW	191	88	54	38	28	21	16	11	6
L	WWNI	286	136	85	61	47	37	30	23	17
C	TW	182	83	51	34	24	17	12	8	5
R	TWNI	271	127	80	57	43	33	26	19	13
	DG	197	86	49	29	17	-	-	-	-
	DGNI	302	141	88	62	46	35	26	19	12
	WW	17.8	14.9	13.4	11.9	10.2	7.9	5.0	2.4	.5
	WWNI	27.5	24.2	21.8	20.7	18.9	16.5	13.4	8.6	3.6
D	TW	16.6	14.3	11.9	9.2	6.9	4.9	3.2	1.8	.7
	TWNI	25.6	22.6	20.7	18.8	16.5	13.2	9.4	5.8	2.3
	DG	17.1	13.5	9.5	5.6	2.6	-	-	-	-
	DGNI	28.1	24.7	22.3	19.8	16.6	12.4	8.6	5.0	1.9

WACO TEXAS — 31.6 N, 2058 DD, T(JAN)=47

		SSF = .1	.2	.3	.4	.5	.6	.7	.8	.9
	WW	263	121	75	53	40	31	24	17	10
L	WWNI	373	176	110	79	61	48	39	30	22
C	TW	245	114	70	47	34	24	18	12	8
R	TWNI	350	164	103	74	56	43	33	25	17
	DG	287	129	77	49	32	21	12	-	-
	DGNI	396	185	116	82	61	47	35	25	16
	WW	24.7	20.5	19.2	17.3	14.9	11.9	8.1	4.4	1.3
	WWNI	36.6	30.8	28.5	27.4	24.8	21.6	17.4	11.3	4.7
D	TW	22.7	19.7	16.7	13.2	10.2	7.5	5.2	3.2	1.4
	TWNI	33.5	29.0	27.0	24.4	21.6	17.3	12.3	7.6	3.1
	DG	25.8	21.2	16.6	11.7	7.6	4.2	1.7	-	-
	DGNI	37.6	32.5	29.7	26.4	22.2	17.0	11.8	7.1	2.8

WICHITA FALLS TEXAS — 34.0 N, 2904 DD, T(JAN)=41

		SSF = .1	.2	.3	.4	.5	.6	.7	.8	.9
	WW	207	96	59	41	31	24	18	13	7
L	WWNI	304	145	91	65	50	40	32	25	18
C	TW	197	90	55	37	26	19	13	9	5
R	TWNI	288	136	85	61	46	36	28	21	14
	DG	217	96	56	34	21	11	-	-	-
	DGNI	322	151	94	67	50	38	28	20	13
	WW	19.6	16.3	14.6	13.4	11.4	9.0	5.8	2.9	.7
	WWNI	29.2	26.1	23.2	22.1	20.6	17.9	14.6	9.4	3.9
D	TW	18.0	15.6	13.1	10.1	7.7	5.5	3.7	2.2	.9
	TWNI	27.4	24.2	22.1	20.2	17.7	14.3	10.2	6.3	2.5
	DG	19.1	15.4	11.2	7.1	3.9	1.3	-	-	-
	DGNI	30.3	26.5	24.2	21.4	18.0	13.6	9.4	5.6	2.1

BRYCE CANYON UTAH — 37.7 N, 9133 DD, T(JAN)=20

		SSF = .1	.2	.3	.4	.5	.6	.7	.8	.9
	WW	98	47	29	20	15	11	7	3	-
L	WWNI	171	85	55	40	30	24	19	15	11
C	TW	99	46	28	19	12	8	5	3	-
R	TWNI	163	80	52	37	28	22	17	12	8
	DG	80	32	-	-	-	-	-	-	-
	DGNI	175	85	54	39	29	22	16	11	7
	WW	9.7	8.3	7.1	6.0	4.7	3.0	1.4	.2	-
	WWNI	17.1	16.2	15.0	13.7	12.2	11.0	8.6	5.5	2.3
D	TW	9.5	8.1	6.4	4.6	3.1	1.9	1.0	.3	-
	TWNI	16.3	15.2	13.9	12.5	10.9	8.6	6.1	3.7	1.5
	DG	6.9	3.7	-	-	-	-	-	-	-
	DGNI	17.4	15.9	14.1	12.5	10.1	7.3	4.9	2.7	.9

CEDAR CITY UTAH — 37.7 N, 6137 DD, T(JAN)=29

		SSF = .1	.2	.3	.4	.5	.6	.7	.8	.9
	WW	139	64	39	27	20	15	11	7	-
L	WWNI	223	106	68	49	37	29	23	18	13
C	TW	135	62	38	25	17	12	8	5	2
R	TWNI	211	100	64	45	34	27	20	15	10
	DG	130	55	29	12	-	-	-	-	-
	DGNI	231	109	68	48	36	27	20	14	9
	WW	12.9	11.1	9.6	8.4	6.9	5.0	2.8	1.1	-
	WWNI	22.1	19.0	18.3	16.3	14.9	13.2	10.4	6.7	2.8
D	TW	12.5	10.6	8.6	6.4	4.7	3.2	1.9	.9	.3
	TWNI	20.3	18.2	16.5	15.0	13.1	10.5	7.4	4.5	1.8
	DG	11.3	8.0	4.3	1.0	-	-	-	-	-
	DGNI	21.8	19.7	17.4	15.4	12.7	9.4	6.4	3.6	1.3

SALT LAKE CITY UTAH — 40.8 N, 5983 DD, T(JAN)=28

		SSF = .1	.2	.3	.4	.5	.6	.7	.8	.9
	WW	131	59	35	23	16	11	7	3	-
L	WWNI	215	101	64	45	33	26	20	15	11
C	TW	127	57	34	22	14	9	6	3	-
R	TWNI	203	95	59	42	31	23	18	13	8
	DG	118	45	18	-	-	-	-	-	-
	DGNI	221	102	63	44	32	23	17	11	6
	WW	11.7	9.9	7.9	6.4	4.6	3.1	1.4	.2	-
	WWNI	20.9	17.8	16.2	14.2	12.4	10.5	7.8	5.2	2.1
D	TW	11.6	9.3	7.1	5.2	3.5	2.1	1.0	.3	-
	TWNI	19.1	16.9	14.8	13.0	10.8	8.6	6.0	3.6	1.4
	DG	9.5	5.5	1.5	-	-	-	-	-	-
	DGNI	20.5	17.7	15.2	12.9	10.2	7.5	4.9	2.6	.8

BURLINGTON VERMONT — 44.5 N, 7876 DD, T(JAN)=17

		SSF = .1	.2	.3	.4	.5	.6	.7	.8	.9
	WW	42	14	-	-	-	-	-	-	-
L	WWNI	115	51	32	22	16	12	9	7	5
C	TW	50	18	8	-	-	-	-	-	-
R	TWNI	109	49	30	21	15	11	8	6	4
	DG	-	-	-	-	-	-	-	-	-
	DGNI	109	47	28	19	13	9	6	3	-
	WW	3.1	1.5	-	-	-	-	-	-	-
	WWNI	10.0	8.8	7.8	6.6	5.6	4.5	3.3	2.2	.8
D	TW	3.8	2.3	.9	-	-	-	-	-	-
	TWNI	9.6	8.4	7.2	6.1	5.0	3.9	2.7	1.5	.5
	DG	-	-	-	-	-	-	-	-	-
	DGNI	9.2	7.6	6.2	4.8	3.5	2.2	1.1	.3	-

NORFOLK VIRGINIA — 36.9 N, 3488 DD, T(JAN)=40

		SSF = .1	.2	.3	.4	.5	.6	.7	.8	.9
	WW	162	75	46	32	23	17	13	8	-
L	WWNI	251	120	76	54	41	32	26	20	14
C	TW	156	72	43	29	20	14	9	6	3
R	TWNI	237	113	71	50	38	29	22	17	11
	DG	160	68	37	20	-	-	-	-	-
	DGNI	263	123	77	54	40	30	22	16	10
	WW	15.2	12.6	11.1	9.5	7.9	6.0	3.6	1.6	-
	WWNI	24.5	21.6	19.5	17.9	15.9	13.9	11.6	7.3	3.1
D	TW	14.3	12.1	10.0	7.5	5.4	3.8	2.3	1.2	.4
	TWNI	22.7	20.1	18.2	16.2	14.2	11.4	8.1	4.9	2.0
	DG	13.9	10.2	6.3	2.9	-	-	-	-	-
	DGNI	24.9	21.9	19.3	16.8	14.1	10.4	7.1	4.1	1.5

RICHMOND VIRGINIA		SSF =	.1	.2	.3	.4	.5	.6	.7	.8	.9
		WW	132	61	37	25	18	13	9	5	-
37.5 N	L	WWNI	216	103	65	46	35	27	22	17	12
3939 DD	C	TW	130	59	35	23	16	11	7	4	2
T(JAN)=37	R	TWNI	205	97	61	43	32	25	19	14	9
		DG	122	49	24	-	-	-	-	-	-
		DGNI	223	105	65	45	34	25	18	13	8
		WW	12.1	10.3	8.5	7.3	5.7	4.1	2.2	.7	-
		WWNI	20.5	19.0	16.9	14.9	13.6	11.5	9.4	6.1	2.5
	D	TW	11.7	9.8	7.9	5.8	4.1	2.6	1.4	.6	.1
		TWNI	19.2	17.5	15.5	13.9	11.9	9.6	6.7	4.1	1.6
		DG	10.1	6.7	2.9	-	-	-	-	-	-
		DGNI	20.8	18.5	16.3	13.9	11.5	8.4	5.6	3.2	1.1

ROANOKE VIRGINIA		SSF =	.1	.2	.3	.4	.5	.6	.7	.8	.9
		WW	124	58	35	24	18	13	9	5	-
37.3 N	L	WWNI	206	99	63	45	34	27	21	17	12
4307 DD	C	TW	123	56	34	22	15	10	7	4	1
T(JAN)=36	R	TWNI	195	93	59	42	32	24	19	14	9
		DG	113	46	21	-	-	-	-	-	-
		DGNI	212	100	63	44	33	24	18	13	7
		WW	11.6	9.9	8.3	7.1	5.6	3.9	2.1	.6	-
		WWNI	19.3	18.5	16.4	14.7	13.5	11.5	9.4	6.0	2.5
	D	TW	11.3	9.5	7.6	5.6	3.9	2.5	1.4	.6	.1
		TWNI	18.5	17.0	15.3	13.7	11.8	9.5	6.7	4.1	1.6
		DG	9.5	6.2	2.4	-	-	-	-	-	-
		DGNI	20.1	18.0	15.9	13.7	11.3	8.3	5.5	3.1	1.0

OLYMPIA WASHINGTON		SSF =	.1	.2	.3	.4	.5	.6	.7	.8	.9
		WW	112	48	28	17	10	5	-	-	-
47.0 N	L	WWNI	191	90	55	38	28	21	16	11	7
5530 DD	C	TW	109	48	27	16	10	5	-	-	-
T(JAN)=37	R	TWNI	181	84	52	36	26	19	14	10	6
		DG	90	23	-	-	-	-	-	-	-
		DGNI	195	89	54	36	25	18	12	8	3
		WW	10.1	7.3	5.6	3.6	1.9	.5	-	-	-
		WWNI	19.1	15.6	13.3	11.7	9.5	7.2	5.0	3.2	1.2
	D	TW	9.8	7.3	4.9	3.2	1.8	.7	-	-	-
		TWNI	17.3	14.5	12.4	10.3	8.2	6.1	4.2	2.4	.8
		DG	6.6	1.1	-	-	-	-	-	-	-
		DGNI	18.3	14.9	12.3	9.6	7.1	5.0	3.0	1.4	.2

SEATTLE-TACOMA WASHINGTON		SSF =	.1	.2	.3	.4	.5	.6	.7	.8	.9
		WW	128	54	30	19	11	6	-	-	-
47.4 N	L	WWNI	213	98	59	41	30	22	16	11	7
5185 DD	C	TW	123	53	29	18	11	6	-	-	-
T(JAN)=38	R	TWNI	199	91	55	38	27	20	14	10	6
		DG	108	33	-	-	-	-	-	-	-
		DGNI	217	97	58	38	27	19	13	8	3
		WW	11.1	8.0	6.0	3.9	2.1	.7	-	-	-
		WWNI	20.3	16.5	14.0	12.1	9.7	7.3	5.1	3.2	1.2
	D	TW	10.8	7.9	5.4	3.6	2.0	.8	-	-	-
		TWNI	18.6	15.3	13.0	10.7	8.4	6.2	4.3	2.4	.8
		DG	8.0	2.5	-	-	-	-	-	-	-
		DGNI	19.9	15.9	12.9	10.0	7.4	5.1	3.1	1.4	.2

SPOKANE WASHINGTON		SSF =	.1	.2	.3	.4	.5	.6	.7	.8	.9
		WW	91	37	20	10	-	-	-	-	-
47.6 N	L	WWNI	169	78	47	32	23	17	12	9	6
6835 DD	C	TW	92	38	20	11	5	-	-	-	-
T(JAN)=25	R	TWNI	161	73	44	30	21	15	11	8	4
		DG	56	-	-	-	-	-	-	-	-
		DGNI	171	75	45	29	20	13	9	5	-
		WW	7.6	5.3	3.3	1.4	-	-	-	-	-
		WWNI	15.4	13.0	11.2	9.3	7.2	5.5	3.7	2.5	.9
	D	TW	7.6	5.3	3.3	1.7	.5	-	-	-	-
		TWNI	14.7	12.3	10.3	8.2	6.4	4.7	3.3	1.8	.6
		DG	2.9	-	-	-	-	-	-	-	-
		DGNI	15.0	12.3	9.7	7.3	5.2	3.5	1.9	.7	-

YAKIMA WASHINGTON		SSF =	.1	.2	.3	.4	.5	.6	.7	.8	.9
		WW	117	48	26	15	8	-	-	-	-
46.6 N	L	WWNI	201	90	55	38	27	20	14	10	7
6009 DD	C	TW	114	47	25	15	8	4	-	-	-
T(JAN)=28	R	TWNI	190	85	51	35	25	18	13	9	5
		DG	92	-	-	-	-	-	-	-	-
		DGNI	205	89	53	35	24	16	11	7	3
		WW	9.6	7.0	4.9	2.8	1.2	-	-	-	-
		WWNI	18.3	14.9	13.0	10.8	8.4	6.5	4.5	3.1	1.2
	D	TW	9.5	6.8	4.5	2.8	1.4	.2	-	-	-
		TWNI	16.9	14.0	11.8	9.5	7.5	5.5	3.9	2.2	.8
		DG	6.1	-	-	-	-	-	-	-	-
		DGNI	17.7	14.4	11.5	8.8	6.4	4.5	2.6	1.1	.2

CHARLESTON WEST VIRGINIA		SSF =	.1	.2	.3	.4	.5	.6	.7	.8	.9
		WW	93	41	23	15	10	7	3	-	-
38.4 N	L	WWNI	170	80	50	35	26	20	16	12	9
4590 DD	C	TW	94	41	24	15	9	5	2	-	-
T(JAN)=34	R	TWNI	161	75	47	33	24	19	14	10	7
		DG	66	-	-	-	-	-	-	-	-
		DGNI	172	79	48	33	24	18	13	8	5
		WW	8.2	6.4	4.9	3.7	2.6	1.3	.1	-	-
		WWNI	15.5	14.5	12.6	10.9	9.9	8.2	6.6	4.3	1.7
	D	TW	8.2	6.4	4.8	3.1	1.9	.9	.2	-	-
		TWNI	14.9	13.3	11.6	10.3	8.6	7.0	4.9	2.9	1.1
		DG	4.5	-	-	-	-	-	-	-	-
		DGNI	15.7	13.5	11.5	9.7	7.8	5.5	3.5	1.8	.5

HUNTINGTON WEST VIRGINIA		SSF =	.1	.2	.3	.4	.5	.6	.7	.8	.9
		WW	99	44	26	17	12	8	4	-	-
38.4 N	L	WWNI	178	84	53	37	28	21	17	13	9
4624 DD	C	TW	101	44	26	16	10	6	3	-	-
T(JAN)=34	R	TWNI	169	79	49	34	26	20	15	11	7
		DG	76	22	-	-	-	-	-	-	-
		DGNI	181	83	51	35	26	19	14	9	5
		WW	8.9	7.1	5.4	4.3	3.1	1.7	.5	-	-
		WWNI	16.3	15.2	13.2	11.5	10.5	8.7	7.0	4.5	1.8
	D	TW	8.8	7.0	5.3	3.6	2.2	1.1	.4	-	-
		TWNI	15.6	14.0	12.2	10.8	9.1	7.3	5.1	3.1	1.2
		DG	5.5	1.3	-	-	-	-	-	-	-
		DGNI	16.6	14.3	12.3	10.3	8.4	6.0	3.9	2.0	.6

EAU CLAIRE WISCONSIN		SSF =	.1	.2	.3	.4	.5	.6	.7	.8	.9
		WW	50	19	8	-	-	-	-	-	-
44.9 N	L	WWNI	121	56	34	24	17	13	10	8	5
8388 DD	C	TW	56	22	11	4	-	-	-	-	-
T(JAN)=12	R	TWNI	115	53	32	22	16	12	9	6	4
		DG	-	-	-	-	-	-	-	-	-
		DGNI	117	52	31	20	14	10	7	4	-
		WW	3.9	2.2	.9	-	-	-	-	-	-
		WWNI	10.9	9.7	8.4	7.0	6.1	5.1	3.7	2.4	.9
	D	TW	4.5	2.9	1.4	.1	-	-	-	-	-
		TWNI	10.4	9.1	7.7	6.6	5.4	4.2	2.9	1.7	.6
		DG	-	-	-	-	-	-	-	-	-
		DGNI	10.2	8.5	6.9	5.4	4.0	2.6	1.4	.5	-

GREEN BAY WISCONSIN		SSF =	.1	.2	.3	.4	.5	.6	.7	.8	.9
		WW	57	22	11	5	-	-	-	-	-
44.5 N	L	WWNI	132	59	37	26	19	14	11	8	5
8098 DD	C	TW	63	25	13	6	-	-	-	-	-
T(JAN)=15	R	TWNI	125	56	35	24	17	13	10	7	4
		DG	-	-	-	-	-	-	-	-	-
		DGNI	127	56	34	22	16	11	7	4	-
		WW	4.4	2.9	1.6	.3	-	-	-	-	-
		WWNI	11.8	10.1	9.1	7.7	6.6	5.4	4.0	2.6	1.0
	D	TW	5.1	3.4	1.9	.7	-	-	-	-	-
		TWNI	11.2	9.7	8.3	7.1	5.9	4.6	3.2	1.8	.7
		DG	-	-	-	-	-	-	-	-	-
		DGNI	11.1	9.2	7.5	6.0	4.5	3.0	1.7	.7	-

LA CROSSE WISCONSIN		SSF =	.1	.2	.3	.4	.5	.6	.7	.8	.9
		WW	59	23	12	6	-	-	-	-	-
43.9 N	L	WWNI	133	61	37	26	19	15	11	9	6
7417 DD	C	TW	64	26	13	7	-	-	-	-	-
T(JAN)=16	R	TWNI	126	57	35	24	18	13	10	7	5
		DG	-	-	-	-	-	-	-	-	-
		DGNI	129	57	34	23	16	11	8	5	-
		WW	4.8	3.0	1.9	.7	-	-	-	-	-
		WWNI	12.0	10.5	9.1	7.8	6.9	5.8	4.3	2.8	1.1
	D	TW	5.2	3.6	2.1	.9	-	-	-	-	-
		TWNI	11.2	9.9	8.4	7.3	6.1	4.8	3.3	1.9	.7
		DG	-	-	-	-	-	-	-	-	-
		DGNI	11.3	9.4	7.8	6.3	4.8	3.2	1.9	.8	-

MADISON WISCONSIN		SSF =	.1	.2	.3	.4	.5	.6	.7	.8	.9
		WW.	62	25	13	7	-	-	-	-	-
43.1 N	L	WWNI	137	62	39	27	20	15	12	9	6
7730 DD	C	TW	67	27	15	8	3	-	-	-	-
T(JAN)=17	R	TWNI	130	59	36	25	19	14	11	8	5
		DG	-	-	-	-	-	-	-	-	-
		DGNI	133	59	36	24	17	12	8	5	2
		WW	4.9	3.4	2.3	1.1	-	-	-	-	-
		WWNI	12.3	10.7	9.6	8.3	7.3	6.2	4.6	3.0	1.1
	D	TW	5.5	3.9	2.4	1.2	.1	-	-	-	-
		TWNI	11.7	10.2	8.9	7.7	6.5	5.1	3.5	2.1	.8
		DG	-	-	-	-	-	-	-	-	-
		DGNI	11.7	9.8	8.2	6.8	5.2	3.5	2.1	.9	.1

MILWAUKEE WISCONSIN		SSF =	.1	.2	.3	.4	.5	.6	.7	.8	.9
		WW	64	26	14	8	-	-	-	-	-
42.9 N	L	WWNI	140	63	39	27	20	16	12	9	6
7444 DD	C	TW	69	28	15	8	4	-	-	-	-
T(JAN)=19	R	TWNI	132	60	37	26	19	14	11	8	5
		DG	-	-	-	-	-	-	-	-	-
		DGNI	136	60	36	25	17	12	9	5	2
		WW	5.0	3.6	2.4	1.2	-	-	-	-	-
		WWNI	12.9	10.7	9.8	8.5	7.4	6.1	4.6	3.0	1.1
	D	TW	5.6	4.0	2.5	1.3	.2	-	-	-	-
		TWNI	11.8	10.4	9.0	7.8	6.5	5.1	3.6	2.1	.8
		DG	-	-	-	-	-	-	-	-	-
		DGNI	11.8	10.0	8.3	6.8	5.2	3.6	2.1	.9	.1

CASPER WYOMING		SSF =	.1	.2	.3	.4	.5	.6	.7	.8	.9
		WW	107	49	30	20	14	10	7	-	-
42.9 N	L	WWNI	184	90	57	41	31	24	19	15	10
7555 DD	C	TW	107	49	29	19	12	8	5	2	-
T(JAN)=23	R	TWNI	175	84	53	38	28	22	17	12	8
		DG	90	34	-	-	-	-	-	-	-
		DGNI	190	90	56	39	29	21	16	11	6
		WW	10.3	8.4	6.9	5.7	4.3	2.8	1.2	-	-
		WWNI	17.7	16.4	14.9	13.3	11.8	10.4	7.9	5.2	2.1
	D	TW	9.8	8.2	6.4	4.6	3.0	1.8	.8	.2	-
		TWNI	16.9	15.3	13.7	12.2	10.5	8.4	5.9	3.5	1.4
		DG	7.3	3.8	-	-	-	-	-	-	-
		DGNI	18.1	16.1	14.0	12.1	9.7	7.1	4.7	2.5	.8

CHEYENNE WYOMING		SSF =	.1	.2	.3	.4	.5	.6	.7	.8	.9
		WW	111	52	32	22	16	12	8	5	-
41.1 N	L	WWNI	190	91	58	42	32	25	21	16	11
7255 DD	C	TW	112	51	31	20	14	9	6	3	1
T(JAN)=27	R	TWNI	181	86	55	39	30	23	18	13	9
		DG	96	38	16	-	-	-	-	-	-
		DGNI	195	92	58	41	31	23	17	12	7
		WW	10.3	8.9	7.7	6.6	5.4	3.6	1.9	.5	-
		WWNI	17.7	17.2	15.5	14.3	12.9	11.7	9.3	5.9	2.4
	D	TW	10.1	8.7	7.0	5.1	3.5	2.3	1.3	.5	.1
		TWNI	16.9	15.8	14.5	13.1	11.6	9.2	6.5	4.0	1.6
		DG	7.9	5.0	1.2	-	-	-	-	-	-
		DGNI	18.3	16.7	14.9	13.3	10.8	7.9	5.3	3.0	1.0

ROCK SPRINGS WYOMING		SSF =	.1	.2	.3	.4	.5	.6	.7	.8	.9
		WW	98	45	28	19	13	9	6	-	-
41.6 N	L	WWNI	176	84	54	38	29	23	18	14	10
8410 DD	C	TW	100	45	27	17	11	7	4	2	-
T(JAN)=19	R	TWNI	167	79	50	36	27	21	16	12	8
		DG	78	28	-	-	-	-	-	-	-
		DGNI	179	84	53	37	27	20	15	10	6
		WW	9.1	7.7	6.3	5.2	3.8	2.4	.9	-	-
		WWNI	16.4	15.7	14.1	12.8	11.4	9.9	7.6	5.0	2.0
	D	TW	9.0	7.5	5.8	4.1	2.6	1.5	.7	.1	-
		TWNI	15.7	14.4	13.1	11.7	10.0	8.0	5.6	3.4	1.3
		DG	6.2	2.6	-	-	-	-	-	-	-
		DGNI	16.8	15.1	13.3	11.5	9.2	6.7	4.4	2.3	.7

		SSF =	.1	.2	.3	.4	.5	.6	.7	.8	.9
SHERIDAN											
WYOMING		WW	86	38	22	14	9	5	-	-	-
44.8 N	L	WWNI	161	77	48	34	25	19	15	12	8
7708 DD	C	TW	88	38	22	14	8	4	-	-	-
T(JAN)=21	R	TWNI	153	72	45	32	23	18	13	10	6
		DG	57	-	-	-	-	-	-	-	-
		DGNI	163	75	46	32	23	17	12	8	4
		WW	8.0	6.2	4.5	3.3	1.9	.8	-	-	-
		WWNI	14.9	13.8	12.4	10.6	9.1	7.8	5.8	3.9	1.5
	D	TW	7.7	6.0	4.4	2.8	1.5	.6	-	-	-
		TWNI	14.2	12.8	11.2	9.7	8.2	6.4	4.5	2.7	1.0
		DG	3.7	-	-	-	-	-	-	-	-
		DGNI	15.1	13.1	11.0	9.2	7.1	5.1	3.2	1.6	.4

Appendix C
ANALYSIS METHODS

The following tables provide an overview of the methods available for analyzing the performance of solar designs. This information has been extracted directly from a publication by the Solar Energy Research Institute (SERI) entitled Analysis Methods for Solar Heating and Cooling Applications (see BIBLIOGRAPHY, reference no. 2).

The tables provide very general information concerning availability and applications of common solar analysis methods. They are not intended as statements of advocacy of any of the methods described.

Special attention should be paid, in the table on pages 311 and 312, to the column entitled "Tools Required." The categories listed follow the same organization as the "Energy Use Analysis" description (pages 18-20) in Passive Solar Fundamentals.

For example, the column entitled "Tables/ Graphs" indicates what are referred to as Manual Methods on page 18. If such a method is desired, simply read down this column to find a variety of possible options. The next three columns ("4 Function Calculator," "Scientific Calculator," and "Programmable Calculator") cover the Programmable Calculator Programs discussed in the text. The final column, missing its title, indicates those methods that are designed to be used with a microcomputer (in Section 2.2100 under Microcomputer Methods).

The table on page 313 is devoted exclusively to the methods discussed under Computer Programs on page 19.

For further information on a specific analysis method, contact the Reference/ Source listed. For general information, contact:

Public Information Office
Solar Energy Research Institute
1617 Cole Boulevard
Golden, Colorado 80401
(303) 321-1000

or:

National Solar Heating and Cooling
Information Center
P.O. Box 1607
Rockville, Maryland 20850
(800) 523-2929

Description	Author	Availability: Cost ($)	Availability: Date	Availability: Reference/Source	Applications, Active: Space Heating	Applications, Active: Domestic Hot Water	Applications, Active: Liquid Collector	Applications, Active: Air Collector	Applications, Passive: Direct Gain	Applications, Passive: Thermal Storage Wall	Applications, Passive: Attached Sunspace	Applications, Passive: Thermal Storage Roof	Applications, Passive: Convective Loop	Tools Required: Tables Graphs	Tools Required: 4 Function Calculator	Tools Required: Scientific Calculator	Tools Required: Programmable Calculator	Tools Required:	Output: Solar Fraction	Output: Optimum Area	Output: Economic Analysis	Output: Other
A Simplified Method for Calculating Solar Collector Array Size for Space Heating	J D Balcomb J C Hedstrom		197€	Sharing the Sun: Solar Technology in the Seventies, Vol 4, American Section International Solar Energy Society, 1976, pp 281-284	•	•	•	•						•	•					•		
Passive Solar Design Handbook SLR Method	J D Balcomb Bruce Anderson	14	1980	NTIS 5285 Port Royal Road Springfield, VA 22161					•	•	△			•	•				•	•	•	
Predicting the performance of Solar Energy Systems	U S Army Construction Engineering Research Lab		1977	Rept No AD-A035 608 9 ST (NTIS)	•	•	•	•						•	•				•	•	•	
Copper Brass Bronze Design Handbook —Solar Energy Systems	Copper Development Association	3	1978	Copper Development Association, Inc 1011 High Ridge Road Stamford, CN 06905	•		•	•						•	•				•			
Copper Brass Bronze Technical Report: How to Design and Build a Swimming Pool Heater	Cooper Development Association	Free	1976	Copper Development Association, Inc			•							•	•				•		•	
PEGFIX and PEGFLOAT	W Glennie	75 both	1978	Princeton Energy Group 729 Alexander Road Princeton, NJ 08540 (609) 452-8235					•		•						•					•
IMPSLR interactive SLR	PEG	250	1980	Princeton Energy Group					•	•								•	•			
NEATWORK thermal network	PEG	250	1980	Princeton Energy Group					•	•								•	•			•
PEGSOL two-zone	PEG	400	1980	Princeton Energy Group					•	•	•							•	•			•
Solarcon Programs for Comprehensive Active System Design (5 programs & 5 sets, weather data) (5 Insolation models)	R W Graeff	383 data	1977	Solarcon, Inc 607 Church Street Ann Arbor, MI 48104 (313) 769-6588	•	•	•	•								•	•	•	•		•	
Solarcon Programs for Passive System Design (5 programs & 5 sets, weather data)	R W Graeff	495	1979	Solarcon, Inc					•	•							•	•	•			
Solar Heating Systems Design Manuai	ITT Corporation Fluid Handling Division	2 50	1977	Bulletin TESE-576, Rev 1 ITT Training & Education Dept Fluid Handling Division Morton Grove, IL 60053	•	•	•	•						•	•				•			
A General Design Method for Closed-Loop Solar Energy Systems	S A Klein W A Beckman		1977	Proceedings of the 1977 Annual Meeting, Vol 1, American Section International Solar Energy Society, 1977, pp 8 1-8 5			•							•		•			•			
Solar Heating Design by the F-CHART Method	S A Klein, W A Beckman, J A Duffie	10	1977	John Wiley and Sons, New York, N Y 1977 (Publisher)	•	•	•	•						•	•				•		•	
TEANET thermal network model	J T Kohler P W Sullivan	95	1978	Total Environmental Action, Inc Church Hill Harrisville, NH 03450 (603) 827-3374					•	•	•	•	•				•					•
The GFL Method for Sizing Solar Energy Space and Water Heating Systems	G F Lameiro P Bendt		1978	Rept No SERI-30 Solar Energy Research Institute 1617 Cole Boulevard Golden, CO 80401	•	•	•	•						•		•			•			
A Rational Procedure for Predicting the Long-Term Average Performance of Flat-Plate Solar Energy Collectors	B Y H Liu R C Jordan		1963	Solar Energy, Vol 7, No 2, 1963, pp 53-70	•	•	•	•						•		•			•			
Pacific Regional Solar Heating Handbook	Los Alamos Scientific Lab		1976	Rept No TID-27630 (NTIS)	•	•	•	•						•		•			•			

(continued on next page)

Description	Author	Availability: Cost ($)	Availability: Date	Availability: Reference/Source	Applications: Active: Space Heating	Applications: Active: Domestic Hot Water	Applications: Active: Liquid Collector	Applications: Active: Air Collector	Applications: Passive: Direct Gain	Applications: Passive: Thermal Storage Wall	Applications: Passive: Attached Sunspace	Applications: Passive: Thermal Storage Roof	Applications: Passive: Convective Loop	Tools Required: Tables Graphs	Tools Required: 4 Function Calculator	Tools Required: Scientific Calculator	Tools Required: Programmable Calculator	Tools Required:	Output: Solar Fraction	Output: Optimum Area	Output: Economic Analysis	Output: Other
McClintock Corp	R McClintock	195 Therm alone 175 econ anal alone 75	1977	McClintock Corp P O Box 430980 Miami FL 33143 (305) 666-1300	●	●	●	●									●	●	●	●	●	
Minimum Cost Sizing of Solar Heating Systems	J C Ward		1976	Sharing the Sun Vol 4 1976 pp 336-348	●	●	●	●						●		●			●	●	●	
Designing and Building a Solar House Your Place in the Sun	D Watson	9	1977	Garden Way Publishing Charlotte VT 05445	●	●	●	●						●	●				●		●	
Optimal Sizing of Solar Collectors by the Method of Relative Areas	C D Barley C B Winn	95	1978	Solar Environmental Engr Co Inc	●	●	●	●						●		●				●	●	
SEEC I—Heat Load Monthly Solar Fraction Economics (F-CHART)	C B Winn	125	1976	Solar Environmental Engineering Co Inc 2524 East Vine Drive Fort Collins CO 80524 (303) 221-5166	●	●	●	●									●	●	●		●	
SEEC II—Collector Optimization Annual Solar Fraction Economics	C B Winn D Barley G Johnson J Leflar	95	1978	Solar Environmental Engineering Co Inc	●	●	●	●									●		●	●	●	
SEEC III-SEEC II Plus Insulation Optimization	C B Winn D Barley G Johnson J Leflar	125 (Calculator version)	1978	Solar Environmental Engineering Co Inc	●	●	●	●									●		●	●	●	
SEEC IV—Collector Heat Exchanger Sizing and Analysis	C Byron Winn J Leflar	95	1978	Solar Environmental Engineering Co Inc	●	●	●										●					●
SEEC V—Duct design and Air Flow Analysis	J Leflar	95	1978	Solar Environmental Engineering Co Inc	●			●									●					●
SEEC VI—Passive Solar Heating	K Sharp	125	1978	Solar Environmental Engineering Co Inc					●	●							●	●	●	●	●	
SEEC VII—Sun Angles and Shading	K Sharp	75	1979	Solar Environmental Engineering Co Inc													●					●
SEEC VIII—Swimming Pools	P Jacobs K Sharp	125	1980	Solar Environmental Engineering Co Inc			●										●		●	●	●	●
Sunshine Power Programs for Modeling Solar Energy Components and Systems Solar Energy Library (9 programs)	G Shramek		1977	Sunshine Power Co 1018 Lancer Drive San Jose CA 95129 (408) 446-2446	●	●	●	●									●		●			
Mazria Design Patterns (Rule-of-Thumb) in The Passive Solar Energy Book	Edward Mazria	11	1979	Rodale Press Emmaus PA 18049					●	●	●	●	●	●					●	●		
PASCALC Los Alamos SLR	J T Kohler	PASCALC therm 75 econ 20	1979	Total Environmental Action Inc Church Hill Harrisville NH 03450					●	●	●						●		●	●	●	
SOLARGRAF/DHW nomographs	W Wright	10	1980	Northeast Solar Energy Center 470 Atlantic Ave Boston MA 02110		●	●	●						●	●				●			
SOLARGRAF/SLR nomographs	W Wright	10	1980	Northeast Solar Energy Center					●	●				●	●				●			
SUN-PULSE II	G Tully	100	1980	McGraw-Hill Book Company P O Box 400 Hightstown NJ 08520	●	●	●	●	●	●							●		●			●
G-CHART	SEDCLA	25 + tax	1979	Solar Energy Design Corp of America	●	●	●	●						●		●			●		●	
P-CHART SLR nomograph	SEDCLA	20 + tax	1980	Solar Energy Design Corp of America					●	●				●	●				●	●	●	
Solar Energy Programs	S A Klein W A Beckman	40	1980	F-CHART P O Box 5562 Madison WI 53705	●	●	●	●	●								●		●		●	●
PASODEI SLR	L-P-M	300	1980	Londe-Parker-Michels Inc Mr Steve Andes 7438 Forsyth Suite 202 St Louis MO (314) 725-5501					●	●	●							●	●	●	●	
OVERHANG shading	L-P-M	350	1980	Londe-Parker-Michels Inc														●				●
EXFIND hour sim Trombe	L-P-M	300	1980	Londe-Parker-Michels Inc						●	●							●				●
INSULATE econ optimized	L-P-M	300	1980	Londe-Parker-Michels Inc														●				●
LPMTZ two-zone passive	L-P-M	1500	1980	Londe-Parker-Michels Inc					●	●	●							●	●			

Δ McFarland, R. D. and Jones, R. W., "Performance Estimates for Attached-Sunspace Passive Solar Heated Buildings," Proc. 1980 AS/ISES, Vol. 3.2, pp. 784-788, Phoenix, AZ, June 1980.

Program Name	Latest Version	Availability: Purchase ($)	Availability: Time Share	Availability: Special Arrangements	Availability: Comments	Application: User Manual	Application: Service Hot Water	Application: Space Heating	Application: Space Cooling	Application: Process Heat	Application: Active System	Application: Passive System	Intended Users: Research Engineers	Intended Users: Architect/Engineers	Intended Users: Builders	Computation Interval: Hour	Computation Interval: Month	Computer Versions Available	Economic Analysis	Sponsor
BLAST*	1980	Nom.	●		Training available	●	●	●	●		●	●	●	●		●		CDC	●	USAF, USA, GSA
DEROB	1979	200				●		●	●			●	●	●		●		CDC		NSF, ERDA, DOE
DOE-2*	1980	400	●			●	●	●	●		●	●	●	●		●		CDC	●	LASL, DOE
EMPSS	1978	500		●	Consulting with ADL	●	●	●	●		●	△	●	●		●		IBM	●	EPRI
F-CHART	1978	100	●		Training available	●	●	●			●			●	●		●	CDC, IBM UNIVAC †	●	DOE
FREHEAT	1979	150			Limited documentation			●				●	●			●		CDC	●	DOE
HISPER	1978	Avail. on request			Limited documentation	△	●	●	●		●		●			●		UNIVAC PDP		NASA, MSFC
HUD-RSVP/2	1979	175	●		Based on F-CHART	●	●	●			●			●	●		●	CDC UNIVAC	●	HUD
PACE	1980	100	●		Based on F-CHART, SLR	●	●	●			●	●		●	●		●	CDC UNIVAC	●	DOE, SERI
SHASP	1978	Avail. on request				●	●	●	●		●		●			●		UNIVAC	●	DOE
SOLAR-5	1979		●		Not Distributed			●	●			●		●	●	●	●	CDC		UCLA, DOE
SOLCOST	1979	300	●			●	●	●			●	●	●	●	●		●	CDC, IBM UNIVAC	●	DOE
SOLFIN 2	1980	Nom.		●	Documentation Cost $6.	●	●	●	●	●	●	●	●	●	●	Annual		IBM, CDC	●	California Energy Commission
SOLOPT	1978	20				●	●	●			●		●				●	AMDAHL	●	Texas A&M Unlv.
SOLTES	1979	175			No Cost for DOE	●	●	●	●	●	●		●			●		CDC		DOE, Sandia
SUNCAT	1979	Nom.			Limited documentation	●.		●				●	●	●		●		Data General Eclipse	●	NCAT
SUNSYM®	1979		●	●	Offered as service only	●	●	●	●		●			●		●		IBM	●	Sunworks Comp. Systems
SYRSOL	1978	Nom.			Avail. but not actively marketed		●	●	●		●		●	●	●	●		IBM	●	ERDA, NSF, DOE
TRACE SOLAR*	1980		●	●	Offered as service only	●	●	●	●	●	●		●	●		●	●	IBM	●	The Trane Co.
TRNSYS	1979	200	●		Training required	●	●	●	●	●	●	●	●			●		CDC, IBM UNIVAC †	△	DOE
TWO ZONE	1977	No Charge		●		●		●	●		●	●	●			●		CDC	●	LBL
UWENSOL	1980	200				●		●				●	●			●		CDC		State of Wash.
WATSUN II, III	1980	Contact Author				●	●	●			●		●			●		IBM	●	Nat'l Research Center of Can.

*Programs are primarily developed for large-scale, multi-zone applications

△ Being added

† ANSI 1966 Std. Fortran

INDEX

Page numbers in italic indicate photos. Page numbers set in boldface and followed by t indicate tables.

B

D

E

F

U

V